Systemic Pathology of Fish

Contributors

THOMAS W. DUKES, D.V.M., M.S., *Animal Disease Research Institute, Agriculture Canada, Ottawa.*

MICHAEL ANTHONY HAYES, B.V.Sc., Ph.D., *Diplomate American College of Veterinary Pathology, Associate Professor, Department of Pathology, Ontario Veterinary College, University of Guelph.*

JOHN LEATHERLAND, B.Sc., Ph.D., *Professor, Department of Zoology, College of Biological Science, University of Guelph.*

BRIAN WILCOCK, B.A., D.V.M., M.S., Ph.D., *Associate Professor, Department of Pathology, Ontario Veterinary College, University of Guelph.*

Systemic Pathology of Fish

A Text and Atlas of Comparative Tissue Responses in Diseases of Teleosts

HUGH W. FERGUSON, B.V.M.&S., Ph.D., M.R.C.V.S.

Iowa State University Press / Ames

To Pam, Jamie, and Katie

Hugh W. Ferguson is director of the Fish Pathology Laboratory and is an associate professor in the Department of Pathology, Ontario Veterinary College, University of Guelph. He is a Diplomate of the American College of Veterinary Pathologists and a pathologist recognized by the American Fisheries Society and the Canadian Veterinary Medical Association.

Manufactured in the United States of America

First edition, 1989

Library of Congress Cataloging-in-Publication Data

Ferguson, Hugh W. (Hugh William), 1949–
Systemic pathology of fish.

Bibliography: p.
Includes index.
1. Fishes—Disease. I. Dukes, Thomas W. II. Title.
SH171.F42 1989 597′.02 88–13739
ISBN 0–8138–0147–8

Contents

Preface

THIS MONOGRAPH is the result of requests from graduate and undergraduate students for a set of notes introducing the comparative pathology of fish. I firmly believe that to be a good fish pathologist one must first of all become fully conversant with comparative pathology and that there are very few, if any, shortcuts. The bacteria, viruses, and parasites associated with fish diseases may differ from their mammalian counterparts, and an understanding of water chemistry, limnology, and fish histology is required, but the general principles of vertebrate pathology still apply, and lesion interpretation is essentially the same in fish as in "higher" vertebrates.

These notes therefore tend to compare the responses of fish tissues with those of mammals and, where possible, to highlight the differences. They are not intended to be an exhaustive list of etiologies, but rather should be viewed as an attempt to illustrate some of the mechanisms of disease and pathophysiology in fish organs and tissues.

Accordingly, the subject has been tackled from the standpoint of the fish, not the disease-causing agent. I trust that experts in the fields of fish parasitology and microbiology will bear with me if I fail to include what they see from their viewpoints as essential information or if the material is at times overly simplistic.

There are enormous gaps in our knowledge and understanding of fish pathology, and with fifteen years' exposure to the subject, I am just starting to feel experienced enough to say, at times, I don't know! This is rarely good enough for students, however, who want to grasp the whole picture, not just selected segments, before starting to focus on the details. In trying to accommodate this dilemma there will be times when we have been too sweeping, too provocative, or just simply wrong. I do hope that those who care about the subject will let me know.

The text assumes a reasonable working knowledge of pathology on the part of the reader, particularly general pathology, and I make no apologies therefore for using the jargon appropriate to the subject. Anybody needing a refamiliarization with general pathology is directed to the bibliography at the end of the book.

Most examples are drawn from diagnostic or experimental material processed by the Fish Pathology Laboratory, Ontario Veterinary College, and as a result restrict themselves almost exclusively to teleost fish. In addition, many are from species commercially important in a North American context. Inevitably, the examples that have been selected to illustrate various disease processes reflect the authors' interests.

Acknowledgments

I first want to record my thanks for all the help I receive from my colleagues in the Department of Pathology at Ontario Veterinary College. Not only do they tolerate my interest in fish and comparative pathology, but they encourage me in its pursuit, based, I know, on a genuine appreciation for its inherent value. A more benign yet stimulating work environment would be hard to imagine. In particular I am very grateful for the chapters on eyes, on neoplasia, and on the endocrine and reproductive systems by Drs. Brian Wilcock and Tony Hayes, by Dr. Tom Dukes of A.D.R.I., Ottawa, and by Dr. John Leatherland, Zoology Department.

The technical and secretarial staff are indeed the hub around which any department circles. They in no small part helped with this monograph. Sandy Brown succeeded in deciphering my scrawl and with her usual good humor and competence produced the final typed copy. In this task she was ably helped by Jean Bagg. Michael Bakerpearce, Helga Hunter, and Sofie Tatarski were responsible for the histological preparation; Ted Eaton helped with some of the photographs; and Tim Sullivan produced many of the best gross pictures.

I would like to acknowledge the following journals for permission to publish micrographs: *Canadian Journal of Zoology,* Figure 4.1 A and B; *Diseases of Aquatic Organisms,* Figures 12.11, 12.16; *Journal of the American Veterinary Medical Association,* Figure 7.8; *Journal of Comparative Pathology,* Figures 5.3B, 5.6, 6.4 A–C; *Journal of Fish Diseases,* Figures 2.17 D and E, 4.19 B, E and F, 7.7, 7.8; *Veterinary Pathology,* Figures 4.7, 9.7 C and D; *The Veterinary Record,* Figures 7.4, 7.10 A and C.

The Fish Pathology Laboratory has had many excellent research assistants, graduate students, and postdoctoral fellows over the years. Apart from at times carrying the lion's share of the diagnostic load, they have been dedicated and enthusiastic experts, and it has been a privilege to work with and learn from them. During my tenure, they include Robert Armstrong, Pierre-Yves Daoust, Sally Goldes, Brad Hicks, Richard Moccia, Vaughn Ostland, Ian Smith, David Speare, Betty Wilkie, and Wolfgang Zenker.

The Fish Pathology Laboratory, Ontario Veterinary College, receives funding from several sources, including the Ontario Ministry of Agriculture and Food.

Systemic Pathology of Fish

1 Introduction

Postmortem Techniques

Sample Submission and Processing. Unlike mammals, which tend to cool down after death, thereby slowing autolysis, fish out of water often warm up. Tissues must therefore be placed into fixative as quickly as possible or artifacts can start to hinder interpretation. This is especially true with gills, and a time lag of even five minutes (not long if one is doing a necropsy!) may produce lamellar epithelial swelling and lifting from the basement membrane. Sick but live fish are probably the most useful samples to receive for diagnostic purposes. Freshly dead fish, kept cool but not frozen, are the next most useful. Ten percent buffered formalin is the fixative of choice for most situations and 24 hours on a shaker is the routine in this laboratory, prior to trimming. Bouin's, however, is especially useful for eyes and gills, for skin, or for small fish. Its demineralizing properties minimize distortion and tearing artifacts from scales and other bony structures and, in the case of small fish, permit whole-body sections without the need to dissect out (and often destroy) the various organs. Fixation for 24–48 hours, followed by storage in 70% alcohol until processing, is usually adequate. For ultrastructural work, 2–2.5% glutaraldehyde in a cacodylate or phosphate buffer with a total osmolality of roughly 320 mOsm, and pH 7.2 has been found acceptable for primary fixation under most circumstances. This is usually followed by secondary fixation with 1% osmium tetroxide in a phosphate buffer, followed by dehydration in graded ethanol prior to resin embedding.

The routine processing and staining methods used in any pathology laboratory are employed for the preparation of paraffin sections. Blood smears and tissue imprints stained with an automatic slide stainer (Hema-Tek Slide Stainer, Ames Company, Division Miles Laboratories Inc., Elkhart, IN 46514) using a polychrome methylene blue stain give good results for most purposes.

Any discussion of fixative at this stage is possibly premature because the importance of routinely employing tissue scrapes, smears, and squashes from live and anesthetized or from freshly dead fish cannot be overemphasized. This is especially true for skin and gill diseases and where protozoan or other parasitic diseases are suspect. Some of the protozoa do not survive processing very well, and there is little substitute for seeing the cell move when trying to decide if it is of host origin or not!

Blood Sampling and Euthanasia. Blood smears often provide diagnostic clues, but generally speaking they are not as useful as in mammalian work. Similarly, blood sampling for serology or clinical chemistry is not routine in many fish disease laboratories. This is sometimes a consequence of the small size of the animals (and hence small blood samples), but modern analytical equipment often requires only microliter samples, and with time, a data bank can be established, allowing useful comparisons to be made. Caudal vein is the preferred sampling site when possible because cardiac puncture may damage an organ sometimes crucial to a diagnosis. The dorsal aorta, running along the roof of the mouth, is another possible sampling site, but its use is limited to certain species, notably salmonids.

For euthanasia, severance of the spinal cord just behind the opercula is preferred to a sharp blow on the head, due to problems of intracranial hemorrhage and brain trauma with the latter method. Even spinal severance is not without problems, however, as the frequency of branchial lamellar aneurysms is greatly increased. Aqueous anesthesia therefore is the most commonly employed method of euthanasia in this laboratory. Anesthetics employed include tricaine methane sulphonate or MS-222 (1 in 1000 is lethal within a few minutes for most species), benzocaine, quinaldine, 2-phenoxyethanol, and less commonly, carbon dioxide.

Necropsy Procedure. As with mammals there are a great many suitable approaches to performing a necropsy on a fish. As has been said so often, probably the most important aspect is consistency, thereby allowing comparisons to be made from one animal to the next. Small fish up to about 4 cm may be fixed and processed whole. It is only important in these cases to remove an operculum and incise the abdominal cavity to allow proper penetration of fixative. If the fish are bigger, the following technique may be useful:

1. Fish lie on their right side (except flatfish, which are approached from the top) on a nonabsorbent surface. Similarly, avoid the use of paper towels or any other material designed to reduce slipperiness because pathogens are easily removed, as is the superficial epidermis. Carefully examine any external lesions, and make skin scrapings.
2. Make an incision just anterior to the pelvic girdle to minimize the chances of inadvertently incising the gut. In salmonids, the organ closest to this location is the spleen, but a bit of blunt dissection to gain access to the peritoneal cavity usually avoids all damage.
3. Make a midline incision progressing anteriorly to the ramus of the mandible and posteriorly to the anus. In some fish, especially old ones, a very stout pair of clippers is needed to get through the pelvic girdle.
4. Remove the uppermost abdominal body wall to expose viscera, as well as the operculum, paying particular attention to pseudobranch, thymus, and thyroid (being diffuse, the thyroid cannot easily be seen in the normal state). Examine gills and prepare gill scrapings and whole mounts. Remove arches for histopathology: unless there are obvious lesions, we routinely sample the second and fourth. *If gill disease is suspected, fix the arches as soon as possible after euthanasia.*
5. Reflect viscera and swimbladder. Remove spleen for microbiological culture if required. Perform any microbiological procedures on kidney at this stage.
6. Examine and remove samples from heart, liver, kidney, and gastrointestinal tract making sure to open stomach (if present) to check for the presence of food.
7. Remove the eye and cut off the head just behind the opercula.
8. Place the head, cut face down, on the chopping board and remove the cranium by slicing vertically downward with a strong knife to expose the brain.
9. Remove a block of skin and muscle, or several if the history suggests it. (We

routinely remove one from immediately anterior to the dorsal fin and a second from the lateral line of the still uppermost [left side] body of the fish.)

General Pathology of Fish

By comparison with mammals, fish do not under most circumstances appear to mount an especially pronounced acute inflammatory response despite possessing the cells and probably most of the systems to do so. The reasons for this are not fully understood, although it must be remembered that the acute inflammatory response exists to localize, destroy, or *dilute* the inciting agent, and inhabitants of the aquatic environment should, at least in some situations, have little trouble accomplishing the latter. Nevertheless, fish do respond, and the cardinal signs of acute inflammation (namely heat, pain, redness, swelling, and loss of function) probably are all present but to varying levels. Although there is a degree of thermoregulation in some large fish, such as tuna, accomplished by means of countercurrent heat-exchange mechanisms, small fish are ectothermic (poikilothermic), and their body temperature is determined by the surrounding water. Thus, the justification for *heat* remaining as a cardinal sign of acute inflammation may be questionable. Behavioral fever, however, has been shown to occur in fish injected with pyrogens, whereby they selected water at a higher than normal temperature. There are, however, recent contradictory studies that failed, using endotoxin and prostaglandin E_1, to confirm these observations (Marx et al. 1984). Endogenous mediators, including complement (both classical and alternative pathways), are present in fish, and although it is probable that their involvement is similar to mammals, much more work needs to be done in this area.

Fish possess neutrophils and other granulocytes, plus monocytes, lymphocytes, and thrombocytes; these are found in different proportions in the blood depending on the species, although lymphocytes are usually the most common cell (Fig. 1.1). Neutrophils are found in acute inflammation particularly where there is tissue destruction. One of the best examples of this is found in diatom-induced branchitis, in which the sharp silica shells and spines seem to cause local damage and foreign-body-type reactions (see Fig.2.9). Often, however, they are not especially prominent, possibly reflecting their relatively low (compared to mammals) numbers in the peripheral circulation. In only some species is their nucleus multilobed (notably present in salmonids and white suckers), and their distinction from macrophages in paraffin sections can therefore be difficult. They possess many of the enzyme systems seen in mammalian neutrophils, such as the oxygen-dependent ones for the generation of reactive oxygen species (e.g., H_2O_2 or superoxide), which have antimicrobial activity and are important for intracellular killing. Despite this, the neutrophils of many species of fish are, at best, only poorly phagocytic, although uptake does vary with the particle involved and with the immune status of the individual animal. This relative inactivity is especially obvious when compared with macrophages, which in the same piece of tissue may be virtually obliterated by the quantity of ingested material, whether antigenic or inert (see Fig. 4.18). Neutrophils may have a more important role to play in extracellular killing accomplished by secreting enzymes and other antimicrobial substances into the immediate vicinity of the invading particle. From a comparative standpoint, it is interesting to consider that the neutrophil's primary role may have been cast for enzyme production and transport rather than for phagocytosis and that this latter facility has been "fine-tuned" only at a later stage in phylogeny. Elucidating the precise role of the fish neutrophil in inflammation therefore remains a worthy challenge.

The formation of frank pus is not a feature of teleost inflammation, although focally intense aggregates of neutrophils are sometimes seen. Fibrinous exudates, however, are not infrequently encountered, often with neutrophilic invasion, suggesting that as in mammals, fish fibrin or its degradation products are chemotactic.

While fish often fail to mount acute inflammatory responses as impressive as those seen in mammals under similar circumstances, there should be no disappointment among pathologists over chronic, especially granulomatous, inflammation. Many bacteria, notably *Mycobacterium* spp., elicit such a response, and the extent of granulomata formation may be so extensive as to virtually obliterate the involved organ, often the reticuloendothelial (and hemopoietic) tissues of spleen and kidney, but also heart and other viscera. Macrophages, epithelioid

Fig. 1.1.

A. Blood smear from rainbow trout showing two neutrophils with nicely segmented nuclei. **B.** Neutrophil from peripheral blood showing fibrillar nature of single-granule type (*arrow*).

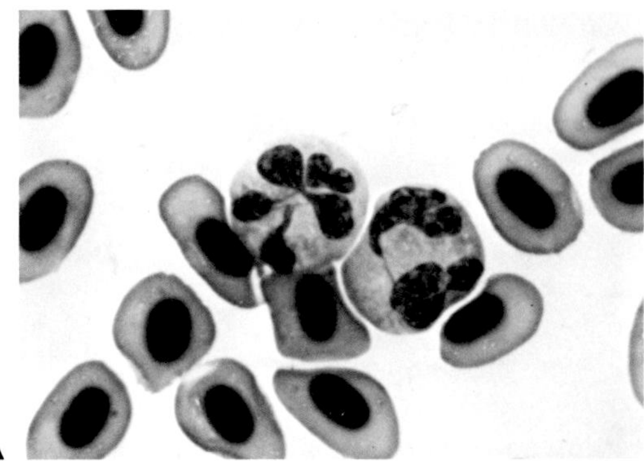

A

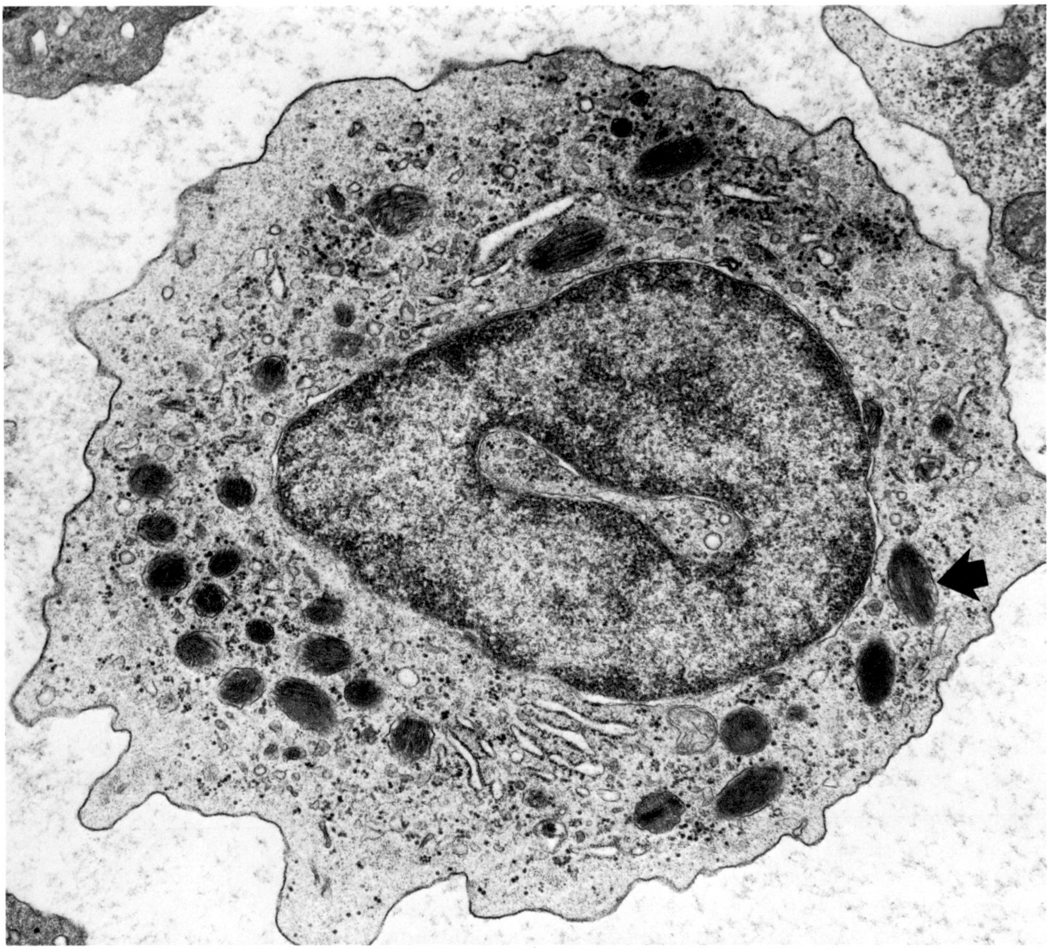

B

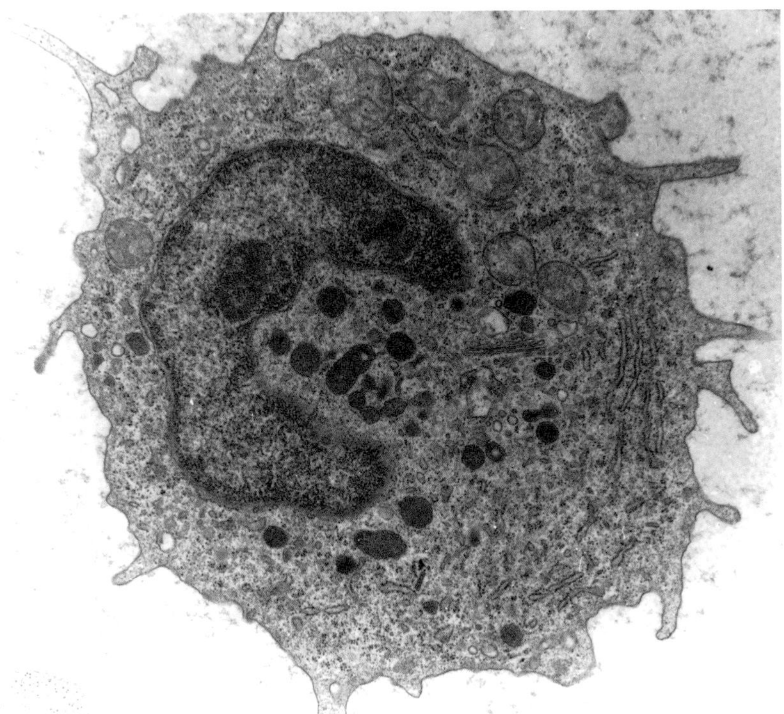
C

C. Monocyte from peripheral blood of plaice showing pseudopodia and large number of lysosomes of varying size.
D. Blood smear from trout showing several thrombocytes (*arrow*).
E. Thrombocyte showing typical crosshatched appearance of nuclear chromatin and cytoplasmic vesicles of cytocavitary network.

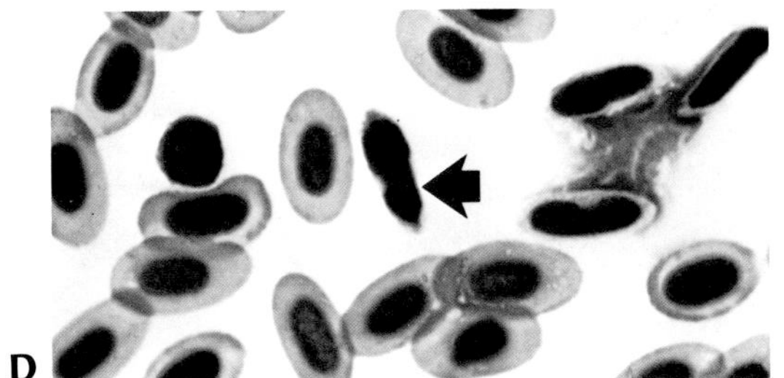
D

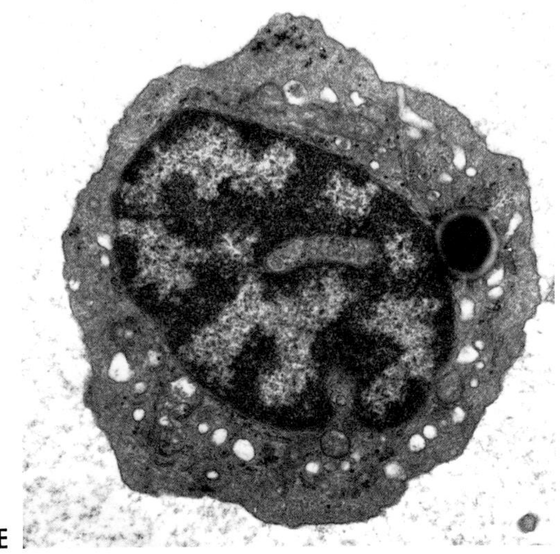
E

cells, lymphocytes, plasma cells, and fibroblasts all participate in the response. Giant cells are also found, particularly in fungal and parasitic infections and foreign-body-type reactions, but they are not common in chronic bacterial diseases; both foreign-body- and Langhans-type giant cells are found, although the latter are rarely described from other than experimental situations.

A prominent feature of chronic inflammatory responses is the presence of melanin or other pigment-containing macrophages, melanomacrophages. These cells not only form discrete aggregates (melanomacrophage centers or MMC) in spleen, kidney, and often liver, to which particulate material is sequestered, but also they may be found within the encapsulating response to many foreign bodies or parasites. This may be so extensive as to render the aggregate grossly black; hence the metacercariae of many digenetic trematodes encysted within muscle or skin achieve the popular name black spot

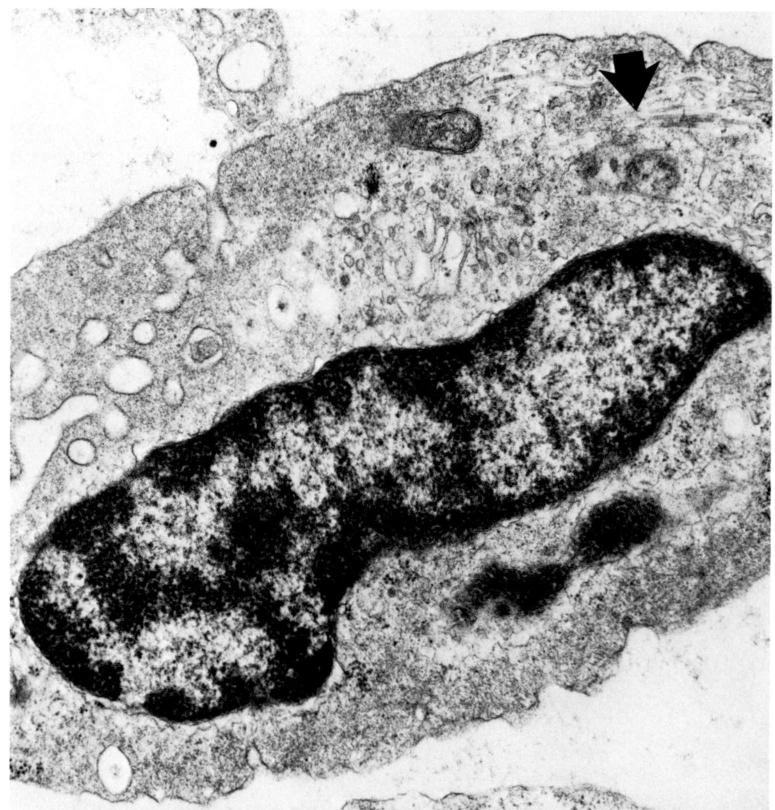

F

F. Thrombocyte showing peripheral microtubules (*arrow*). **G.** Several lymphocytes from plaice showing peripheral scant cytoplasm containing prominent mitochondria.

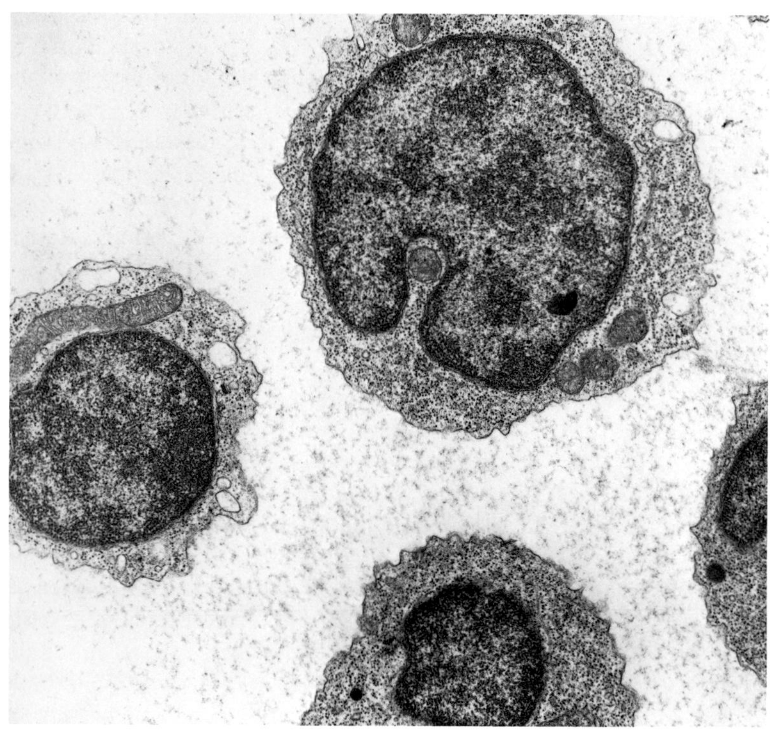

G

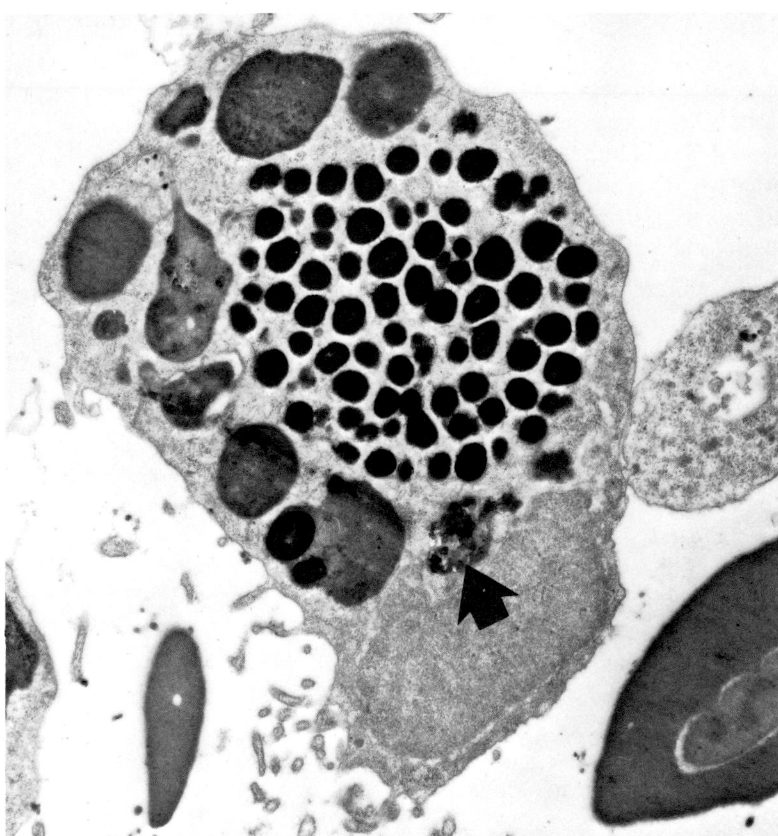

H. Melanomacrophage in circulation of fish injected with colloidal carbon, some of which may be seen within cytoplasm of cell (*arrow*).

(see Fig. 3.25). The reason for this affinity is unclear, but it is felt that the melanin may function to scavenge free radicles produced by the invader and may thereby protect surrounding host tissues from further damage. These cells are occasionally encountered in the peripheral circulation (Fig. 1.1H).

An overriding consideration when questioning the speed of the inflammatory response is the effect of water (and hence body) temperature. A drop of 10°C will roughly halve the metabolic rate of a fish. While it has been shown experimentally that this will decrease the rate of clearance of bacteria from the bloodstream, whether or not this particular aspect has any biological importance is less clear; despite operating at low temperatures, efficiency is still very high. Fish do, however, succumb rapidly to diseases, especially bacterial ones, with rising water temperatures. One of the major effects of temperature is on the specific immune responsiveness, and while a discussion on this whole aspect is beyond the scope of this book (see bibliography), suffice it to say that the lymphocyte subpopulations with B-like and T-like properties would seem to have different temperature sensitivities. The ability to respond may depend on the interaction between the replicative or enzymatic powers of the pathogen (in the case of a microorganism) and the speed of recognition, cooperation, and synthesis within the immune system. The undoubted importance of stressors (handling, crowding, and fright, to name a few) and the influence of corticosteroids on immune responsiveness is poorly understood, but it has been shown experimentally that the administration of physiological doses of cortisol (probably the single most important final common pathway for stress) to brown trout results in a marked lymphocytopenia within 36 hours (Pickering 1984).

Hyperplastic responses are well developed in fish, as is neoplasia. Classification of fish tumors is largely based on mammalian criteria, and while many exhibit histological evidence of malignancy and may show local invasion, with the exception of hepatic carcinomas, relatively few are seen to metastasize. The reasons for this fundamental biological difference from mammals are surely worth investigating. Could it be a reflection of lower immunological or other selection pressures exerted on the altered cells, or some basic enzymatic deficiency? Repair and regenerative mechanisms in fish are also very efficient, and even very severe lesions may heal with little gross evidence of scarring. Once again, however, the process is a temperature-dependent one.

References

Marx, J., R. Hilbig, and H. Rahmann. 1984. Endotoxin and prostaglandin E_1 fail to induce fever in a teleost fish. *Comp. Biochem. Physiol.* 77:483–87.

Pickering, A. D. 1984. Cortisol-induced lymphocytopenia in brown trout, *Salmo trutta* L. *Gen. Comp. Endocrinol.* 53:252–59.

2 Gills and Pseudobranchs

Gills

The gills have a very large surface area, up to ten times that of the rest of the body, and the respiratory surface on the lamellae is covered by a thin epithelium. The gills therefore have a necessarily intimate interface with their environment, providing for gaseous exchange as well as acid-base balance, osmoregulation, and the excretion of nitrogenous waste products; in teleosts this is largely in the form of ammonia. These functions are accomplished by means of Na^+/NH_4^+, Na^+/H^+, and CL^-/HCO_3^- coupled exchange mechanisms. Gill tissues derive much of their oxygen directly from the water, and together with the epidermis, which also draws directly, these tissues may account for 20% to 40% of the oxygen used by resting fish.

Normal Structure and Function. Each of the four pairs of gill arches found in most teleosts is supported by a bony skeleton (cartilaginous in young fish). From the arch, diverging rows of filaments are given off, and on both sides of each filament are located the platelike lamellae where gaseous exchange occurs (Fig. 2.1). The efficiency of exchange, which in the case of oxygen is roughly 35% to 75% (depending on species), is largely a function of the countercurrent exchange between blood and water, i.e., the blood flows in a direction opposite to that of the water. By means of this system, blood leaving the lamellae and entering the efferent filamental arteriole that runs down the outer aspect of the filament to the arch has a partial pressure of oxygen approaching that of the incoming water. Water flow over the lamellae is continuous, achieved by means of alternating the buccal pump pushing water to the gills and the opercular pump drawing it out. Some species dispense with one or both of these mechanisms relying on forward movement when swimming to pass water over the gills simply by keeping the mouth open (ram ventilation). Other species have modifications to organs such as the gastrointestinal tract or the swimbladder to facilitate gaseous exchange (accessory breathing apparati); survival in poorly oxygenated waters is thereby enhanced.

Each lamella is best regarded as a thin envelope of cells, the two surfaces of which are supported (or held together) by pillar cells (Figs. 2.2 and 2.3A). Each respiratory surface comprises a single or double layer of epithelial cells joined at their margins by sometimes extensive interdigitations and by desmosomes and tight junctions. Between these two layers is an interstitial space in which inflammatory cells are often found

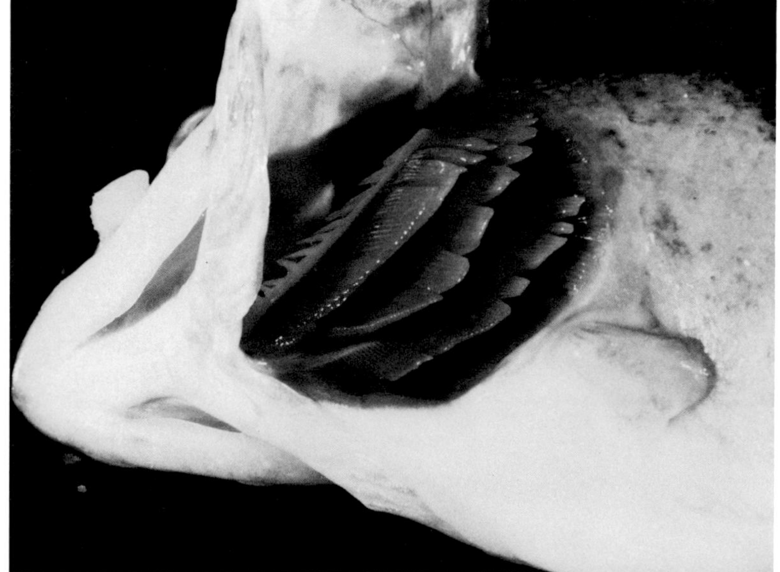
A

B

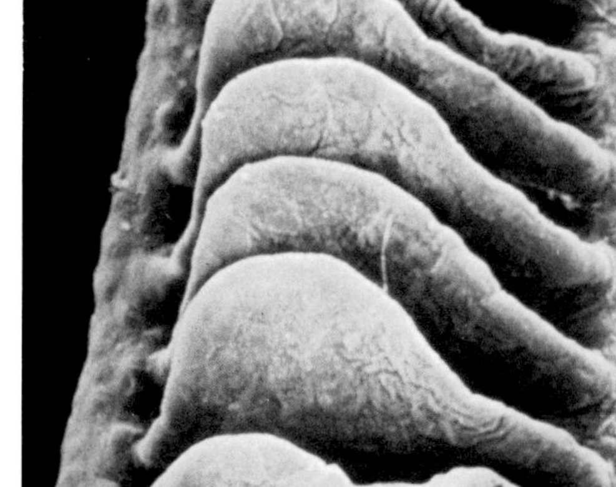
C

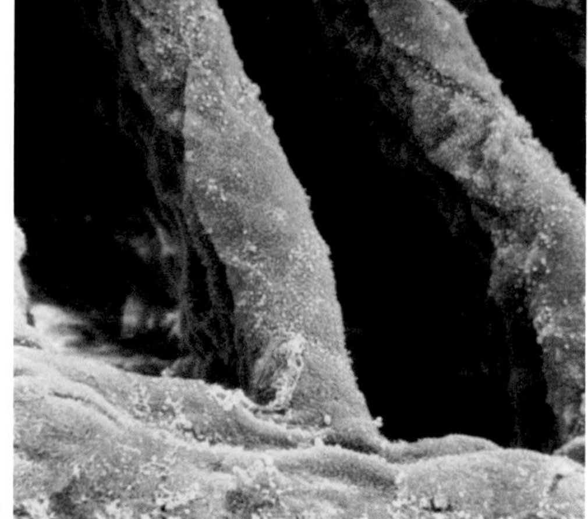
D

Fig. 2.1. A. Normal trout with operculum reflected to show gill arches beneath. **B.** Normal trout gill showing diverging rows of filaments on which lamellae are situated. **C.** Higher power showing lamellae on one side of filament. (*Courtesy of S. Goldes*) **D.** View between adjacent lamellae showing interlamellar spaces through which water flows. Note microridges on the epithelial surfaces. (*Courtesy of S. Goldes*)

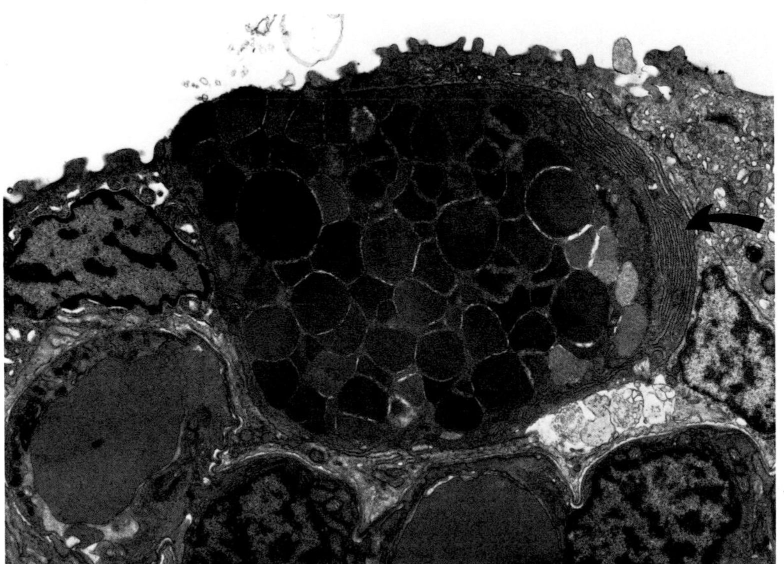

E. Lamellar mucus-producing cell (goblet cell) from rainbow trout. The nucleus is located at the base, above which may be seen numerous stacks of endoplasmic reticulum (*arrow*).

(Fig 2.3B). The inner layer of cells sits on a basement membrane that traverses the apposing inner faces of the lamella in grooves located within the pillar cells, thereby providing additional tensile support for the actomyosin fibrils found within the cytoplasm of the pillar cells. Where the ends of each pillar cell touch the basement membrane, they spread laterally to form wide flanges: thus each pillar cell is shaped like a spool. Where the flanges touch those from adjacent pillar cells, they are joined by desmosomes and tight junctions. Erythrocytes therefore percolate through the spaces created by this arrangement except at the outer margin of the lamella (the marginal channel; see Fig. 2.14D), which has an endothelial lining and may be the preferred channel for blood flow. Blood flow through lamella is pulsatile, but in hypoxia, this increase in pressure is compensated by a synchronous increase in water pressure associated with each breath; hence lamellar thickness remains the same due to synchrony of heart rate and respiration. This arrangement also facilitates the return of extracellular fluid to the central venous sinus.

The outer layer of lamellar epithelium is thrown into convoluted fingerprint-like microridges (Fig. 2.1D). It is suggested that in

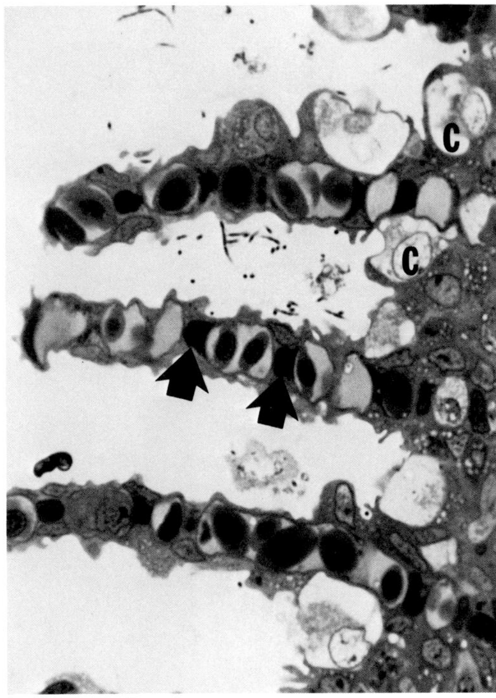

Fig. 2.2. Goldfish with bacterial gill disease. Regularly spaced pillar cells (*arrows*) have erythrocytes in the spaces between. Large pale chloride cells (*C*) may be a specific target in this fish as most are necrotic. Bacteria are present in interlamellar spaces.

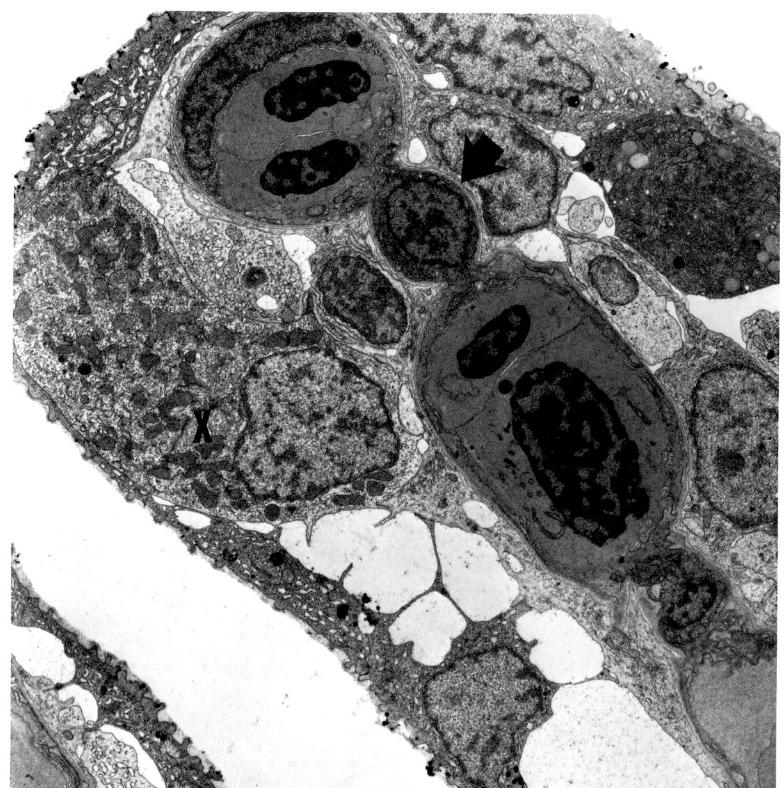

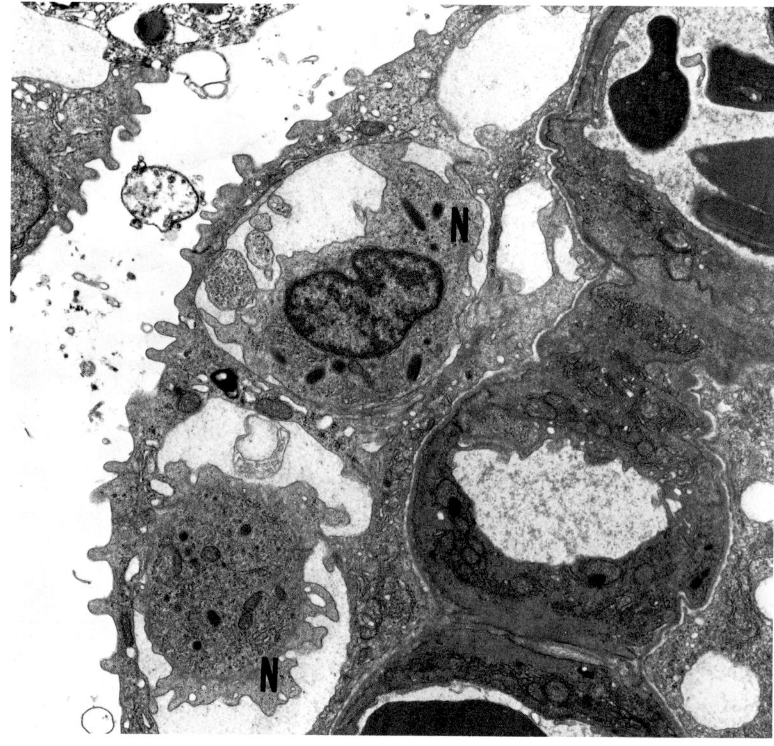

Fig. 2.3. A. Outer margin of normal trout lamella showing epithelial cells, chloride cell (*X*), and pillar cell (*arrow*). (*Courtesy of P.-Y. Daoust*) **B.** Trout with bacterial gill disease showing neutrophils (*N*) beneath epithelial cells. These inflammatory cells are identified by their fibrillar granules and other cytoplasmic characteristics. **C.** Unidentified but very characteristic "X-cells" beneath lamellar epithelium of eelpout. (*Courtesy of S. Desser*)

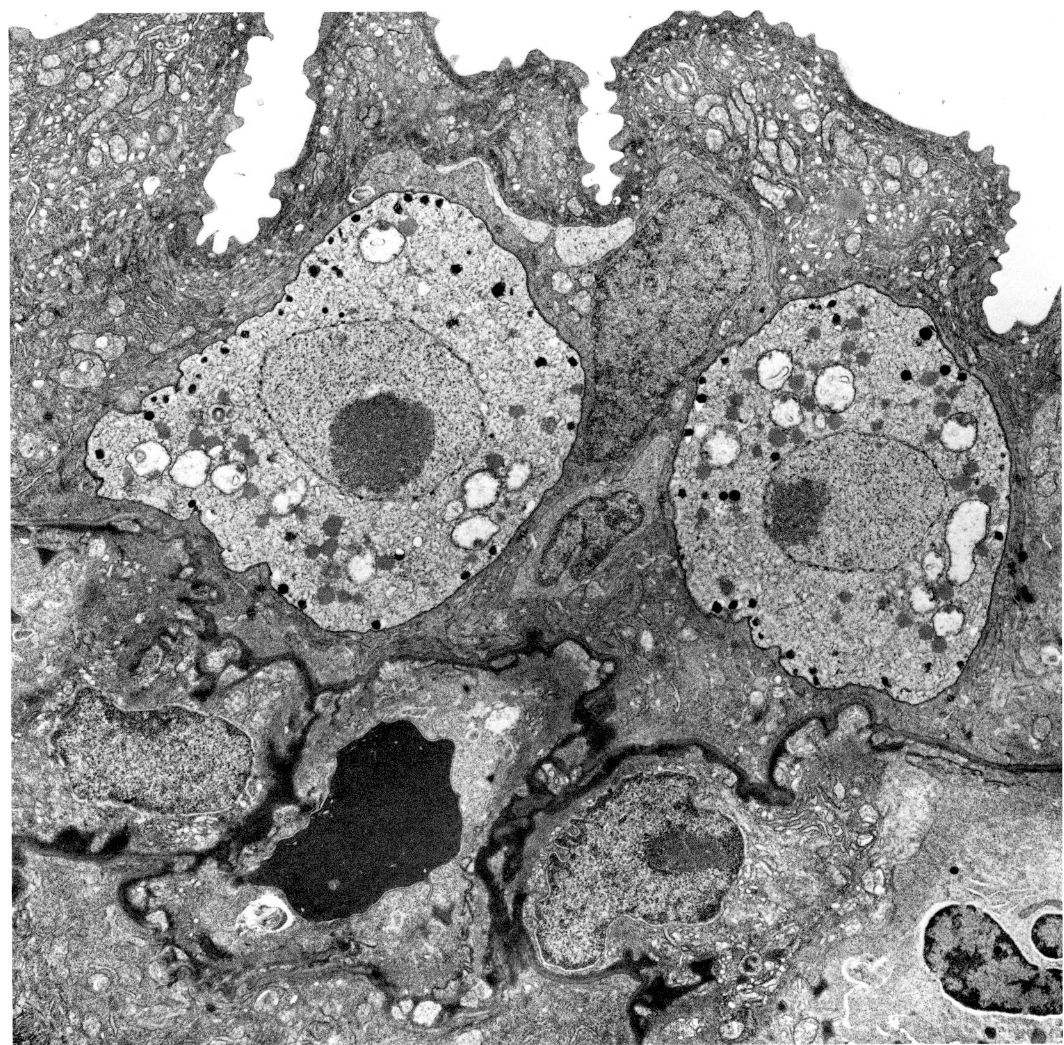

C

addition to greatly increasing the respiratory surface area, these aid in the flow and attachment of mucus. Goblet cells (Fig. 2.1E), eosinophilic osmoregulatory chloride cells, and their associated "accessory" cells are also present in the lamellar epithelium, particularly at the base of the lamellae, and are more numerous in marine than in freshwater species. Under pathological conditions these cells are often found in increasing numbers at the distal ends of the lamellae. The granular eosinophilia of chloride cells is mainly due to the presence of the large numbers of mitochondria needed for their energetically expensive activities of osmoregulation. In seawater they are responsible for the net secretion of sodium chloride.

At the base of the lamellae beneath the epithelium may be found small clumps of progenitor cells giving rise to the epithelial cells that subsequently travel up the lamellae. Transit time from base to tip of lamella in young trout at 12°C is approximately 7 to 8 days.

In the filament may be found the supporting cartilage, afferent and efferent filamental arterioles, and other anastomosing vessels comprising the central venous sinus. Filamental epithelium contains large numbers of goblet cells, especially on the leading and trailing edges and at the base of the lamellae. Production of mucus increases when the gills are irritated by pollutants or parasites. Similarly, quantitative and qualitative variations occur with other changes in water quality such as ammonia and calcium content. The negative charges of the mucus sialic acid groups effectively bind diffusible cations such as those of mercury, zinc, and copper. Thus the selective accumulation (from water) of heavy metals by the gills, reaching levels greatly exceeding those in other tissues, is explained by analysis that shows that approximately 80% of this resides initially within the mucus (Satchell 1984).

Other cells found within the filamental interstitium include variable numbers of lymphocytes, macrophages, eosinophilic granular cells, neuroepithelial cells (containing 5-hydroxytryptamine), and rodlet cells.

Rodlet cells are an enigma. They are found in very many species of teleost and within many tissues, although they are most commonly encountered within the gills, intestine, renal tubules, and bulbus arteriosus. They are viewed by some as parasites, but their distribution in such a variety of species and tissues, their presence in larval fish, and their failure to elicit any inflammatory response prompt most histologists to view them as normal cells. They have a very characteristic flask-shaped appearance with often refractile eosinophilic "rods" in their cytoplasm (see Fig. 4.15D).

While on the subject of mystery cells, it is probably worth mentioning the large ovoid pale-staining "X-cells" that have been described from gills, from pseudobranchs, and from the skin. They are sometimes found in numbers large enough to suggest neoplasia (see also Chapter 13) although once again, whether the cells are of host origin or not is unknown. In the gills, they are found at the base and along the length of the lamellae, situated between the epithelial and pillar cells (Fig. 2.3C).

Movements of branchial arches, by means of the arch muscles, alter water flow and, under situations of increased demand, thereby recruit a higher proportion of filaments, probably the more dorsal ones. Possible consequences of this differential branchial irrigation, which under conditions of low demand creates low flow "backwater" locations, include impaired removal of bacteria and other parasites. This idea is certainly borne out under practical diagnostic conditions when recovery of bacteria from dorsal locations on the arches is frequently more successful than from ventral ones.

In addition to differential irrigation, the lamellae may be differentially perfused with blood. Thus under resting conditions, only the proximal portions of the filaments may be utilized, and the ventral ones in favor of the dorsal. The consequences of this for the distribution of pathogens with differing oxygen requirements are poorly understood but are possibly important. Certainly under conditions of high demand, with optimal irrigation of the respiratory surfaces plus maximum recruitment and utilization of lamellae, there will be increased opportunity for exposure to toxicants and other pathogens. This is backed up by experiments in rainbow trout in which uptake of methylmercury from the water correlates positively with oxygen consumption. Compared to mammals therefore, those very features that facilitate efficient uptake of oxygen from a medium that contains relatively little of it also give toxicants direct and immediate access to the respiratory surfaces. (In mammals of course, alveolar and bronchiolar air provide for a modicum of buffering capacity.)

Response to Injury. Clinical evidence of gill disease is manifested in a variety of ways including increased rate of opercular movement, flaring of opercula (most easily seen from above the fish), and increased production of mucus, which can sometimes be seen streaming from the opercula or creating a foam on top of the water. Affected fish often darken, and they may gasp at the water surface. Exposure to air probably causes collapse of the lamellae and a consequent counterproductive reduction in surface area. Fish may gather at the sides of tanks or at the water inflow, where oxygen levels can be higher. Sometimes oxygen levels are in fact *lower* at the inflow than in the rest of the tank (seen with wellwater supplies), but the fish nevertheless still gather there, suggesting possibly a perceived need for flowing water rather than for oxygen per se. Alternatively, fish may congregate at the outflow, especially in cases of toxic insult (an attempt by the fish to escape). With time, there is erosion of the trailing edges of the opercula, and although the consequences of this lesion for overall efficiency of respiration are probably quite severe (reduced water pressure leading to thicker lamellae and edema are possibilities), neither the pathogenesis of the lesion nor the timing of its development is properly understood.

Correlation of clinical disease with obvious pathological change or the presence of pathogens is also poorly understood, as are the systemic consequences of gill disease. Considering the large surface area involved, the potential for rapid and clinically significant ionic and protein change following acute diffuse damage is great, and as with so much else in pathology, morphological change probably lags far behind functional impairment. Even at the later stage when there may be morphological correlates, interpretation of events is severely hampered by the flushing action of the water, which effectively and immediately removes some of the evidence in the form of cellular and other inflammatory exudates. One study that did attempt to examine the clinical results of toxic gill damage demonstrated, among other parameters, low plasma sodium, bicarbonate, and chloride as well as low pO_2 and high pCO_2; hematocrit, blood glucose, and total protein were also increased (Albassam et al. 1987). These changes suggested osmoregulatory and acid-base imbalances as well as impaired gas exchange and were considered the direct consequences of damage to lamellar epithelium and possibly chloride cells in particular.

DISTRIBUTION OF LESIONS. The delicate structure of the gills responds to injury in a variety of ways depending on the type and severity of causative agent and the length of exposure. In most cases, lesions are diffuse among arches, although specific agents will target particular cells. For example, chloride cell degeneration and necrosis are seen in nitrite and cadmium toxicity, and in some species, polyunsaturated fatty acid deficiency. Alternatively, specific zones of the gill may be targeted, as seen in the massive epithelial hyperplasia and lamellar fusion affecting the distal third of the filaments in pantothenic acid deficiency, the classical "nutritional gill disease" (Fig. 2.4). Rarely, a disease process

Fig. 2.4. Pantothenic acid deficiency or nutritional gill disease in rainbow trout. Note the extensive fusion and obliteration of normal lamellar architecture affecting distal ends of the filaments. The resulting loss of surface area and hence capacity for gaseous exchange probably had little clinically observable effect, as under resting conditions the fish would not normally perfuse the distal filaments. Are these two facts related?

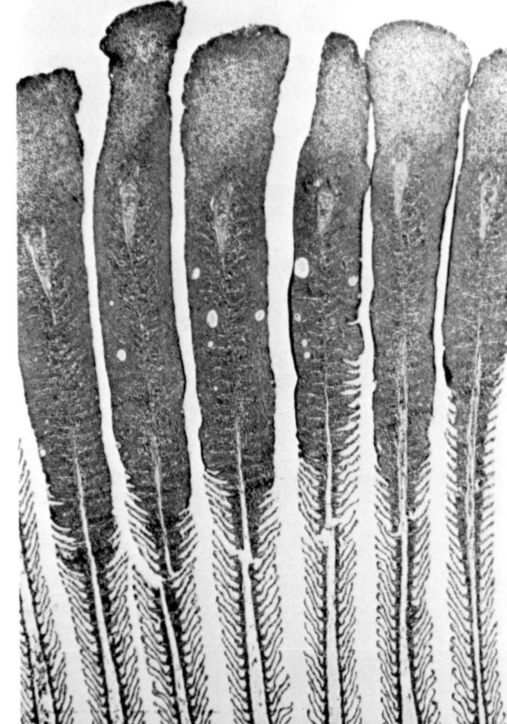

restricts itself to one arch, while adjacent arches remain totally normal. In such cases, it is much easier for the fish to compensate for the loss of respiratory surface by increased perfusion or recruitment of lamellae and increased branchial irrigation. This may be so successful that the fish does not appear to be clinically sick. In pantothenic acid deficiency, for example, there may be little need for compensation, as trout under resting conditions would not normally utilize the respiratory capacity of the distal filaments—that very area obliterated by the tissue response seen in this disease. Respiratory embarrassment only supervenes therefore if the lesions spread to involve more of the gill, when the respiratory reserve becomes exhausted and decompensation occurs. Of course, before this happens, any fish so affected will be less able to cope with additional demands for oxygen required for swimming or for metabolizing food. Overgrowth and increased numbers of lamellae at tips of filaments are long-term compensatory mechanisms noted in rainbow trout and are probably also present in other species.

ACUTE LESIONS. These are most common with chemical pollutants and include vacuolation and swelling of lamellar epithelium (Fig. 2.5); blebbing or loss of microridges; congestion, with probable alteration of vascular pathways due to shunting mechanisms; increased mucus production (often manifested in intensive culture as foaming of the water); edema, with apparent separation of epithelium from the underlying basement membrane; and an increase in the interstitial fluid spaces. There may even be frank necrosis of epithelium, with karyorrhexis and pyknosis of nuclei. It is *absolutely crucial* to appreciate, however, that virtually all these lesions can be easily reproduced as artifacts due to poor fixation. (In our laboratory we find that anesthesia followed by immediate fixation in Bouin's—no longer than 1 to 2 minutes before fixing—reduces artifact to a minimum.)

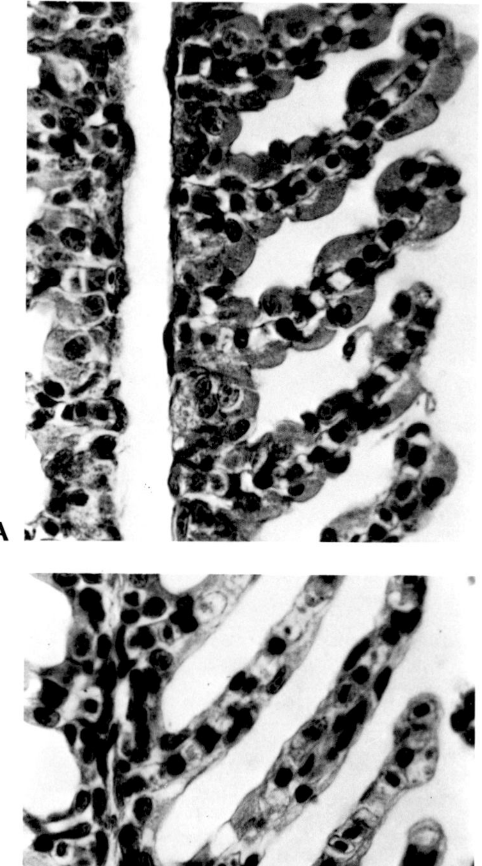

Fig. 2.5. A. Trout gills 24 hours after a 30-minute bath in 7 ppm chloramine-T. There is marked swelling of the epithelial cells and an increased cytoplasmic granularity and eosinophilia. This is so pronounced that the cells resemble chloride cells. Compare with control lamellae in **B.**

A more reliable and obvious lesion may be distortion of lamellae with protrusion of cytoplasmic processes from epithelial cells to touch adjacent lamellae (lamellar synechiae) with subsequent lamellar fusion (Figs. 2.6 and 2.21B) by means of tight junctions. These responses, which can occur within 48 hours, are common in toxicity with heavy metals such as cadmium, mercury, or copper. The mechanism(s) underlying such a response is unknown, although it is proposed that the toxicant alters the glycoprotein in the mucus covering of the cell, thereby affecting the negative charge of the epithelium and favoring adhesion to adjacent lamellae as a direct result. Under hypoxic situations, the act of gasping at the water surface, causing exposure of the lamellae to the air and hence collapse, may be another mechanism promoting fusion. Whatever the reason, consequences of such a response include interference with the flow of water over the

Fig. 2.6. A. Lamellar synechia and fusion in mercury-exposed trout. (*Courtesy of P.-Y. Daoust*) **B.** Lamella of trout exposed to mercury 24 hours previously, showing two large debris-containing cells (macrophages) just beneath the outer epithelial layer. (*Courtesy of P.-Y. Daoust*)

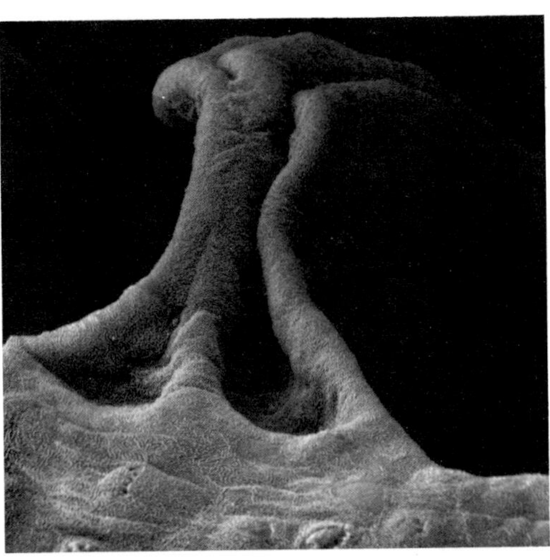

A

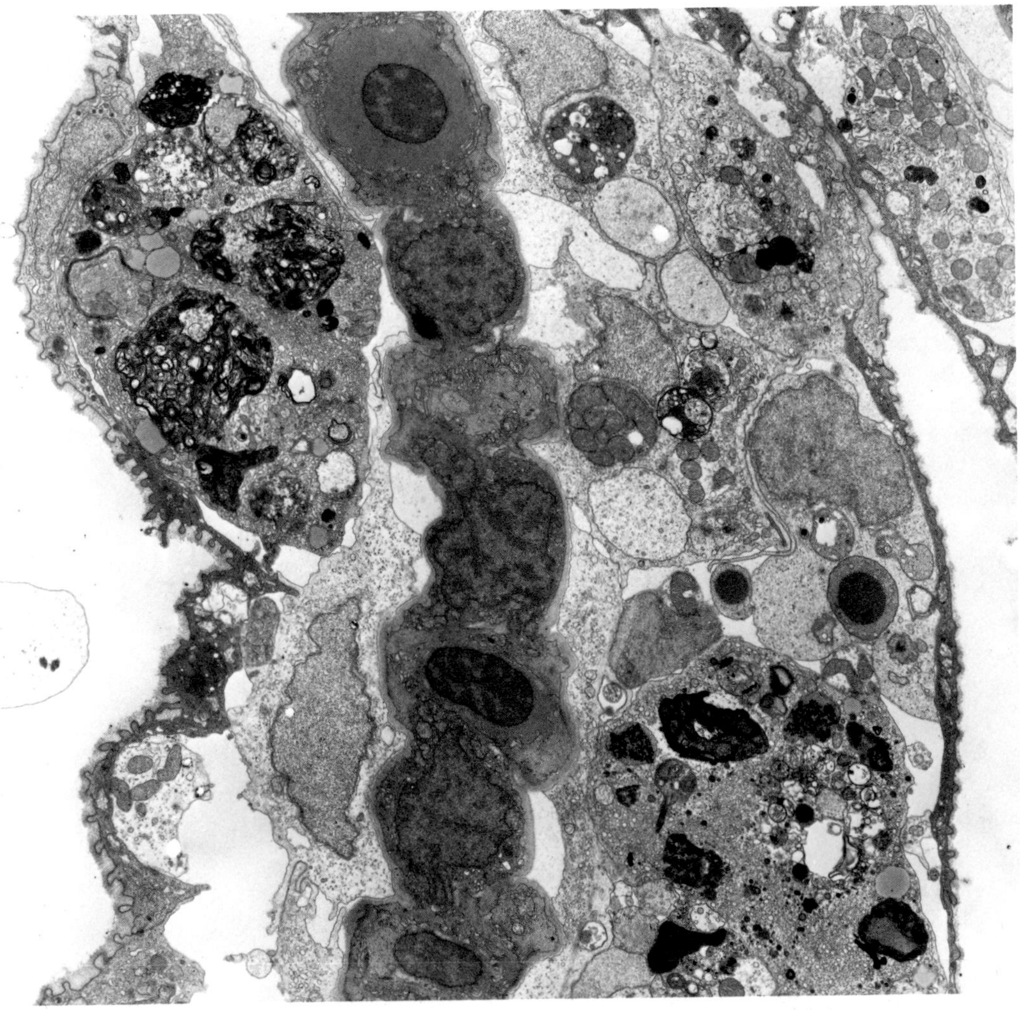

B

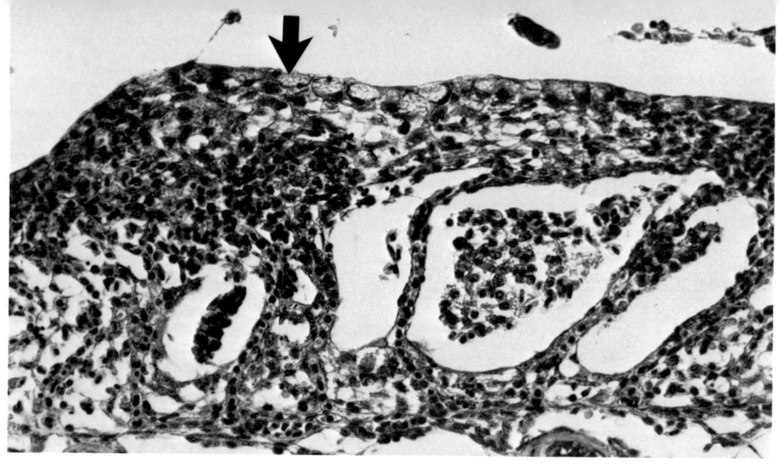

Fig. 2.7. Rainbow trout gill showing pronounced lamellar fusion and hyperplasia, with mucous metaplasia of outermost layer of cells (*arrow*). Debris, including sloughed epithelial cells, is present in the interlamellar space.

lamellar surface combined with a reduction in surface area. In addition, however, the normal route for exfoliation of epithelial cells is probably also disturbed. The piling up of these cells on the inner aspect of the "arch" created by the lamellar synechia may soon obliterate the interlamellar space, thereby compounding the gas-exchange situation even more. Alternatively, the cells may indeed slough but be trapped by the walls of the arch. Such aggregates appear as necrotic debris in the interlamellar space (Fig. 2.7). Thus, the hypercellularity so often described in association with gill disease may, at least in part, be the result of a failure of normal exfoliation, and although an increase in the rate of cell division (hyperplasia) is probably the dominant event seen under most of these circumstances, the distinction between the two processes is nevertheless a useful one to make. Hyperplasia of the progenitor zone is seen, but in addition, mitotic activity may be also encountered in epithelial cells far removed from the base of the lamellae, suggesting that they have retained their capacity for division.

Another acute response to branchial injury, associated with chemical or physical trauma (including rough handling at euthanasia), is aneurysm of lamellae (telangiectasia) in which there is breakdown of vascular integrity due to rupture of the pillar cells and a pooling of blood. Such lesions, which are easily recognized grossly as punctate red spots (Fig. 2.8), often thrombose and fibrose, followed by resorption.

A

Fig. 2.8. A. Rainbow trout gill with nitrite toxicity. Punctate hemorrhages are present throughout. **B.** Cichlid gill with severe multifocal lamellar aneurysms. **C.** Close-up showing early thrombosis in one lamella.

Hemorrhage is a feature of gill disease associated with blooms of some of the *Chaetoceros*-like diatoms and is possibly the result of physical damage caused by their spiny silicate projections. Other responses seen in this condition include a pronounced and dominating neutrophilic infiltration along with foreign-body-type reactions surrounding the organisms that have become engulfed by the hyperplastic lamellar epithelium. Giant cells may occasionally also be seen attempting to invest the silica shells or spines (Fig. 2.9).

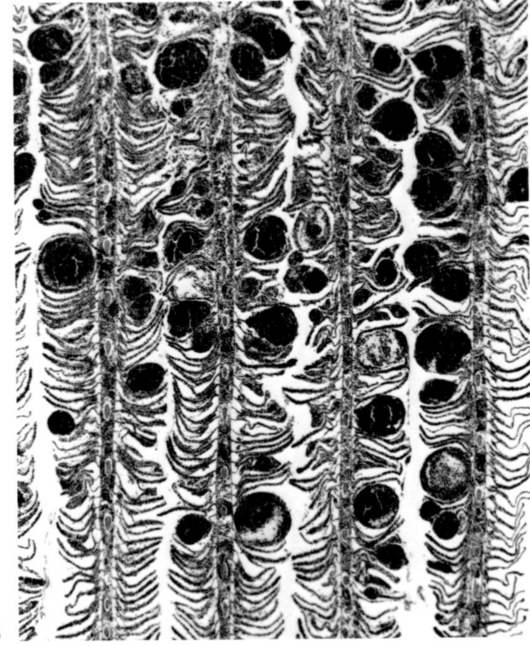
B

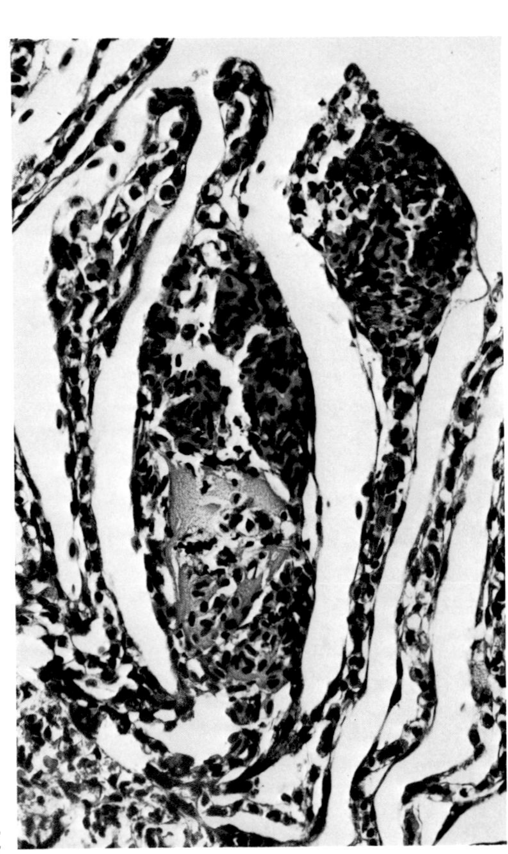
C

Chronic responses, although relatively mild ones, include goblet and chloride cell hyperplasia, often extending out onto the lamellae in large numbers. More severe changes include marked epithelial hyperplasia usually with fusion of adjacent lamellae and obliteration of the interlamellar space. Sometimes adjacent filaments fuse, and rarely this may be so severe as to involve virtually the whole arch (Fig. 2.17B). By contrast with this proliferative gill disease—the term usually employed to denote these nonspecific changes—experimentally induced chronic vitamin A toxicity causes hypoplasia and breakdown of epithelial integrity with lamellar distortion as a result. An influx of inflammatory cells (branchitis) is seen in most of the situations already described (Figs. 2.3B and 2.9), although this may be difficult to appreciate, especially in the lamellae, where they are encountered in

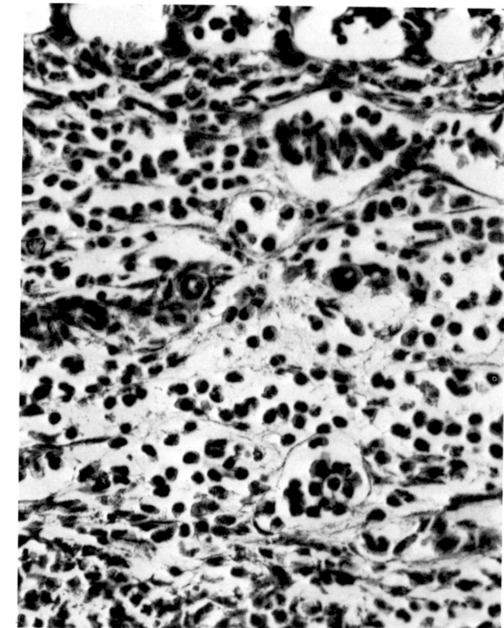

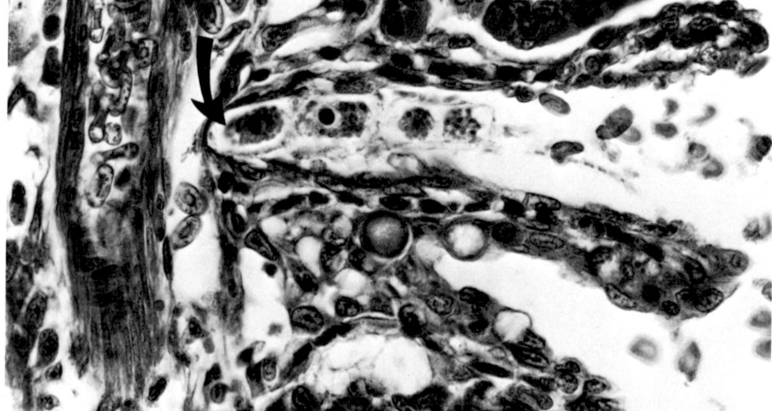

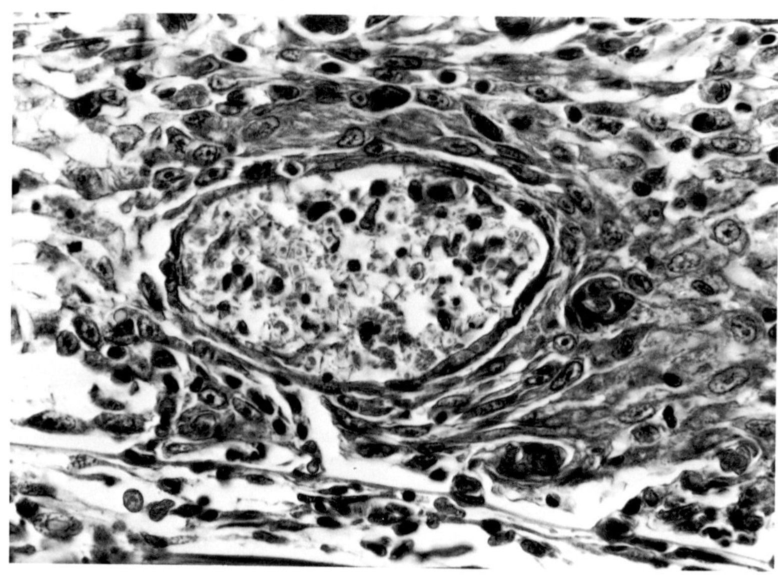

Fig. 2.9. A. Acute suppurative branchitis in rainbow trout. Large numbers of neutrophils may be seen within the central venous sinuses of the filament. This fish had a septic vegetative valvular endocarditis associated with *Lactobacillus* sp., and it is likely that there has been spread to the gills.
B–D. Suppurative and hemorrhagic branchitis associated with *Chaetoceros*-like diatom. Damage is possibly due to trauma from the sharp silica shell and spines of the organisms that become trapped between lamellae (**B,** *arrow*), and are subsequently engulfed, unable to leave the gills because of the spines, which act rather like barbs. **C.** An epithelial-lined vesicle containing mainly refractile diatom spines. This subsequently sees an influx of neutrophils and appears as a microabscess filled with numerous spines and neutrophils. **D.** A foreign-body-type response, with a giant cell (*arrow*) containing diamond-shaped refractile structures; these represent the spines cut in transverse section.

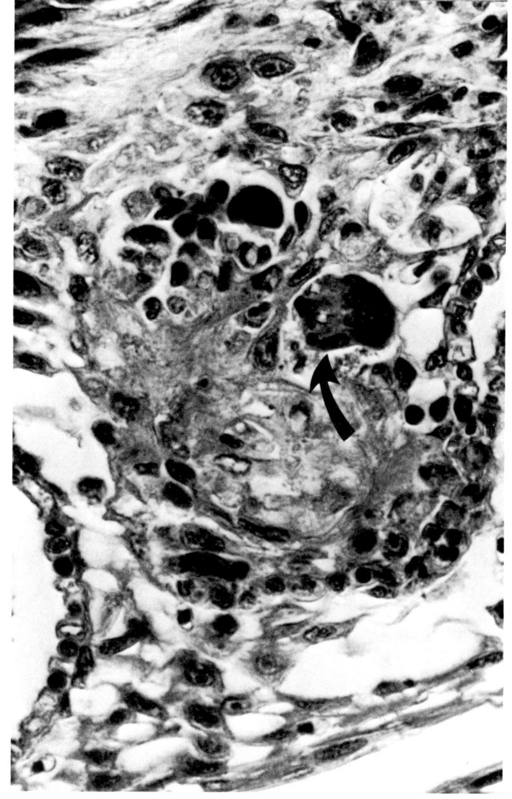

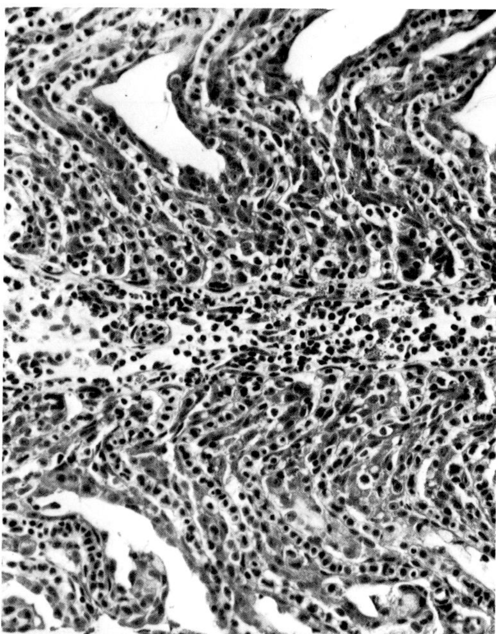

Fig. 2.10. Chronic proliferative branchitis in rainbow trout showing lamellar fusion at both distal and proximal ends and generalized influx of a mixture of inflammatory cells along filament.

the subepithelial nontissue spaces (possibly lymph spaces between the two epithelial cell layers). In chronic infections, especially parasitic disease, there may be large numbers of inflammatory cells along the filaments, including lymphocytes and macrophages (Fig. 2.10). The lamellae may also become diffusely involved in chronic granulomatous diseases such as bacterial kidney disease.

A cell type commonly encountered in many species, including salmonids, is an eosinophilic prominently granular cell, similar histologically to the cell found in the submucosa of the intestinal tract. While its identity is unknown, numbers increase in chronic disease, especially along the filaments, thereby suggesting that they are inflammatory cells (Fig. 2.11).

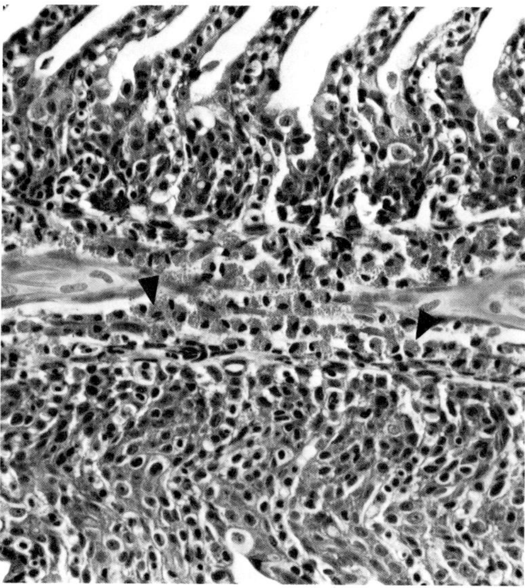

Fig. 2.11. Rainbow trout gill recovering from protozoan branchitis. Note the large numbers of eosinophilic granular cells within the filament (*arrows*).

Although intended as a protective response, any increase in thickness of the respiratory surface, due to increased quantities of mucus or due to edema, hypertrophy, or hyperplasia of the epithelium, results in a greater distance for exchange of gases and metabolites with the water. Respiratory, osmoregulatory, and excretory impairment are possible consequences, particularly if the reaction is diffuse, affecting all the gill. (It should be noted that the thickness of the gill membranes varies from species to species under normal conditions. Thus sluggish fish tend to have thicker membranes than active ones.) Lamellar hyperplasia, causing a decreased efficiency of carbon dioxide exchange, may lead to respiratory acidosis and a counterproductive reduced affinity of hemoglobin for oxygen due to the large Bohr effect seen in fish.

Mechanisms of *repair* are poorly studied although it is well known from clinical observations that even severely damaged gills can make a relatively rapid and often amazingly complete recovery. Acute responses such as epithelial lifting may recover within 7 days. More chronic changes may take several weeks to repair. Whether or not distortion and scarring occur as sequelae probably depends on the lamellar basement membrane becoming destroyed; if intact, epithelial proliferation can proceed normally from the progenitor compartment at the base of the lamellae, while damaged pillar cells probably repair by in situ division of adjacent viable cells.

Specific Conditions. Even though the gills excrete ammonia, it has long been considered that exposure to high levels of this and other metabolic waste products, as well as a variety of other nonspecific agents, may irritate the lamellae, which then respond so as to create a favorable environment for subsequent colonization by more specific pathogens such as bacteria or protozoa. The significance of un-ionized ammonia as a branchial irritant per se has been disputed recently, although there is no doubt that in high enough concentrations it can induce coma, probably by interfering with neurotransmitters in the brain. Similarly, high levels (approximately 6000 ppm) of the inert suspended solid kaolin fail to produce markedly significant change in young rainbow trout gills, although the branchial epithelium does phagocytose some of the particulate. (The uptake of soluble and particulate antigen across the gill surface is important with respect to vaccines, possibly stimulating locally enhanced as well as general immunity.)

Bacterial gill disease (BGD) occurs in several species (probably most, given the correct conditions), but it is a particularly serious problem of intensive salmonid culture and in some parts of the world represents the most common diagnosis. The precise conditions predisposing to the disease are not yet known, although overcrowding appears to be a preeminent factor. It is characterized by the presence of filamentous bacteria on the gill surface of fish suffering clinically from respiratory distress. A mixed population of bacteria is usually isolated, although the condition has been experimentally reproduced with *Flavobacterium* spp. The disease is most commonly seen in fry and fingerlings, but outbreaks do occasionally occur in populations of older fish with frequently serious results. Clinically, the fish have flared opercula (Fig. 2.12), often with erosion of the trailing edges, and the gills have clubbed filaments covered in excess mucus (Fig. 2.13). In acute cases, the gills may be hyperemic, whereas in the chronic response, when hyperplasia and mucus production are pronounced, they are usually pale. Histologically, the fine threadlike bacteria can be seen attached to the lamellar epithelium (Fig. 2.14), often accompanied by debris trapped by the mucus. With time, there is lamellar fusion with entrapment of debris, obliteration of interlamellar spaces, and, frequently, mucous metaplasia (Fig. 2.15). Severe necrosis is not usually a feature of BGD at the light micro-

A

B

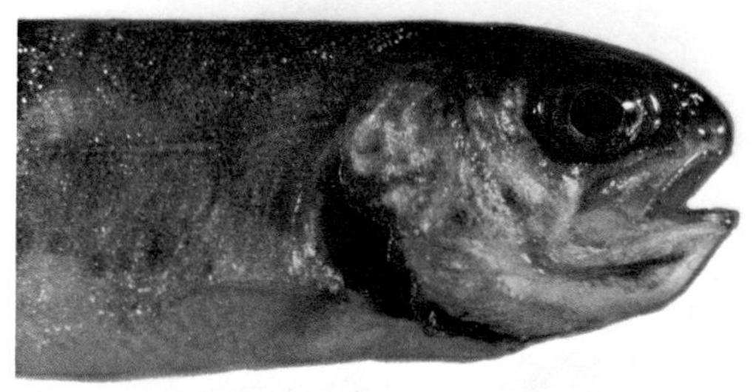

C

Fig. 2.12. Bacterial gill disease in young rainbow trout typified by flared opercula (**A** and **B**). With time, the trailing edges of the opercula become eroded (**A** and **C**) to expose the underlying gills. The full impact of impaired opercular function on respiration as a whole has yet to be properly assessed. (*Courtesy of D. Speare*)

Fig. 2.13. Pallor and glistening appearance of trout gills due to increased quantities of mucus. In intensive culture situations, production of large quantities of mucus from gills (or skin) may result in increased foaming of the water.

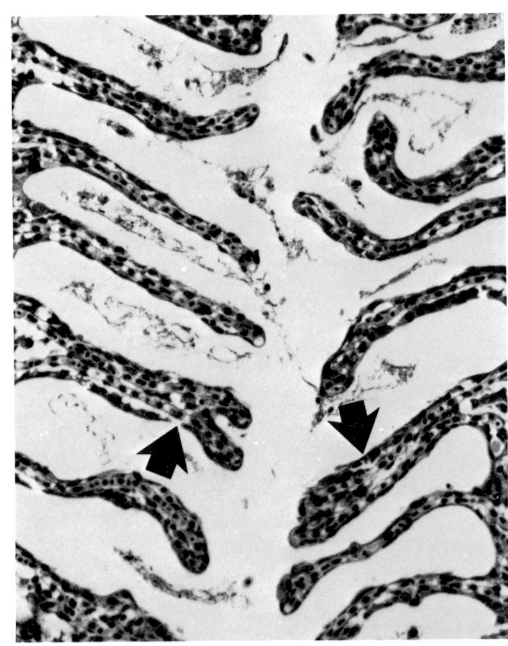

Fig. 2.14. Acute bacterial gill disease in rainbow trout. **A.** Lamellar apposition (synechiae) along full length (*arrows*). Clumps of filamentous bacteria are present in the interlamellar spaces. Their separation from the lamellar surface is partly artifactual as may be seen in **B.**
B. A scanning electron micrograph of lamellae covered with thick lacy mat of bacteria. **C** and **D.** Base and tip respectively of lamella with large numbers of bacteria on surface. Most features of lamellae are well demonstrated in these pictures, including marginal channel (*M*), chloride cells (*C*), and pillar cells (*arrows*).

A

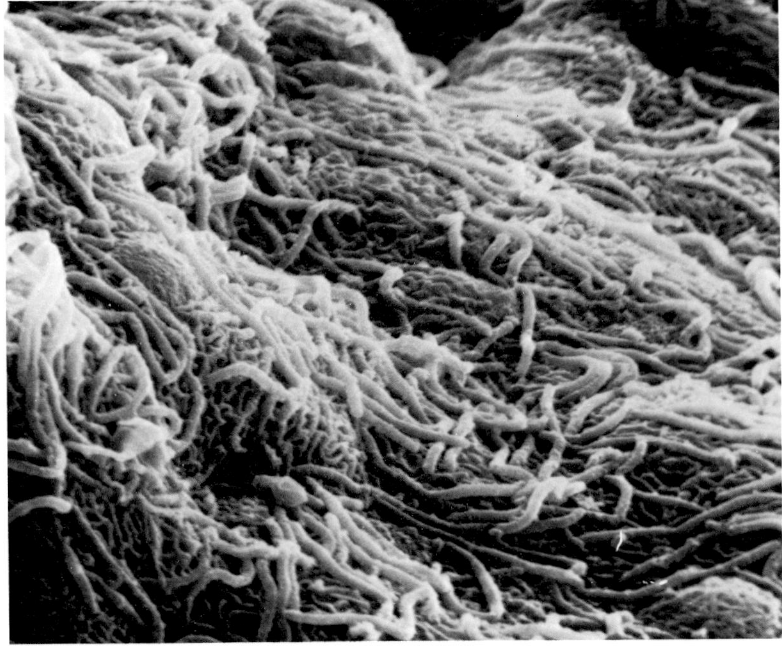

B

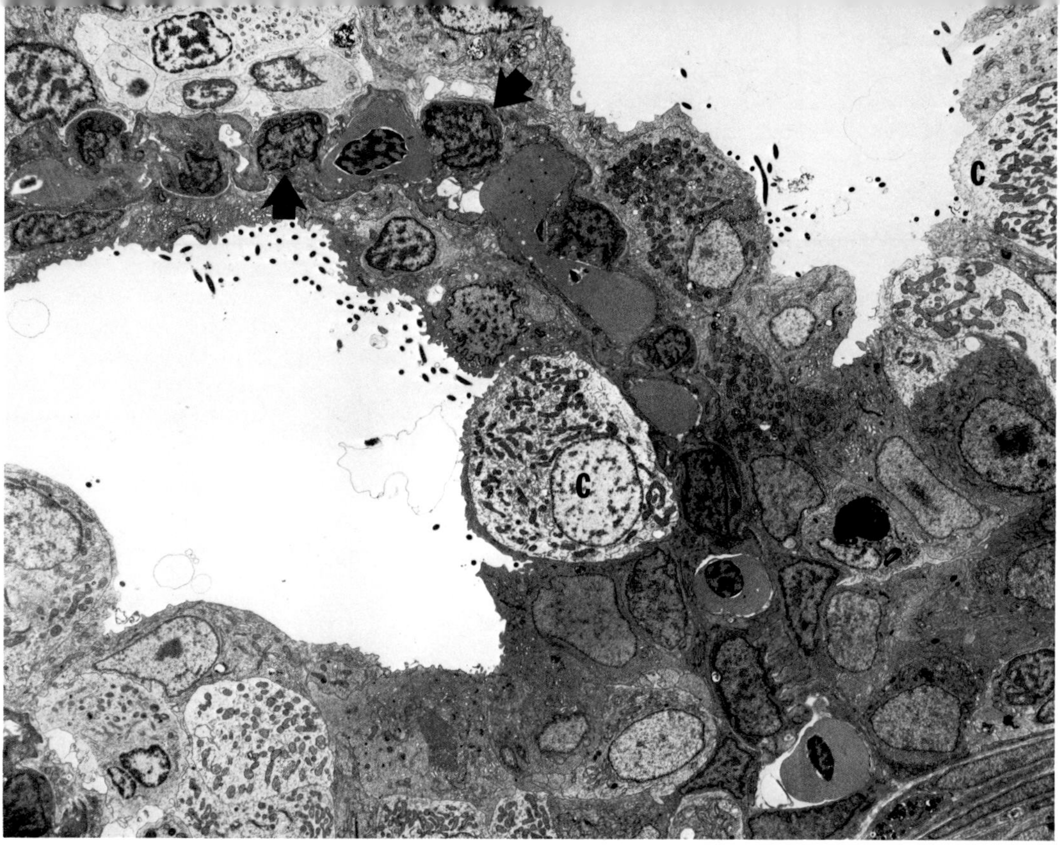

C

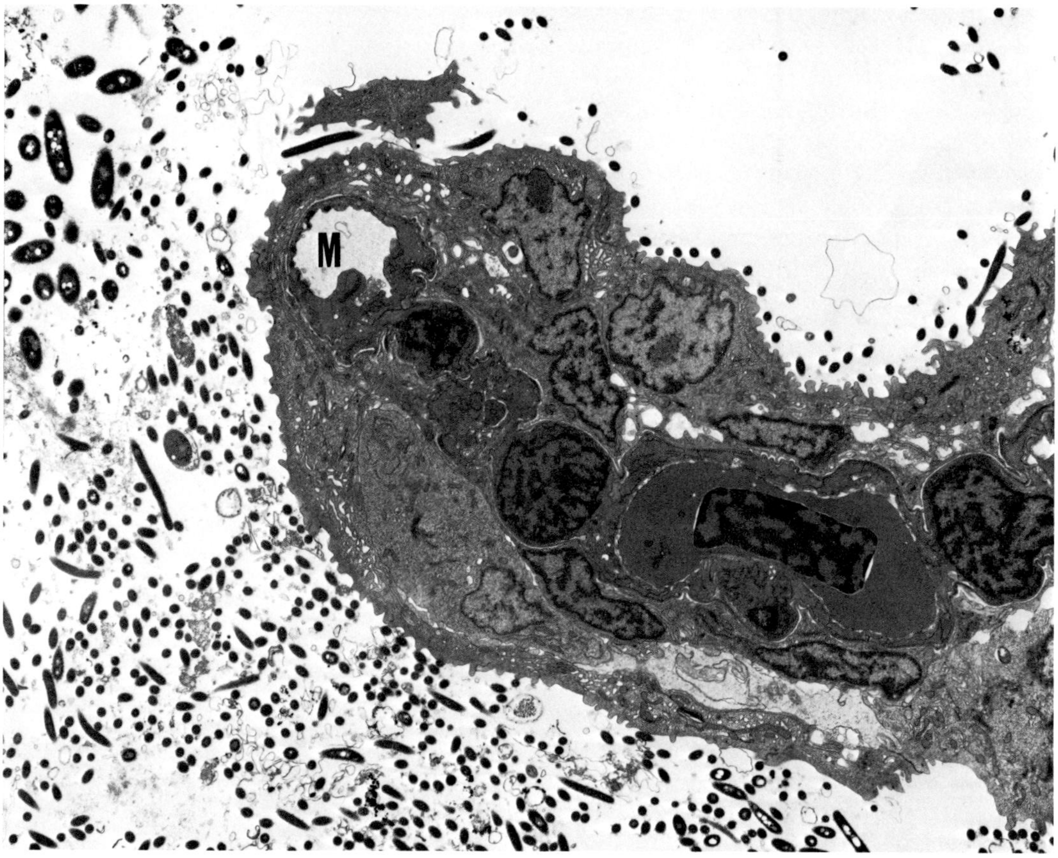

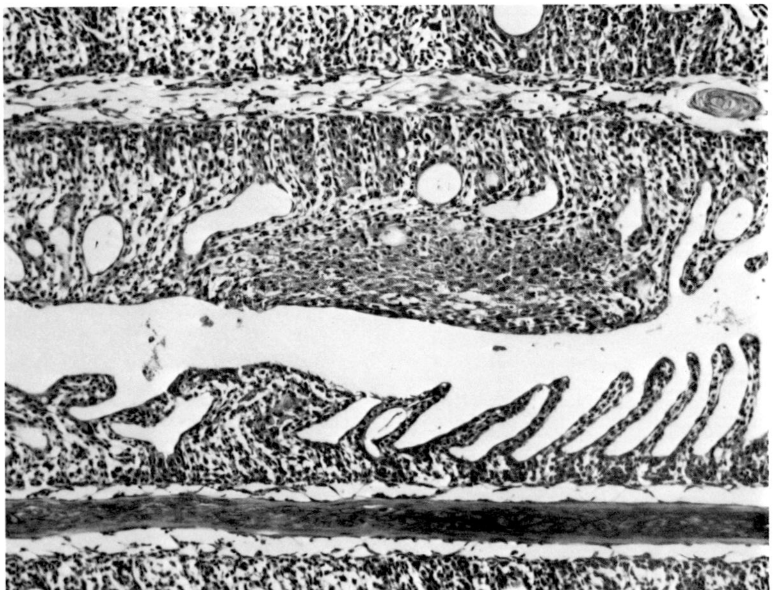

Fig. 2.15. Extensive lamellar fusion and hyperplasia, with obliteration of interlamellar spaces, as shown by this case of chronic bacterial gill disease in trout. Extensive lesions cause respiratory distress.

scopic level, although individual epithelial cells do show swelling, increased eosinophilia, or pyknosis, and, in some cases, chloride cells may be targeted (Fig. 2.2).

By contrast, columnaris disease caused by *Flexibacter columnaris* can also attack the gills (Fig. 2.16), causing severe widespread necrosis of all elements, although lesions in this condition are more commonly found on the epidermis or mouth and at temperatures above those normally associated with BGD.

Nodular gill disease is the name given to an emerging condition of salmonids and nonsalmonids in which there is massive epithelial proliferation leading often to almost total obliteration of an affected arch (Fig. 2.17) and sometimes adhesion of the arch to the operculum. The grossly visible pearly white nodular proliferations are associated with the presence of distinctive densely staining cells ("A cells") with vacuolated cytoplasm arrayed in sheets over the surface of affected tissue. The significance of these cells is unknown, and although they have been identified as amoeba, the possibility that they are transforming epithelial cells cannot be discounted. They frequently are asso-

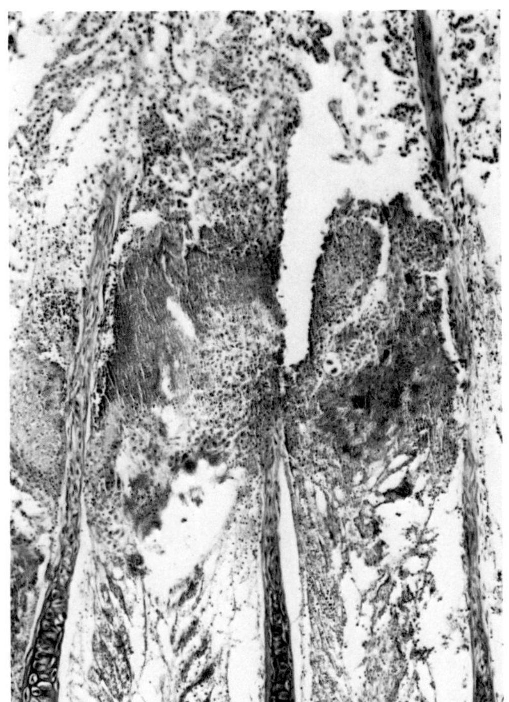

Fig. 2.16. Columnaris disease (*Flexibacter columnaris*) is more frequently associated with dermal or oral lesions. In this yellow perch, however, the gills are involved, and the necrotic lesion contains large numbers of bacteria, seen here occupying most of the central area of the picture.

A

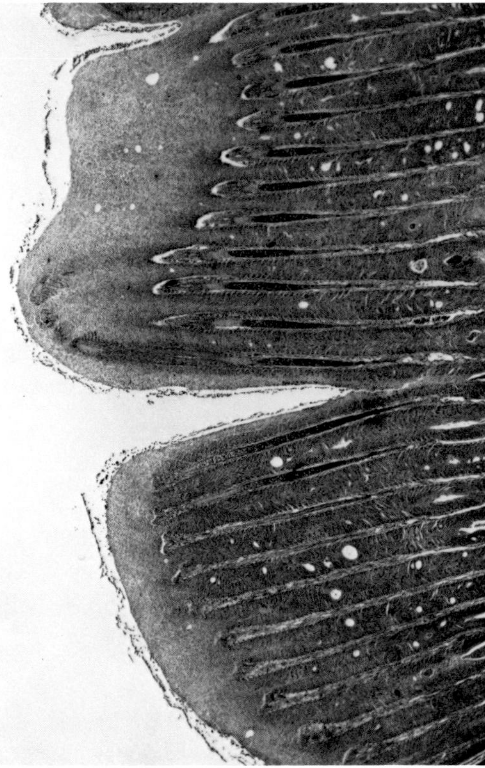

B

Fig. 2.17. Nodular gill disease of rainbow trout. **A.** Multifocal pearl gray nodular lesions. Unaffected tissue remains absolutely normal. Hence acute mortality is not high, and fish continue to eat. The disease is, however, difficult to treat. **B.** Whole arch fused into single nonfunctioning entity. The characteristic but unidentified cells ("A cells") may be seen on the surface. **C.** High power of **B.** Mitotic activity in epithelium is high, and there is a mild diffuse infiltrate of inflammatory cells.

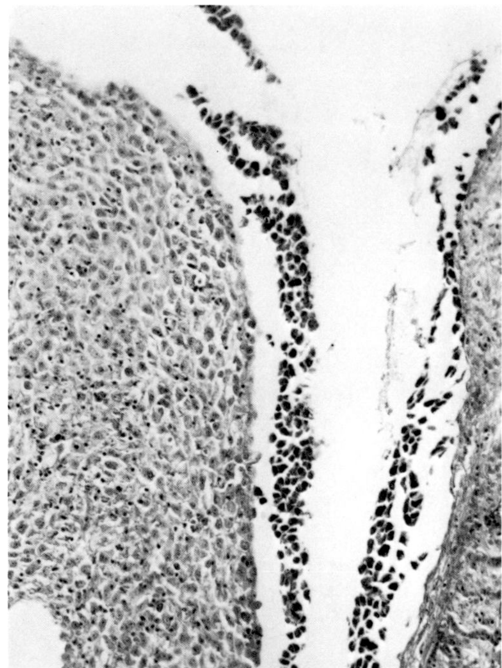

C

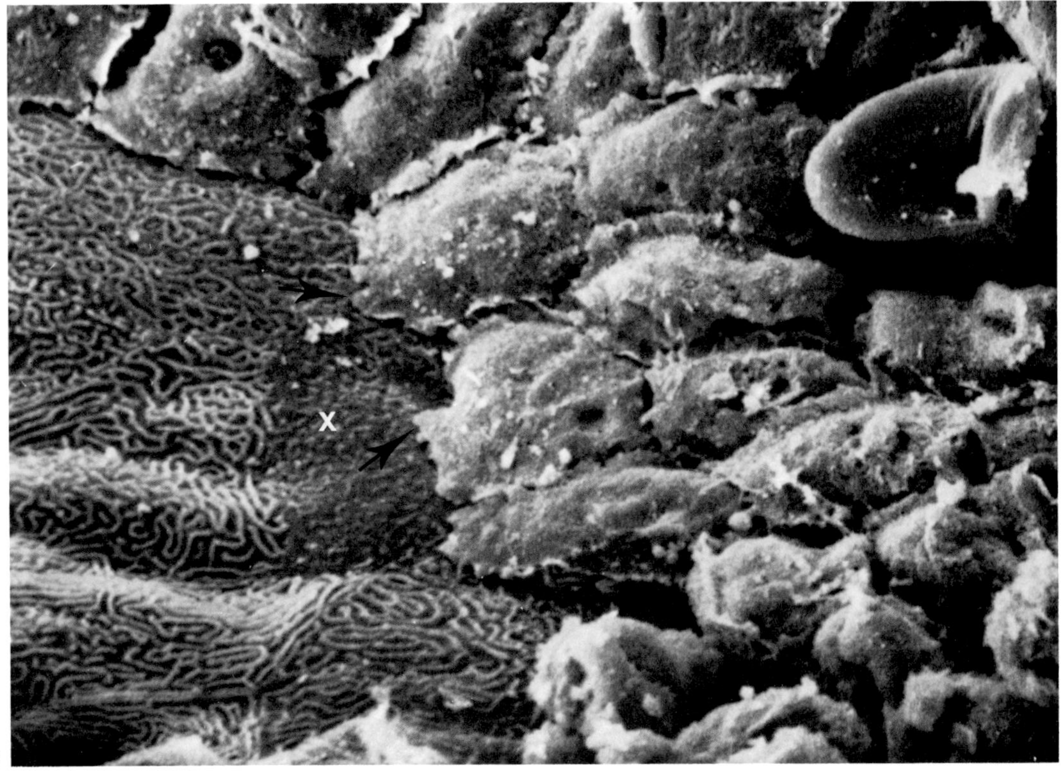

Fig. 2.17. D. Scanning electron micrographs showing a sheet of "A cells" largely covering fingerprint-like microridges of lamellar epithelium. Ruffled edges of the cells (*arrows*) appear to interdigitate with one another. In one place (*X*) it would appear that an "A cell" has been dislodged to reveal erosion of microridges beneath. **E.** Transmission electron micrograph of lamellar surface showing several "A cells" overlying an epithelial cell (*E*). Note crescent-shaped nuclear chromatin, lipid droplets (*L*), vacuoles (*V*), and filamentous coat (*FC*) on outer surface.

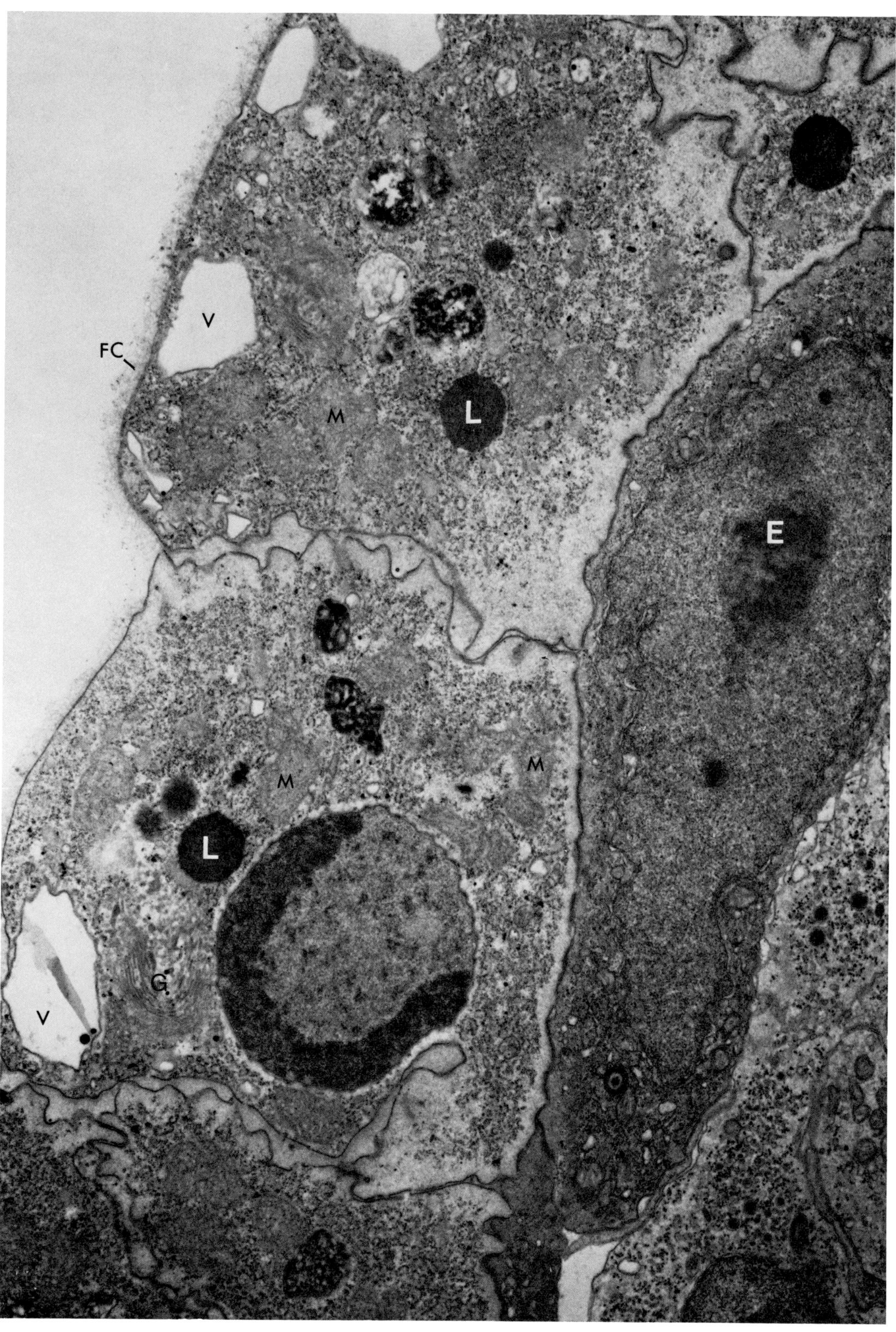
V
FC
M
L
E
M
M
L
G
V

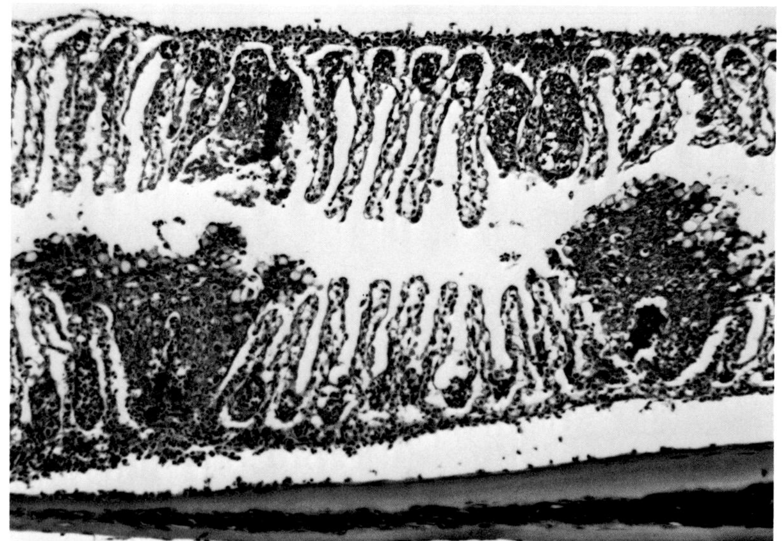

Fig. 2.18. Furunculosis in rainbow trout frequently demonstrates presence of bacterial colonies in gills, here associated with hyperplasia of surrounding lamellar epithelium. The extensive fine vascular bed of the gills provides ample opportunity for trapping septic emboli from the heart (myocardium is a target tissue in this disease; see Fig. 6.11A) or for localization during bacteremia.

ciated with and often contain filamentous bacteria and are also found elsewhere in the branchial cavity, within the nares, and even on top of the head. The precise relationship between this condition and bacterial gill disease remains to be elucidated.

Considering the extensive vascular network of the gills, it is hardly surprising to find their involvement in gas embolism ("gas bubble disease") associated with supersaturated water and in bacteremias and viremias. In furunculosis, for example, colonies of bacteria are frequently encountered within lamellae and filaments (Fig. 2.18), sometimes causing ischemic lesions. In bacterial kidney disease, the gills may contain large numbers of bacteria within pillar cells, macrophages, or epithelial cells (Fig. 2.19); the latter often may be seen exfoliating from the lamellar surface. In viral hemorrhagic septicemia (VHS) too, the pillar cells have been shown to actively phagocytose the rhabdovirus (Chilmonczyk and Monge 1980). Lymphocystis virus may also localize in the gill, especially in pleuronectids, the resulting lesion being indistinguishable from its more common skin counterpart. Cytomegaly is seen in the lamellar epithelium in herpesvirus infection of marine flatfish, and in eels there is seen a proliferative branchitis associated with a virus similar to that of infectious pancreatic necrosis (IPN) (other lesions in this latter disease include a proliferative glomerulonephritis).

Secondary fungal invasion with *Saprolegnia parasitica* (*diclina*) can be superimposed on any primary gill disease, sometimes with severe consequences due to infarction (Fig. 2.20). If the branchial cavity is extensively affected, the thymus lying at the dorsal commissure may also become in-

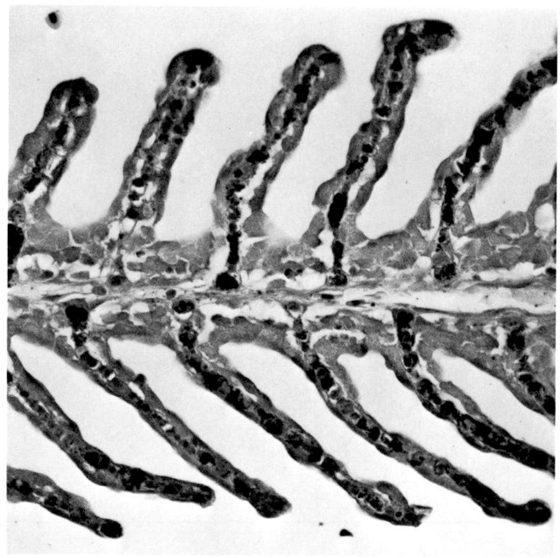

Fig. 2.19. Bacterial kidney disease in coho salmon. The gram-positive *Renibacterium* are present in large numbers throughout the gills.

Fig. 2.20. Multifocal necrotic branchitis due to mycotic invasion, in this case *Saprolegnia. Flexibacter columnaris* can appear grossly similar.

volved, with possible serious immunological consequences. This is relatively common in young fish under intensive culture conditions when feeding practices are poor. A specific mycotic syndrome of the gills, branchiomycosis or gill rot, was first described in 1912 by the German veterinarian Marianne Plehn (as were so many other fish diseases). It is particularly significant as a disease of young cultured carp in which gill lesions start as red hemorrhagic swellings, progressing to white then brown necrotic areas due to infarction from colonization of branchial vessels by the oxygen-seeking nonseptate fungus.

Epitheliocystis is a chlamydia- or rickettsia-like organism infecting the gills and skin of a variety of species in both fresh and saltwater. Although infections are usually benign (Fig 2.21A), proliferative gill lesions associated with mass mortality have been reported in carp fry. All major types of surface epithelial cells are infected including chloride and goblet cells, which become hypertrophic. Mucophilosis is a condition first described in carp by Marianne Plehn, and although once considered to be due to an alga or fungus (*Mucophilus cyprini*), it was shown by Molnar and Boros (1981) to be probably identical to epitheliocystis.

A variety of protozoa are encountered in and on the gills, including amoeba (Fig. 2.21B), coccidia, microsporidia, and myxosporea. Most myxosporea seem to evoke little response and are often encountered as incidental findings in routine diagnostic work (Fig. 2.22). A few, such as *Myxobolus exiguus,* however, are reported to cause serious lesions of gills and epidermis in fish of the genus *Mugil* on the Atlantic, Mediterranean, and Black Sea coasts. Filaments from affected fish may be virtually obliterated by the large number of cysts present. The apicomplexan, *Dermocystidium* (*Perkinsus*) is reported from a variety of commercially important species, the protozoan parasite usually causing grossly visible cysts in skin or gills and resulting in a hyperplastic and granulomatous response (systemic infections are also occasionally encountered); although often insignificant, disease and mortality may be severe. Ciliates such as *Ichthyophthirius multifiliis* (Fig. 2.23), trichodinids, and *Chilodonella* spp. are also seen, as are the dinoflagellates and flagellates such as *Ichthyobodo* (*Costia*) *necator.* If extensive, many of these infections result in diffuse hyperplasia and inflammation, which can often be severe enough to cause clinical disease. Within the gill parenchyma may be found metacercariae of digeneans, glochidia of mollusks, or the eggs of sanguinicolids.

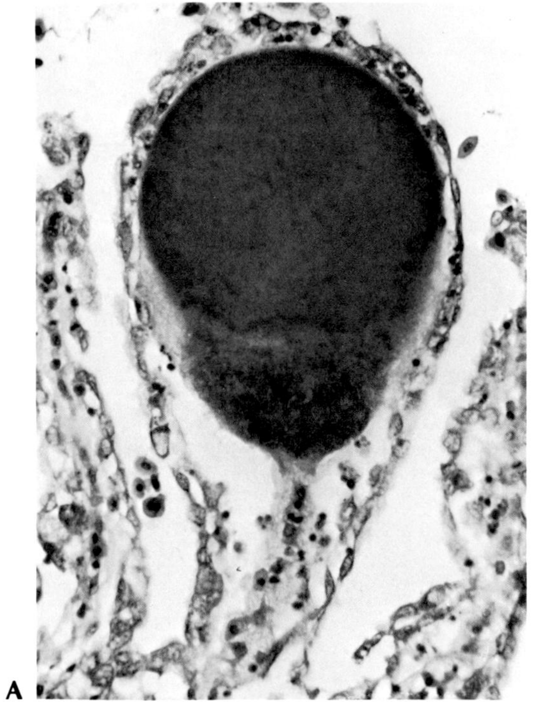

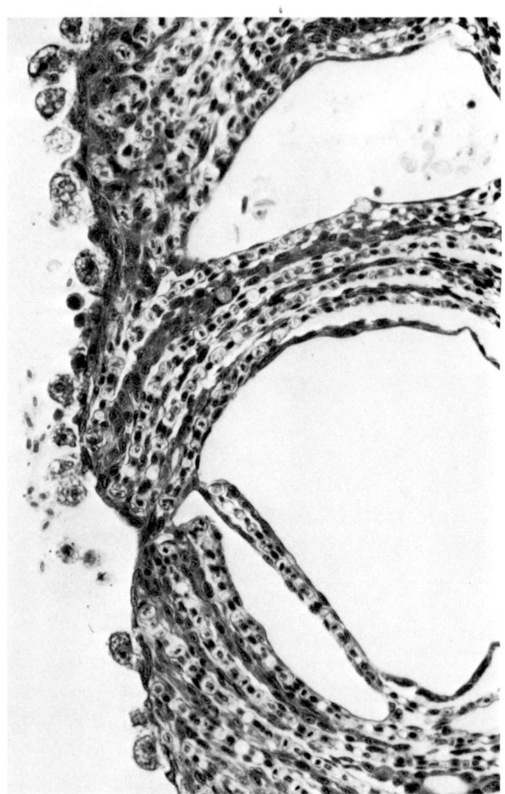

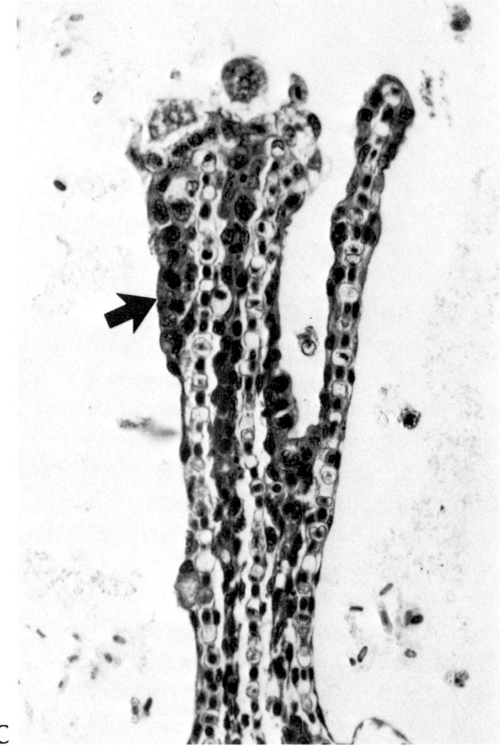

Fig. 2.21. A. Procaryote infection of goldfish, showing aggregate within lamella, typical of epitheliocystis. **B** and **C.** Parasitic branchitis in sea-caged rainbow trout associated with amoebae (*Paramoeba*). The organisms are resting on the fused and metaplastic lamellar surface. **C** shows hypertrophied epithelial cells (*arrow*). (*Material courtesy of B. Munday*)

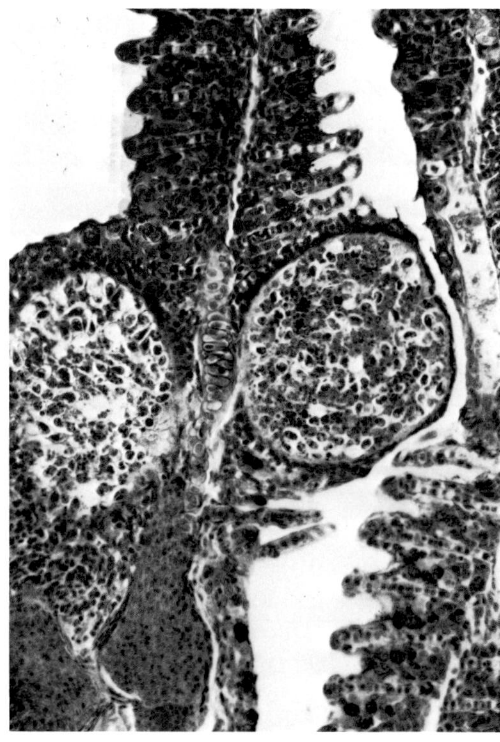

Fig. 2.22. Two myxosporean cysts within the gills of minnow, eliciting little inflammatory response.

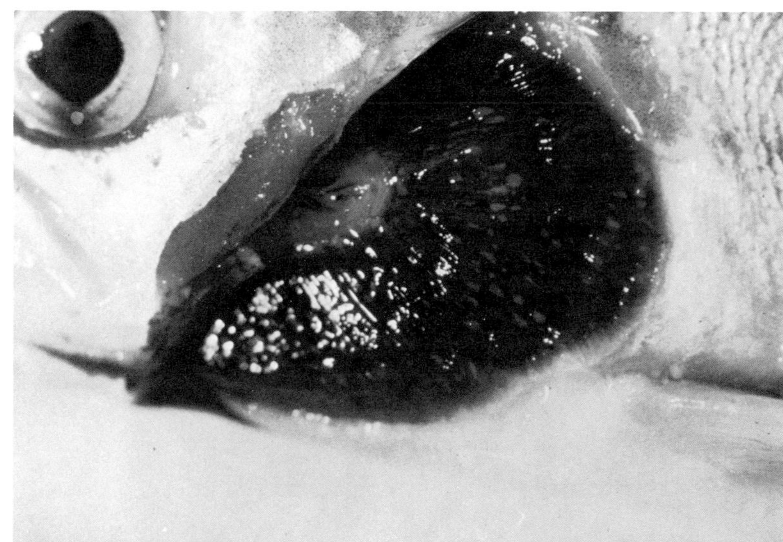

A

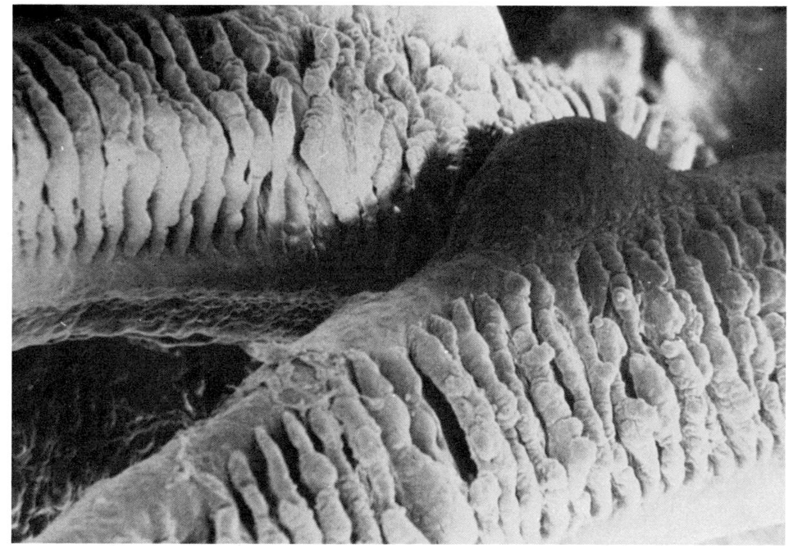

B

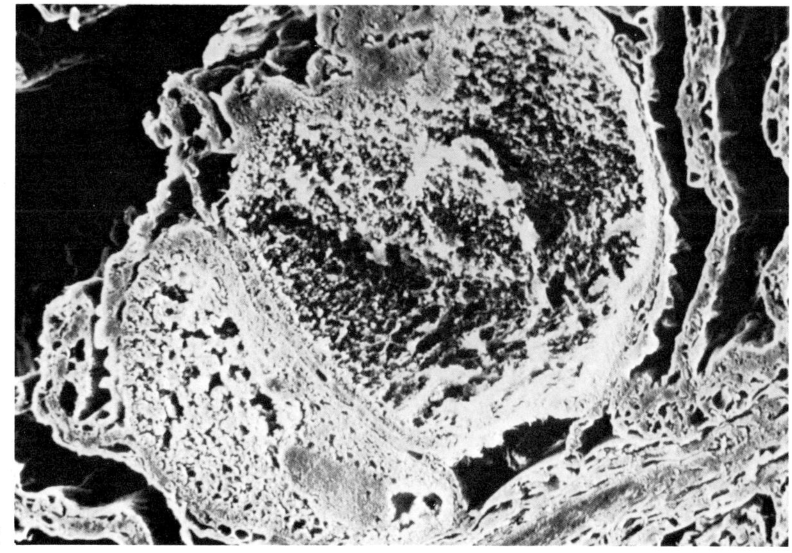

C

Fig. 2.23. *Ichthyophthirius multifiliis* in trout. **A.** Grossly visible white foci throughout gills representing parasites lying *beneath* epithelial surface. **B.** Scanning electron micrograph of **A.** **C.** Freeze-fracture preparation showing protozoa within lamellae. (*Courtesy of B. D. Hicks*)

Ectoparasitic metazoa, especially monogeneans, are also frequently encountered. Many of these may cause few detectable effects on the host except for local damage imposed by the opisthaptor (hold-fast). Large numbers, however, can cause emaciation and anemia. Gyrodactylus is reported to have a synergistic action with the ciliate Trichodina so that together they cause significant damage.

Ergasilid copepods have extrabuccal digestion that leads to lysis of tissue and hyperplasia at the point of attachment. *Lernaeocera branchialis* initially grasps the filament, causing blanching and eventual infarction. In addition to eliciting a host inflammatory response comprising lymphocytes, granulocytes, and macrophages, the sheer physical presence of these and other crustacean parasites may interfere with efficient irrigation of the gills or may cause pressure atrophy (so-called crypting).

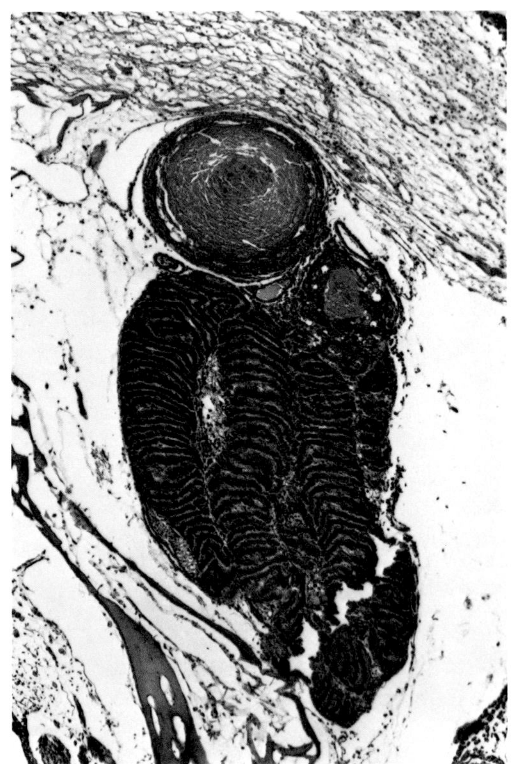

Fig. 2.24. Pseudobranch from aquarium cyprinid with mycobacteriosis showing granulomatous involvement.

The Pseudobranchs

Located on the dorsal part of the inner aspect of each operculum, these structures are a hemibranch, i.e., a gill arch with a single row of filaments, each of which supports lamellae. In some species such as cyprinids, they are deeply embedded and difficult to see grossly, whereas in others they are superficial and appear as small red fleshy elevations. In many marine species the filaments erupt through the overlying epithelium to freely contact the outside water. Only a few species lack pseudobranchs, notably some eels and silurids.

Pseudobranch epithelial cells are similar in both covered and free arrangements, however, and in both marine and freshwater species. The specific pseudobranchial cells are also similar to the chloride cells of the gills in that they possess numerous mitochondria and an extensive network of cytoplasmic tubules. It should be noted that both chloride and pseudobranch cells may be found on free pseudobranchs. The cells are located on a basal lamina with close proximity to the extensive network of blood vessels that supplies the organs.

The function of the pseudobranchs is unknown, although they are unlikely to have any respiratory value as they are already supplied by saturated arterial blood. They do have high carbonic anhydrase levels, however, and there is one theory linking them to the ocular choroid rete (choroid gland) for the purpose of supplying a well-oxygenated blood supply to the retina. There is little evidence to support the apparently attractive idea of an involvement in osmoregulation, despite the superficial similarity of pseudobranchial epithelial cells to chloride cells. Possibly the most tenable hypothesis to date is that of a sensory role.

This is suggested by the presence of a large afferent innervation reaching the brain via the glossopharyngeal nerve, which does respond to pressure changes in the opercular chamber (Laurent and Dunel-Erb 1984).

The pseudobranchs are frequently involved in bacteremias, and there can be extensive mineralization and destruction of the organs in intensively reared salmonids, often, but not invariably, associated with nephrocalcinosis (Fig. 2.24). Degeneration and necrosis of the epithelium are seen in columnaris disease and with exposure to a variety of toxicants especially herbicides, heavy metals, and pesticides. Gas bubbles are also frequently seen in the pseudobranch in conditions of supersaturation.

References

Albassam, M., J. Moore, and A. Sharma. 1987. Ultrastructural and clinicopathological studies on the toxicity of cationic acrylamide-based flocculant to rainbow trout. *Vet. Pathol.* 24:34–43.

Al-Kadhomiy, N. K. 1984. Vascular pathways in the gill filaments of the flounder, *Platichthys flesus* L. *J. Fish Biol.* 24:105–14.

Arillo, A., C. Margiocco, and F. Melodia. 1979. The gill sialic acid content as an index of environmental stress in rainbow trout, *Salmo gairdneri,* Richardson. *J. Fish Biol.* 15:405–10.

Barker, D. L., and J. E. Mills Westermann. 1986. Rodlet cells in teleost fish: some aspects of ecology and histochemistry. In *Pathology in Marine Aquaculture,* ed. C. P. Vivares, J.-R. Bonami, and E. Jaspers, 247–58. European Aquaculture Society, Special Publ. no. 9.

Bell, M. V., R. J. Henderson, B. J. S. Pirie, and J. R. Sargent. 1985. Effects of dietary polyunsaturated fatty acid deficiencies on mortality, growth and gill structure in the turbot, *Scophthalmus maximus. J. Fish Biol.* 26:181–91.

Blazer, V. S., and J. B. Gratzek. 1985. Cartilage proliferation in response to metacercarial infections of fish gills. *J. Comp. Pathol.* 95:273–80.

Bowers, A., and J. B. Alexander. 1981. Hyperosmotic infiltration: immunological demonstration of infiltrating bacteria in brown trout, *Salmo trutta* L. *J. Fish Biol.* 18:9–13.

______. 1982. *In vitro* and *in vivo* passage of bacteria across restricted areas and isolated tissues of trout, *Salmo trutta* L. and *S. gairdneri* Richardson. *J. Fish Dis.* 5:145–51.

Brittelli, M. R., H. H. Chen, and C. F. Muska. 1985. Induction of branchial (gill) neoplasms in the medaka fish (*Oryzias latipes*) by N-methyl-N′-nitro-N-nitrosoguanidine. *Cancer Res.* 45:3209–14.

Burreson, E. M., and J. P. Sypek. 1981. *Cryptobia* sp. (Mastigophora:Kinetoplastida) from the gills of marine fishes in the Chesapeake Bay. *J. Fish Dis.* 4:519–22.

Chilmonczyk, S., and D. Monge. 1980. Rainbow trout gill pillar cells: demonstration of inert particle phagocytosis and involvement in viral infection. *J. Reticuloendothel. Soc.* 28:327–32.

Coughlan, D. J., and S. P. Gloss. 1984. Early morphological development of gills in smallmouth bass (*Micropterus dolomieui*). *Can. J. Zool.* 62:951–58.

Daoust, P.-Y., and H. W. Ferguson. 1983. Gill diseases of cultured salmonids in Ontario. *Can. J. Comp. Med.* 47:358–62.

______. 1985. Nodular gill disease: a unique form of proliferative gill disease in rainbow trout, *Salmo gairdneri* Richardson. *J. Fish Dis.* 8:511–22.

______. 1986. Potential for recovery in nodular gill disease of rainbow trout, *Salmo gairdneri* Richardson. *J. Fish Dis.* 9:313–18.

Daoust, P.-Y., G. Wobeser, and J. D. Newstead. 1984. Acute pathological effects of inorganic mercury and copper in gills of rainbow trout. *Vet. Pathol.* 21:93–101.

Davis, H. S., G. L. Hoffman, and E. W. Surber. 1961. Notes on *Sanguinicola davisi* (Trematoda:Sanguinicolidae) in the gills of trout. *J. Parasitol.* 47:512–14.

Desser, S. S., and R. A. Khan. 1982. Light and electron microscope observations on pathological changes in the gills of the marine fish *Lycodes lavalaei* Vladykov and Tremblay associated with the proliferation of an unidentified cell. *J. Fish Dis.* 5:351–64.

Duhamel, G. E., M. L. Kent, N. O. Dybdal, and R. P. Hedrick. 1986. *Henneguya exilis* Kudo associated with granulomatous branchitis of channel catfish *Ictalurus punctatus* (Rafinesque). *Vet. Pathol.* 23:354–61.

Dunel-Erb, S., Y. Bailly, and P. Laurent. 1982. Neuroepithelial cells in fish gill primary lamellae. *J. Appl. Physiol.: Respir. Environ. Exercise Physiol.* 53:1342–53.

Dykova, I., and J. Lom. 1978. Histopathological changes in fish gills infected with myxosporidian parasites of the genus *Henneguya*. *J. Fish Biol.* 12:197–202.

Dykova, I., J. Lom, and G. Grupcheva. 1983. *Eimeria branchiphila* sp.nov. sporulating in the gill filaments of roach, *Rutilus rutilus* L. *J. Fish Dis.* 6:13–18.

Egidius, E. C., J. V. Johannessen, and E. Lange. 1981. Pseudobranchial tumours in Atlantic cod, *Gadus morhua* L., from the Barents Sea. *J. Fish Dis.* 4:527–32.

Farkas, J. 1985. Filamentous *Flavobacterium* sp. isolated from fish with gill diseases in cold water. *Aquaculture* 44:1–10.

Fischer-Scherl, T., and R. Hoffmann. 1986. Light- and electron-microscope studies on the pseudobranch of the golden orfe, *Leuciscus idus* L. *J. Fish Biol.* 29:699–709.

Foskett, J. K., H. A. Bern, T. E. Machen, and M. Conner. 1983. Chloride cells and the hormonal control of teleost fish osmoregulation. *J. Exp. Biol.* 106:255–81.

Gaino, E., A. Arillo, and P. Mensi. 1984. Involvement of the gill chloride cells of trout under acute nitrite intoxication. *Comp. Biochem. Physiol.* 77:611–17.

Goldes, S. A., H. W. Ferguson, P.-Y. Daoust, and R. D. Moccia. 1986. Phagocytosis of the inert suspended clay kaolin by the gills of rainbow trout, *Salmo gairdneri* Richardson. *J. Fish Dis.* 9:147–51.

Hoffman, G. L., B. Fried, and J. E. Harvey. 1985. *Sanguinicola fontinalis* sp.nov. (Digenea:Sanguinicolidae): a blood parasite of brook trout, *Salvelinus fontinalis* (Mitchill), and longnose dace, *Rhinichthys cataractae* (Valenciennes). *J. Fish Dis.* 8:529–38.

Hughes, G. M., and K. Nyholm. 1979. Ventilation in rainbow trout (*Salmo gairdneri,* Richardson) with damaged gills. *J. Fish Biol.* 14:285–88.

Hughes, G. M., and S. F. Perry. 1976. Morphometric study of trout gills: a light-microscopic method suitable for the evaluation of pollutant action. *J. Exp. Biol.* 64:447–60.

Hughes, G. M., and E. R. Weibel. 1972. Similarity of supporting tissue in fish gills and the mammalian reticuloendothelium. *J. Ultrastruct. Res.* 39:106–14.

Karges, R. G., and B. Woodward. 1984. Development of lamellar epithelial hyperplasia in gills of pantothenic acid-deficient rainbow trout, *Salmo gairdneri* Richardson. *J. Fish Biol.* 25:57–62.

Karlsson, L. 1983. Gill morphology in the zebrafish, *Brachydanio rerio* (Hamilton-Buchanan). *J. Fish Biol.* 23: 511–24.

Karlsson-Norrgren, L., P. Runn, C. Haux, and L. Forlin. 1985. Cadmium-induced changes in gill morphology of zebrafish, *Brachydanio rerio* (Hamilton-Buchanan), and rainbow trout, *Salmo gairdneri* Richardson. *J. Fish Biol.* 27:81–95.

Karlsson-Norrgren, L., I. Bjorklund, O. Ljungberg, and P. Runn. 1986. Acid water and aluminium exposure: experimentally induced gill lesions in brown trout, *Salmo trutta* L. *J. Fish Dis.* 9: 11–25.

Kudo, S., and N. Kimura. 1983. Transmission electron microscopic studies on bacterial gill disease in rainbow trout fingerlings. *Jpn. J. Ichthyol.* 30:247–60.

Kumaraguru, A. K., F. W. H. Beamish, and H. W. Ferguson. 1982. Direct and circulatory paths of Permethrin (NRDC-143) causing histopathological changes in the gills of rainbow trout, *Salmo gairdneri* Richardson. *J. Fish Biol.* 20:87–91.

Langdon, J. S., J. E. Thorpe, and R. J. Roberts. 1984. Effects of cortisol and ACTH on gill Na^+/K^+=ATPase, SDH and chloride cells in juvenile Atlantic salmon *Salmo salar* L. *Comp. Biochem. Physiol.* 77:9–12.

Langdon, J. S., N. Gudkovs, J. D. Humphrey, and E. C. Saxon. 1985. Deaths in Australian freshwater fishes associated with *Chilodonella hexasticha* infection. *Aust. Vet. J.* 62:409–13.

Laurent, P., and S. Dunel-Erb. 1984. The pseudobranch: morphology and function. In *Fish Physiology,* vol. 10B, ed. W. S. Hoar and D. J. Randall, 285–323. Orlando, Fla.: Academic Press.

Liu, K.-C., L.-M. Mai, and C.-H. Chien. 1985. Electron microscopic study of the reproductive stages of the gill rot pathogen, *Branchiomyces* sp., of fish. *Fish Pathol.* 20:267–72.

Mallatt, J. 1985. Fish gill structural changes induced by toxicants and other irritants: a statistical review. *Can. J. Fish. Aquat. Sci.* 42:630–48.

Mathias, J. A., and J. Barica. 1985. Gas supersaturation as a cause of early spring mortality of stocked trout. *Can. J. Fish. Aquat. Sci.* 42:268–79.

Mattey, D. L., M. Morgan, and D. E. Wright. 1980. A scanning electron microscope study of the pseudobranchs of two marine teleosts. *J. Fish. Biol.* 16:331–43.

Meakins, R. H., and J. Kawooya. 1973. The effects and distribution of metacercaria on the gills of the fish *Tilapia zilli* (Gervais) 1848. *Z. Parasitenkd.* 43:25–31.

Molnar, K. 1972. Studies on gill parasitosis of the grasscarp (*Ctenopharyngodon idella*) caused by *Dactylogyrus lamellatus* Achmerov, 1952. IV. Histopathological changes. *Acta Vet. Acad. Sci. Hung.* 22:9–24.

______. 1979. Gill sphaerosporosis in the common carp and grasscarp. *Acta Vet. Acad. Sci. Hung.* 27:99–113.

Molnar, K., and G. Boros. 1981. A light and electron microscopic study of the agent of carp mucophilosis. *J. Fish Dis.* 4:325–34.

Morrison, C. M. 1979. A dense cell in the epithelium of the gill lamellae of the brook trout, *Salvelinus fontinalis* (Mitchill). *J. Fish Biol.* 15:601–5.

Morrison, C. M., and V. Marryatt. 1986. Further observations on *Loma morhua* Morrison and Sprague, 1981. *J. Fish Dis.* 9:63–67.

Morrison, C. M., and V. Sprague. 1981. Electron microscopical study of a new genus and new species of microsporida in the gills of Atlantic cod *Gadus morhua* L. *J. Fish Dis.* 4:15–32.

______. 1981. Microsporidian parasites in the gills of salmonid fishes. *J. Fish Dis.* 4:371–86.

Mulcahy, D., R. J. Pascho, and C. K. Jenes. 1983. Detection of infectious haematopoietic necrosis virus in river water and demonstration of waterborne transmission. *J. Fish Dis.* 6:321–30.

Munshi, J. S. D., and G. M. Hughes. 1981. Gross and fine structure of the pseudobranch of the climbing perch, *Anabas testudineus* (Bloch). *J. Fish Biol.* 19:427–38.

Naito, N., and H. Ishikawa. 1980. Reconstruction of the gill from single-cell suspensions of the eel, *Anguilla japonica*. *Am. J. Physiol.* 238:R165–R170.

Neukirch, M. 1984. An experimental study of the entry and multiplication of viral haemorrhagic septicaemia virus in rainbow trout, *Salmo gairdneri* Richardson, after waterborne infection. *J. Fish Dis.* 7:231–34.

Nieto, T. P., A. E. Toranzo, and J. L. Barja. 1984. Comparison between the bacterial flora associated with fingerling rainbow trout cultured in two different hatcheries in the northwest of Spain. *Aquaculture* 42:193–206.

Niimi, A. J., and S. L. Morgan. 1980. Morphometric examination of the gills of walleye, *Stizostedion vitreum vitreum* (Mitchill) and rainbow trout, *Salmo gairdneri* Richardson. *J. Fish Biol.* 16:685–92.

Oronsaye, J. A. O., and A. E. Brafield. 1984. The effect of dissolved cadmium on the chloride cells of the gills of the stickleback, *Gasterosteus aculeatus* L. *J. Fish Biol.* 25:253–58.

Ototake, M., and H. Wakabayashi. 1985. Characteristics of extracellular products of *Flavobacterium* sp., a pathogen of bacterial gill disease. *Fish Pathol.* 20:167–71.

Paperna, I., and A. P. Alves De Matos. 1984. The developmental cycle of epitheliocystis in carp, *Cyprinus carpio* L. *J. Fish Dis.* 7:137–47.

Paperna, I., and J. G. Van As. 1983. The pathology of *Chilodonella hexasticha* (Kiernik). Infections in cichlid fishes. *J. Fish Biol.* 23:441–50.

Part, P., O. Svanberg, and E. Bergstrom. 1985. The influence of surfactants on gill physiology and cadmium uptake in perfused rainbow trout gills. *Ecotox. Environ. Safety* 9:134–44.

Peters, G., and L. Q. Hong. 1985. Gill structure and blood electrolyte levels of European eels under stress. In *Fish and Shellfish Pathology,* ed. A. E. Ellis. London: Academic Press.

Peters, G., R. Hoffmann, and H. Klinger. 1984. Environment-induced gill disease of cultured rainbow trout (*Salmo gairdneri*). *Aquaculture* 38:105–26.

Philpott, C. W. 1980. Tubular system membranes of teleost chloride cells: osmotic response and transport sites. *Am. J. Physiol.* 238:R171–R184.

Pic, P., and P. Lahitette. 1981. Effects of cytochalasin B on water, Na^+ and Cl^- exchanges in the gill of seawater adapted mullet *Mugil capito*. *J. Comp. Physiol.* 141:523–29.

Poston, H. A., and J. W. Page. 1982. Gross and histological signs of dietary deficiencies of biotin and pantothenic acid in Lake trout, *Salvelinus namaycush*. *Cornell Vet.* 72:242–61.

Randall, D. J., S. F. Perry, and T. A. Heming. 1982. Gas transfer and acid/base regulation in salmonids. *Comp. Biochem. Physiol.* 73B:93–103.

Roberts, R. J., A. M. Bullock, M. Turner, K. Jones, and P. Tett. 1983. Mortalities of *Salmo gairdneri* exposed to cultures of *Gyrodinium aureolum*. *J. Mar. Biol. Ass. U.K.* 63:741–43.

Rodgers, L. J., and J. B. Burke. 1981. Seasonal variation in the prevalence of "red spot" disease in estuarine fish with particular reference to the sea mullet, *Mugil cephalus* L. *J. Fish Dis.* 4:297–307.

Roubal, F. R. 1987. Gill surface area and its components in the yellowfin bream, *Acanthopagrus australis* (Gunther)(Pisces:Sparidae). *Aust. J. Zool.* 35:25–34.

Rourke, A. W., R. W. Davis, and T. M. Bradley. 1984. A light and electron microscope study of epitheliocystis in juvenile steelhead trout, *Salmo gairdneri* Richardson. *J. Fish Dis.* 7:301–9.

Sano, T., N. Okamoto, and T. Nishimura. 1981. A new viral epizootic of *Anguilla japonica* Temminck and Schlegel. *J. Fish Dis.* 4:127–39.

Satchell, G. H. 1984. Respiratory toxicology of fishes. In *Aquatic Toxicology*, vol. 2, ed. L. J. Weber, 1–50. New York: Raven Press.

Shariff, M. 1982. *Henenguya shaharini* sp.nov. (Protozoa:Myxozoa), a parasite of marble goby, *Oxyeleotris marmoratus* (Bleeker). *J. Fish Dis.* 5:37–45.

Smart, G. 1976. The effect of ammonia exposure on gill structure of the rainbow trout (*Salmo gairdneri*). *J. Fish Biol.* 8:471–75.

Smith, C. E., and T. Inslee. 1980. Interlamellar *Henneguya* infestation in adult channel catfish *Ictalurus punctatus* (Rafinesque). *J. Fish Dis.* 3:257–60.

Smith, D. G., and J. Chamley-Campbell. 1981. Localization of smooth-muscle myosin in branchial pillar cells of snapper (*Chrysophys auratus*) by immunofluorescence histochemistry. *J. Exp. Zool.* 215:121–24.

Snieszko, S. F. 1981. Bacterial gill disease of freshwater fishes. In Fish Disease Leaflet no. 62, U.S. Dept. of the Interior. Washington, D.C.

Soderberg, R. W. 1985. Histopathology of rainbow trout, *Salmo gairdneri* Richardson, exposed to diurnally fluctuating un-ionized ammonia levels in static-water ponds. *J. Fish Dis.* 8:57–64.

Solangi, M. A., and R. M. Overstreet. 1982. Histopathological changes in two estuarine fishes, *Menidia beryllina* (Cope) and *Trinectes maculatus* (Bloch and Schneider), exposed to crude oil and its water-soluble fractions. *J. Fish Dis.* 5:13–35.

Thune, R. L., and W. A. Rogers. 1981. Gill lesions in bluegill, *Lepomis macrochirus* Rafinesque, infested with *Cleidodiscus robustus* Mueller, 1934 (Monogenea:Dactylogyridae). *J. Fish Dis.* 4:277–80.

Trust, T. J. 1975. Bacteria associated with the gills of salmonid fishes in freshwater. *J. Appl. Bacteriol.* 38:225–33.

Wakabayashi, H. 1980. Bacterial gill disease of salmonid fish. *Fish Pathol.* 14:185–89.

Wakabayashi, H., and T. Iwado. 1985. Effects of a bacterial gill disease on the respiratory functions of juvenile rainbow trout. In *Fish and Shellfish Pathology*, ed. A. E. Ellis. London: Academic Press.

Wakabayashi, H., S. Egusa, and J. L. Fryer. 1980. Characteristics of filamentous bacteria isolated from a gill disease of salmonids. *Can. J. Fish. Aquat. Sci.* 37:1499–1504.

Wobeser, G., L. F. Kratt, R. J. F. Smith, and G. Acompanado. 1976. Proliferative branchiitis due to *Tetraonchus rauschi* (Trematoda: Monogenea) in captive Arctic Grayling (*Thymallus arcticus*). *J. Fish. Res. Board Can.* 33:1817–21.

Woodward, B. 1984. Symptoms of severe riboflavin deficiency without ocular opacity in rainbow trout (*Salmo gairdneri*). *Aquaculture* 37:275–81.

Wootten, R., and A. H. McVicar. 1982. *Dermocystidium* from cultured eels, *Anguilla anguilla* L., in Scotland. *J. Fish Dis.* 5:215–22.

3 Skin

SKIN LESIONS in fish are common. This is a reflection firstly of the organ's relative delicacy, compared to that of mammals. Secondly, it functions as an osmotic barrier, and hence minor lesions may rapidly progress to major ones particularly in fresh water. Thirdly, skin lesions are easily seen, especially so when many of them cause pronounced color changes. Examination of wet-mounted skin scrapings for the presence of parasites or bacteria is almost essential to a diagnosis, as many of the protozoa in particular are much more easily seen this way than in paraffin sections. On another practical note, Bouin's fixation is a great help because of its demineralizing properties, which make sectioning easier in scaled species; hence artifact is greatly reduced.

Normal

There are great interspecies differences in teleost skin; some species have no scales while others have numerous large eosinophilic club cells (fright-substance cells or Shreckstoffzellen) within the epidermis (Fig. 3.1). There are also large variations within a species and even within different locations on the same fish. For example, epidermal thickness in salmonids is greatest on top of the head and over the fins where no scales

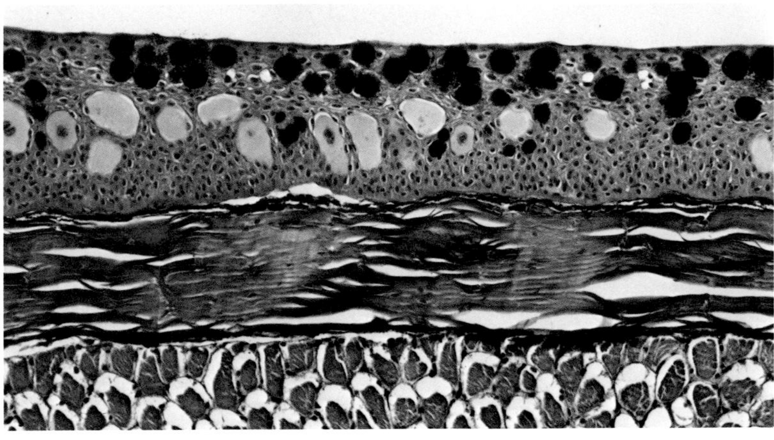

Fig. 3.1. Normal skin from brown bullhead, which has no scales but relatively thick epidermis containing large pale club cells and darker goblet cells above them.

are found. In addition, age, sex, and time of year all exert pronounced influences on the appearance and relative proportions of the different components of the skin. For these reasons, it is important to exercise care when examining pathological changes in this organ. Paired samples taken from the same location on the body of an apparently unaffected fish of the same age, sex, and species, are recommended as an aid to correct interpretation.

The *epidermis* is a nonkeratinizing (in most species) stratified squamous epithelium that varies in thickness from 3 or 4 cells up to approximately 20. A major difference from mammals is that in those teleosts having no keratinized cells, the outermost epidermal cells are viable and retain the capacity to divide. In addition to the filament-containing or malpighian cells that comprise the majority component, other cell types include mucous or goblet cells that are largely responsible for secreting the cuticle (inapparent in routine H & E–stained sections), large eosinophilic club cells (only some species), eosinophilic granule cells (unknown function), wandering leukocytes, and macrophages, which may contain melanin. There are variable numbers of lymphocytes in normal teleost epidermis, and these can greatly increase in number under some disease conditions; they become especially abundant just above the basal layer. Although these are involved in local immune responses (so that, for example, cutaneous antibody levels can rise independent of systemic humoral responses), the full significance of their participation has yet to be established (Lobb 1987). Specialized structures within the epidermis include taste buds and neuromasts.

The *dermis* comprises an upper stratum spongiosum and a deep stratum compactum. Large numbers of a variety of pigment-cell types are present at different levels within the dermis. These include melanophores, xanthophores, and iridophores. The latter contain reflective plates of guanine that render a silvery appearance to the skin when they are orientated at 90° to the oncoming light. *Scales* are a major feature of most species of teleosts (Fig. 3.2), and it is important to appreciate that, where present, they originate in scale pockets in the dermis and become covered by a layer of epidermis as they emerge, often to overlap one another. Thus, scale loss represents a very significant breach in the osmotic defenses of the fish. Scales comprise an outer reticulated osseous part with sawtoothlike ridges and an inner fibrillar layer that is uncalcified in some species and partially calcified in others. The latter consists of parallel collagen fibers embedded in an organic matrix. Scales represent a ready source of calcium, and during periods of starvation or prespawning, they may be resorbed in preference to the skeletal reserves.

Beneath the deep, compact dermal layer, the loose, relatively well-vascularized *hypodermis* provides a frequent avenue for lateral movement of pathogens and inflammatory processes.

Disease

Clinical signs of skin disease include hyperactivity and obvious evidence of irritation such as rubbing against the sides or bottom of tanks or ponds. In so doing, the sides of the fish become visible from above and the behavior is known as flashing. In some cases the fish may even jump out of the water. As with gill disease, there is usually excess mucus production, and this can result in

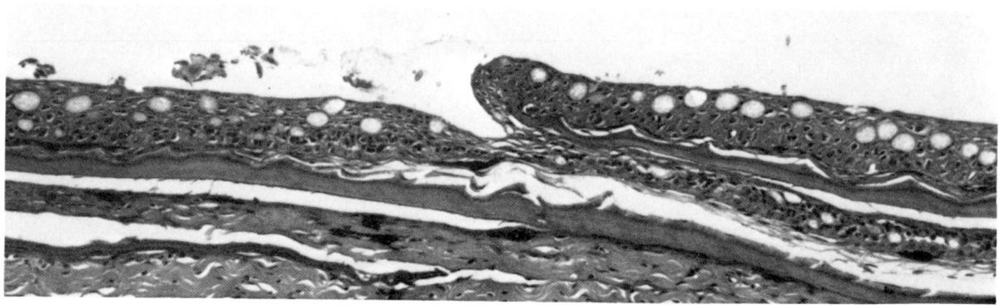

Fig. 3.2. Normal skin showing overlapping distribution of scales. Note ridges on their outer surface. The epidermis contains goblet cells but no club cells in this species.

foam accumulating on top of the water. Severe epidermal hyperplasia may produce a grossly visible white film, although a necrotizing lesion that destroys epidermis as well as the underlying pigment cell layer (as with columnaris disease) can superficially resemble this. Necrosis of the dorsal fin is common in cultured rainbow trout especially in situations where the fish are overstocked. Similar lesions are seen in salmon under the same conditions although in these fish the tail is a more frequent target. "Myxobacteria" are associated with both of these conditions, and although the reason for the different predilection sites is unknown, merely reducing the stocking density can often effect a cure. Fish with severe skin disease may become lethargic and inappetant, although systemic involvement through loss of osmoregulatory control or septicemia from secondary bacterial or fungal infections inevitably complicates the picture.

Although skin lesions are common, the variety of pathological changes tends to be more limited than in mammals. This may be partly explained by the absence of adnexal structures and partly by the extreme and probably dominating osmotic forces that are very rapidly brought to bear once the initial "waterproof" barrier is breached. If or when this occurs, influx of the surrounding water may, in addition, dilute or flush away any antigens present, thereby reducing the need for inflammation. It is of interest to note that in salmonids, the neural lymph duct is blocked and the main lymphatic return to the heart must therefore be via the cutaneous vessels. (Recent detailed anatomical studies suggest that fish may not have true lymphatic vessels, but for the purpose of this discussion, the traditional view will be adhered to.) In plaice (*Pleuronectes platessa*) and presumably most other teleosts, the total lymph volume of the muscle and skin is more than four times that of the circulating blood volume; and more than 80% of that lymph is dermal. The inference is that skin disease, or anything else that alters lymph volume, can profoundly alter the circulating blood picture both in terms of protein and electrolyte content, as well as relative hematocrit.

A listing of some of the agents associated with skin disease is shown in Table 3.1. With our present state of knowledge, however, classifying dermal disease in fish is probably best accomplished according to histological location.

Superficial Disease. The epidermis is avascular. This fact imposes its own restrictions on the variety of potential responses to injury. Acute responses to insult include discharge of goblet and club cells, although this may be difficult to appreciate histologically. Intercellular edema (spongiosis) and intracellular edema (hydropic degeneration) are more easily seen and accompany the early stages of many parasitic and bacterial diseases (Fig. 3.3). If spongiosis is severe, cell-

Fig. 3.3. Acute dermatitis in rainbow trout with monogenetic trematode attached to epidermis. There is pronounced epidermal spongiosis and some hydropic degeneration accompanied by pyknotic debris.

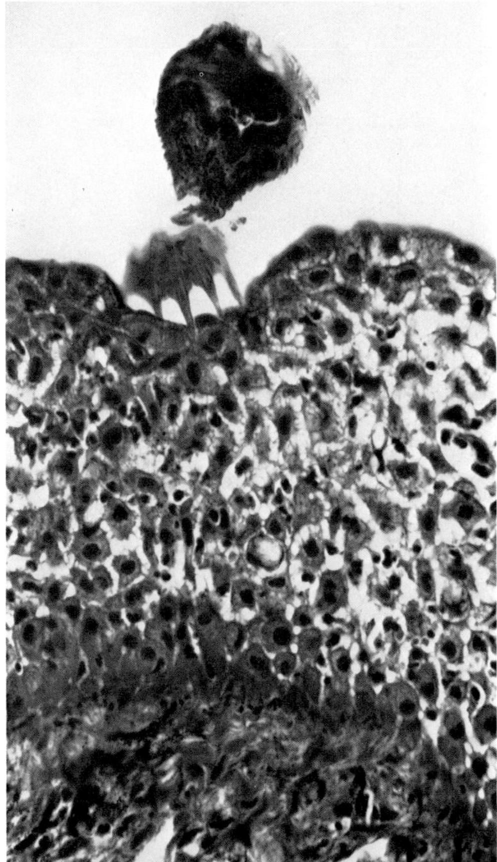

Table 3.1. Some of the agents associated with skin disease and the host response

Agent	Host response
BACTERIA	
Flexibacter columnaris	Acute necrotizing loss of epidermis with ulceration and extension into muscle in severe cases
Cytophaga (*Flexibacter*) *psychrophila*	Acute necrotizing dermatitis with marked epidermal hyperplasia in chronic disease
Flexibacter and *Cytophaga* spp. including saltwater spp.	Combinations of above
Aeromonas salmonicida	Acute or subacute necrotizing myositis or dermatitis with dermal ulceration
A. hydrophila	Similar to *A. salmonicida*
Renibacterium salmoninarum	Dermal or subdermal pyogranulomas, sometimes with cavitation
Vibrio anguillarum, V. ordalii	Acute or subacute ulcerative dermatitis plus myositis
Pseudomonas fluorescens	As for *Vibrio* spp.
Yersinia ruckeri	Acute dermatitis with pronounced hyperemia
Edwardsiella tarda	Subacute ulcerative dermatitis with gaseous bullae
E. ictaluri	Deep dermal ulceration, classically involving frontal bones
Mycobacterium marinum, M. fortuitum and other spp.	Chronic granulomatous dermatitis with ulceration
Nocardia asteroides, N. kampachi	As for mycobacteria
Pasteurella spp.	Acute hyperemic or chronic granulomatous dermatitis
Flavobacterium spp.	As for *Pasteurella* spp.
Clostridium hastiforme and other *Cl.* spp.	Subacute necrotizing peritonitis and myositis with dermal ulceration
VIRUSES	
Lymphocystis disease	Dermal fibroblastic cytomegaly with chronic lymphoid dermatitis
Herpesvirus (cod)—seen ultrastructurally	Epidermal malpighian cytomegaly and hyperplasia, with acute dermatitis
Herpesvirus scophthalmi—seen ultrastructurally	Epidermal malpighian cytomegaly
Cod ulcus syndrome, rhabdo and iridovirus-associated	Dermal vesiculation, erosion, and ulceration, with hyperemia, hemorrhage, and edema
Walleye retrovirus-associated	Dermal fibrosarcoma
Walleye herpesvirus-associated	Epidermal hyperplasia
Carp herpesvirus-associated	Epidermal hyperplasia plus inclusions (so-called pox)
Oncorhynchus masou virus-herpesvirus	Epidermal papillomas
Cod adenovirus—seen ultrastructurally	Epidermal hyperplasia
Herpesvirus salmonis	Dermal hyperemia
Infectious pancreatic necrosis virus (birnavirus)	Dermal hyperemia and darkening
Channel catfish disease virus (herpesvirus)	Dermal hyperemia and hemorrhage
Spring viremia of carp virus (rhabdovirus)	Dermal hyperemia and hemorrhage
Other endotheliotropic viruses	Dermal hyperemia and/or hemorrhage, especially at the base of fins
Chlamydia-like agent of epitheliocystis	Epidermal cytomegaly plus dermatitis
FUNGI AND ALGAE	
Saprolegnia diclina (*parasitica*) and *Achlya* spp.	Acute necrotizing dermatitis with epidermal sloughing and subdermal lateral spreading
Aphanomyces spp.—often systemic granulomas	Presumed necrotic dermatitis or systemic granulomas causing pressure atrophy
Exophiala salmonis plus other spp.—often systemic	Presumed necrotic dermatitis or systemic granulomas causing pressure atrophy
Ochroconis humicola plus other spp. (formerly in genus *Scolecobasidium*)	Necrotic and granulomatous dermatitis with ulceration
Ichthyophonus—systemic granulomas	Dermal granulomas common, causing "sandpaper skin"
Cladophora and *Chlorella* spp.	Frequent bone involvement—cranium and opercula especially

Table 3.1. (*Continued*)

Agent	Host response
PROTOZOA	
Ichthyophthirius multifiliis	Acute to subacute dermatitis with hyperplasia; organisms present *within* epidermis. Reinfection may cause pronounced necrosis
Cryptocaryon spp.	Acute to subacute dermatitis with hyperplasia; organisms present *within* epidermis
Chilodonella spp.	Acute to subacute dermatitis with hyperplasia
Brooklynella spp.	Acute to subacute dermatitis with hyperplasia
Tetrahymena spp.	Disseminated infections with dermal ulceration
Uronema spp.	Disseminated infections with dermal ulceration
Epistylis spp., *Heteropolaria* spp., *Scyphydia* spp.	Necrotic dermatitis with ulceration
Trichodina spp., plus similar spp.	Often mild subacute dermatitis with hyperplasia
Ichthyobodo (*Costia*) *necatrix*	Often severe erosive and ulcerative dermatitis following hyperplasia
Amyloodinium spp. plus other dinoflagellates	Necrotic dermatitis, often severe
Hexamitid protozoa (*Spironucleus* sp.?)	Associated with severe cranial ulcerative dermatitis—"hole in head"
Haemogregarina sachai	Subdermal granulomas
Myxosporean spp.	Dermal or subdermal cysts, often with rupture through skin: granulomas with prespore stages
Microsporidian spp.	Dermal or subdermal cysts, often with rupture through skin
Dermocystidium (*Perkinsus*)	Epidermal cysts with hyperplasia and granulomatous dermatitis
METAZOA	
Gyrodactylus and other spp. monogeneans	Sometimes severe dermatitis with hyperplasia
Digenetic metacercariae	Granulomatous dermatitis, often with melanization—black spot
Philometra spp., *Philometroides* spp.	Chronic active ulcerative dermatitis
Mollusk glochidia larvae	Granulomatous dermatitis
Argulus spp.	Necrotizing and ulcerative dermatitis, often severe
Lernaea spp., *Lepeiophtheirus* spp., *Caligus* spp. plus other copepods	Necrotizing and ulcerative dermatitis, often severe; may progress to granulomatous
Leeches, e.g., *Piscicola* spp.	Ulcerative dermatitis with hyperemia, hemorrhage, and epidermal hyperplasia
Lampreys	Severe ulcerative dermatitis with hyperemia
PHYSICAL AND OTHER AGENTS	
Anoxia	Ulcerative dermatitis
Sunlight	Acute necrotizing dermatitis
Freeze-branding	Pigmentary loss
Nets, hooks, gaffs	Superficial abrasions or deep puncture wounds
Predators, especially birds' beaks	Deep puncture wounds or bilateral "raking"-type lesions
Chemical toxicants	Necrotizing dermatitis; possible hyperplasia in chronic exposure
Fatty acid or thiamin deficiency	Diffuse dermal depigmentation
Vitamin C deficiency	Collagen abnormalities; impaired wound healing
Vitamin A deficiency or toxicity	Epidermal hypoplasia

to-cell contact may totally break down with the consequent formation of vesicles or bullae. These tend to be more common just at or above the basal layer (Fig. 3.4). They may become confluent with lifting and loss of epidermis (Fig. 3.5), leaving sometimes only a single row of cells resting on the basement membrane. Such a dermo-epidermal separation must be carefully differentiated from processing artifact, especially in scaled species. In mammals, *pemphigus* is a group of autoimmune skin diseases characterized by the formation of vesicles, bullae, erosions, and acantholysis and by the production of antibodies directed against epithelial cells. Autoimmune skin diseases have not been

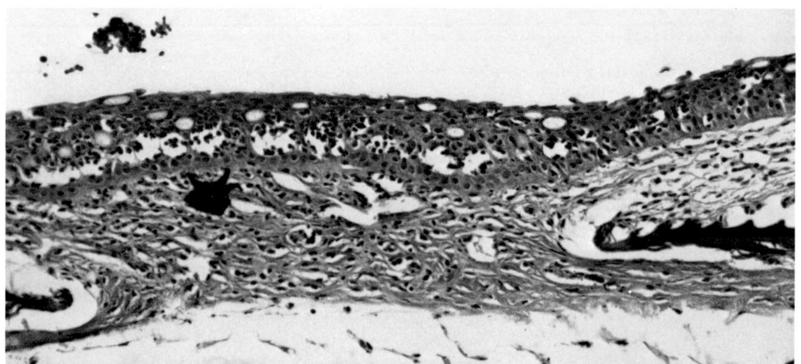

A

B

Fig. 3.4. A. Rainbow trout with multifocal to confluent vesicles just above basal layer. The cause of this lesion was not determined. **B.** Severe epidermal and superficial dermal edema in trout with "myxobacterial" tail rot.

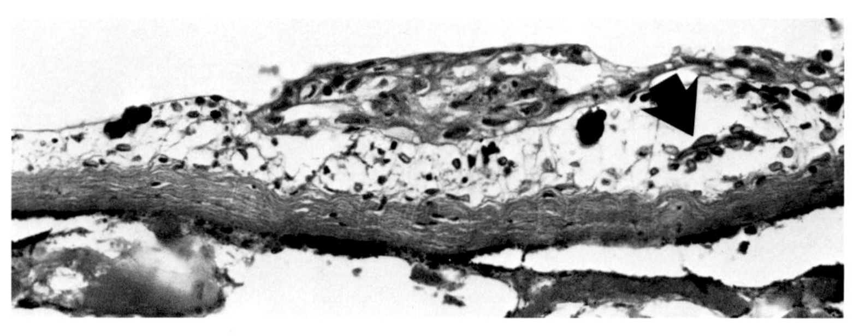

A

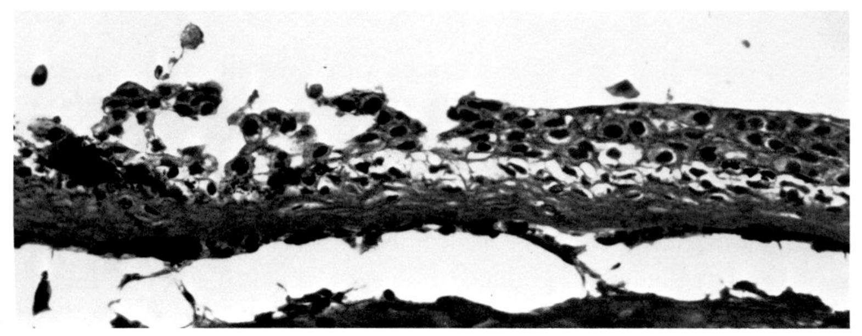

B

Fig. 3.5. A. Mycotic dermatitis (early) showing separation of epidermis from dermis prior to sloughing entirely. Such an underrunning is common with fungi (*arrow*). **B.** Edge of lesion due to *Flexibacter* in trout. Note that cell-to-cell contact has broken down entirely and cells are lifting off in clumps.

described in fish, although the analogy has been drawn between pemphigus and the histological changes seen in the specific entity ulcerative dermal necrosis (UDN) or Irish salmon disease of salmonids (Roberts and Bullock 1976). The condition is seen when Atlantic salmon (other salmonids are also susceptible) enter fresh water on their return migration prior to spawning. Early lesions include breakdown in intercellular junctions, although the skin rapidly becomes secondarily infected, particularly with *Saprolegnia* sp. of fungi. Despite exhaustive research with different hypotheses being favored from time to time, the cause and pathogenesis of this historically important disease remain an almost complete mystery.

Infiltration of vesicles by inflammatory cells to form "pustules" is not commonly reported in fish, but they are quite frequently seen, especially in the early stages of superficial bacterial or fungal diseases.

Individual malpighian cells may die with pyknosis and karyorrhexis or even apoptosis (phagocytosis by adjacent viable cells). A distinctive pattern of such changes is seen in sunburn, when fish in very clear water are subjected to high levels of ultraviolet radiation: dermo-epidermal separation is also seen. Niacin in particular is important in protecting the skin of rainbow trout against the effects of ultraviolet-B irradiation (Poston and Wolfe 1985).

Changes in the dermis that may accompany acute epidermal lesions include superficial dermal edema (Fig. 3.6) and vascular di-

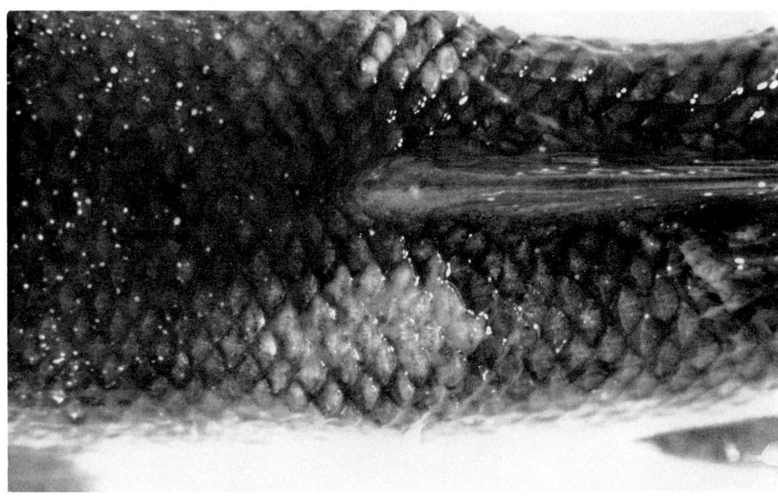

Fig. 3.6. A. White sucker with two large foci of dermal edema on either side of dorsal fin. These were experimentally produced with chemicals. Note pallor and elevation of scales. (*Courtesy of I. Smith*) **B.** Brown bullhead with *Aeromonas hydrophila* bacteremia showing dermal congestion and hemorrhage. The fish was so weak that it was unable to prevent itself from being sucked onto the outflow pipe leading from the tank—hence the large "suckoma" seen ventrally!

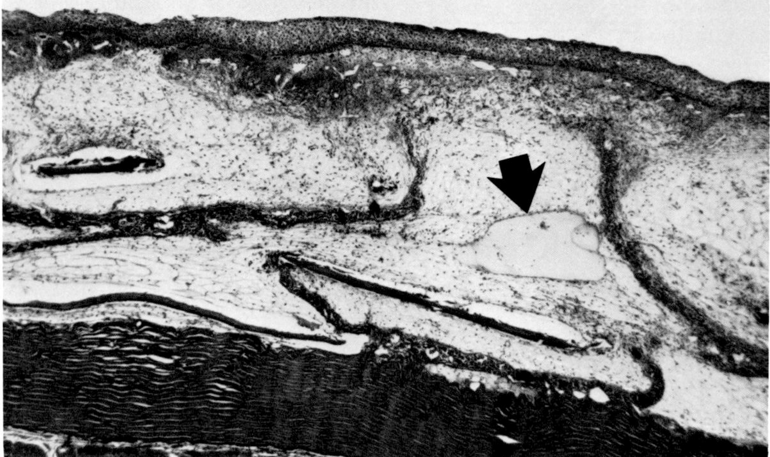

Fig. 3.7. Severe dermal edema in trout used for multiple blood samples. One lymphatic (*arrow*) is prominently dilated.

lation (occasionally hemorrhage) with pavementing and emigration of leukocytes. These may be seen moving across the basal layer into the epidermis. Rupture of pigment cells with dispersion of granules may also be seen. If the osmotic barrier is broken, dermal edema may be pronounced. Such a process has been termed waterlogging in freshwater species, where the osmotic gradient tends to force water into the tissues. Under these circumstances dermal lymphatics become greatly dilated, and the whole dermis may be massively thickened. Diffuse dermal edema may also accompany hypoproteinemia (Fig. 3.7).

If present at all, these changes may rapidly proceed to severe necrosis and sloughing of the epidermis (Fig. 3.8). Examples of this are associated with fungal infection by *Saprolegnia diclina* (Fig. 3.9) and the bacterial conditions peduncle disease (*Flexibacter psychrophila*; Fig. 3.10), columnaris disease (*Flexibacter columnaris*; Fig. 3.11) and carp erythrodermatitis (*Aeromonas salmonicida*). Waterlogging is severe in these acute ulcerative dermatitides, usually extending deep into the hypodermis and even muscle, tracking down the fascial planes. Secondary invaders frequently obscure the primary insult and pathological response.

Fig. 3.8. Severe acute necrotizing dermatitis with hemorrhage and inflammation of underlying muscle. Note the unusual laminar deposition of mineral (*arrow*).

Hypoplasia of epidermis is seen in vitamin A deficiency as well as in vitamin A

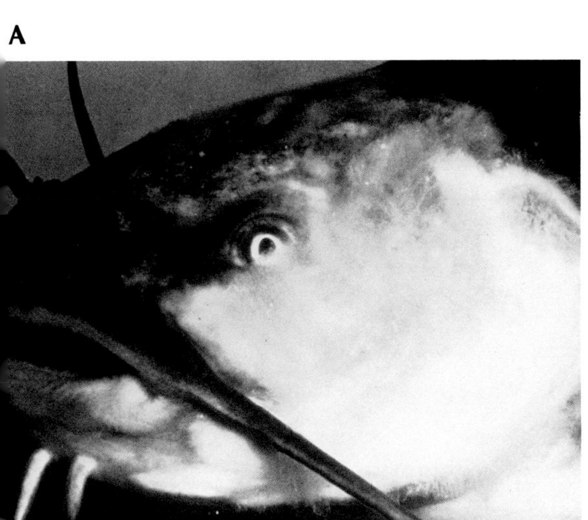

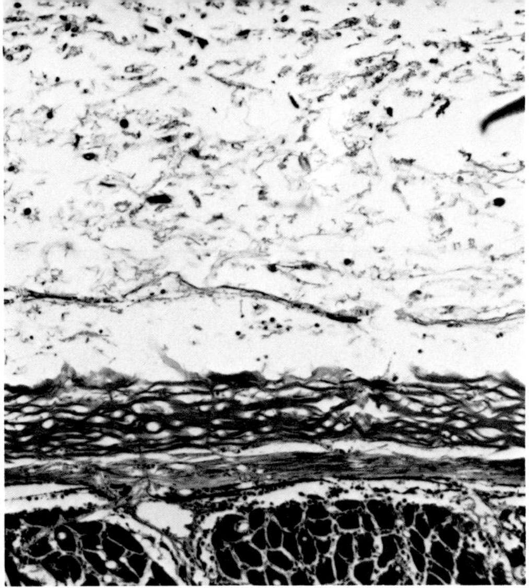

Fig. 3.9. A. Mycotic dermatitis in brown bullhead. (*Courtesy of R. D. Moccia*) **B.** Severe mycotic dermatitis in walleye. Epidermis is entirely replaced by a mat of fungal hyphae. Hyaline change in the collagen is expected in these waterlogging types of lesions.

Fig. 3.10. Tail rot in young rainbow trout. Large numbers of *Flexibacter*-type organisms were isolated.

Fig. 3.11. A. Columnaris disease in young yellow perch showing classical saddlepatch lesion extending as band around dorsal fin. Pallor results from necrosis and sloughing of epidermis, removing in the process some pigment cells. **B.** Columnaris disease in trout. Epidermis is completely missing, and long filamentous bacteria have insinuated themselves among collagen fibers of the dermis.

toxicity, but a much more common chronic response to a large variety of insults is hyperplasia, sometimes with the formation of pronounced dermal papillae and hyperpigmentation (Fig. 3.12). It must be remembered that teleost epidermis is capable of division at all levels and not just in the *stratum spinosum* as is usually the case in mammals. Epidermal hyperplasia accompanies chronic irritation by protozoan and metazoan parasites, bacteria (especially the "myxobacteria"), and chemical or physical agents such as suspended solids, particularly it would seem, at low water temperatures. As reviewed by Robertson (1985), the protozoan parasite *Ichthyobodo necator* induces hyperplasia of the outermost malphigian cells rather than those of the suprabasal layer. The resulting hyperplastic plaque, which is recognized grossly as a blue translucent sheen, may then slough off, leaving only a single layer of cells attached to the basal layer.

Ichthophthirius multifiliis is one of the commonest protozoan infections of freshwater fish in the wild or culture situation, as well as in the display aquarium. The ciliates are large enough to be grossly visible, and their multifocal distribution has earned them the popular name of white spot (Fig. 3.13; see also Fig. 2.23). It has been shown experimentally that invasion of the epidermis (gills are also affected) by moderate numbers of the infective stage (trophont) causes little damage, and that by 40 hours (at 17°C) most of the parasites are situated on the basal lamina (Ventura and Paperna 1985). Heavy infection, however, results in spongiosis and erosion, plus infiltration by neutrophils and lymphocytes; hyperplasia is seen only following subsequent infections. If the hyperplastic tissue is itself parasitized, the trophont may fail to reach the basal lamina, with severe lysis of epidermis, and possibly dermo-epidermal separation, as the consequence. Maturation of the parasite following growth within the epidermis sees its emergence into the surrounding water. If large numbers mature at the same time, the consequent damage to the skin may impose a substantial osmotic stress on the fish.

A

B

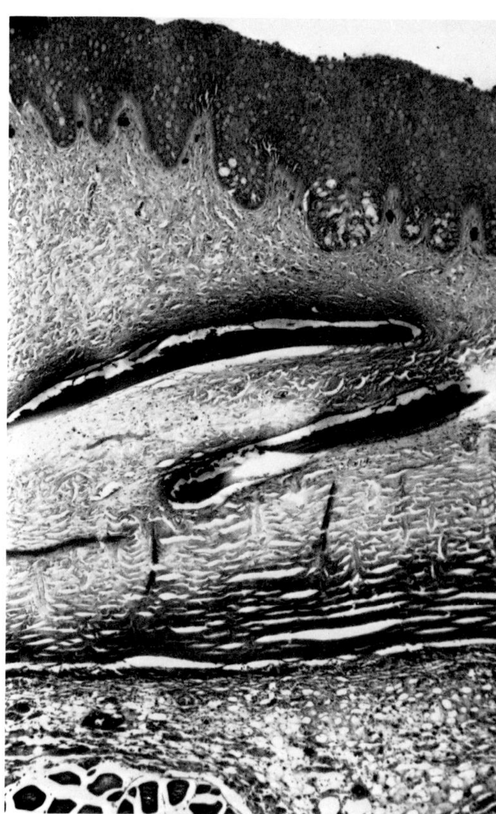

Fig. 3.12. A. Dermal hyperplasia in rainbow trout recovering from tail rot. **B.** Chronic hyperplastic dermatitis showing prominent dermal pegs extending up into thickened epidermis.

Fig. 3.13. White spot disease of rainbow trout. Each spot represents a parasite, visible partly due to its large size (up to roughly 1 mm) and partly due to the accompanying epidermal hyperplasia (see also Fig. 2.23 for gills).

In some species, hyperplasia is also seen as one of the skin changes accompanying sexual maturation, and it can be experimentally reproduced with cortisol injections. Such fish are prone to develop secondary bacterial or fungal skin disease, but the underlying pathogenesis of the lesions must be understood for a correct interpretation. There is much less known about other endocrine dermatoses, but they are probably more important than is presently appreciated.

Pronounced epidermal hyperplasia must of course be differentiated from the diffuse neoplastic proliferations seen in a variety of species. Many of these reactions, termed papillomatosis (Fig. 3.14), are associated with virus or viruslike particles seen ultrastructurally. There is usually little dermal involvement, and this may help with interpretation, although it must be appreciated that chronic insult may predispose to neoplastic transformation, in which case the two types of response can merge together.

Deep Disease. Severe dermal or subdermal lesions may involve the epidermis by extension or may cause pressure atrophy of dermis and epidermis by virtue of their space-occupying nature (Fig. 3.15). There may be loss of the normal complement of mucous and club cells, thinning of the epidermis, and even ulceration. These deep-seated lesions are quite common and are found in diseases such as furunculosis (*Aeromonas salmonicida*; Fig. 3.16) vibriosis, mycobacteriosis (Fig. 3.17), and enteric septicemia of catfish (*Edwardsiella icta-*

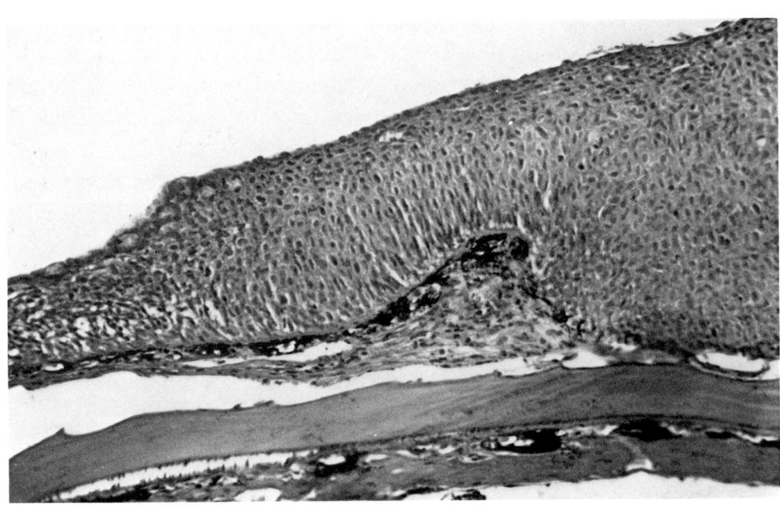

Fig. 3.14. Papillomatosis of Atlantic salmon showing massive hyperplasia of malpighian cells in epidermis (see also Fig. 13.2).

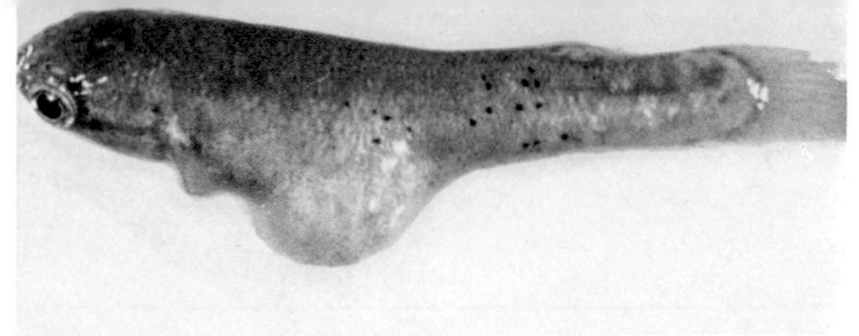

A

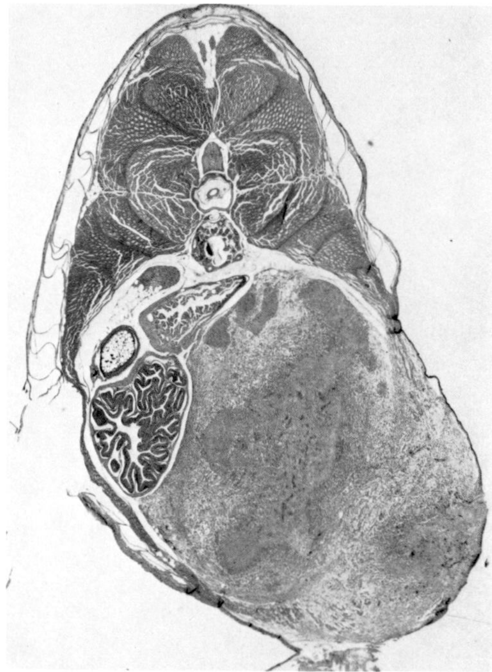

B

Fig. 3.15. A and **B.** *Rivulus marmoratus* with large mycobacterial granuloma grossly distending the peritoneal cavity.

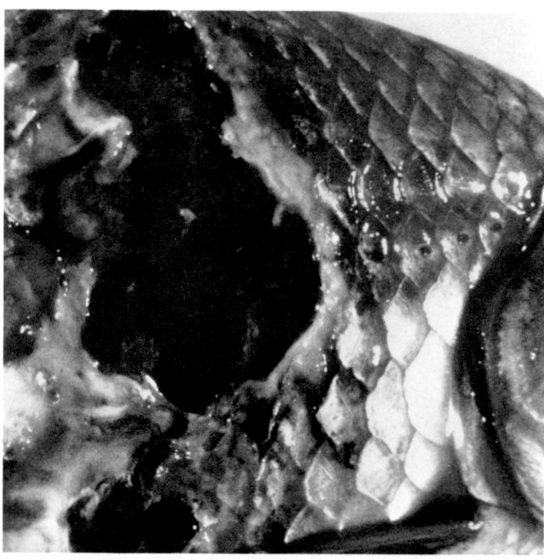

Fig. 3.16. Furunculosis in goldfish causing large ulcers with hemorrhagic base. Loss of osmotic integrity over such a large area poses a serious threat.

luri). (The lesion in this latter disease may be deep enough to involve the cranial bones and cartilage, producing the so-called hole-in-the-head lesion. This should not be confused with another disease of the same name encountered in oscars and related species in which the grossly similar lesions are thought to be caused by hexamitid protozoa or by bacterial infection of the sensory canal system.) *Lactobacillus piscicola* can also cause subdermal blisters in salmonids, often as a postspawning event.

Dermal ulceration has been experimentally produced with hypoxia alone (using water levels of < 1mg/L dissolved oxygen), the resulting lesions a possible consequence of vascular shunting away from the skin, combined with the reduced availability of oxygen for the epidermis, which, being avascular, may partially depend for its supply on the water itself (Callinan 1988). Such a pathogenesis has been proposed to explain the epizootics of dermal ulcerative disease (with numerous secondary invaders as an almost inevitable consequence) seen in feral and cultured fish populations in many parts of the world and which often follow heavy rainfall or flood conditions; due to the high quantities of wastes carried into these waters, oxygen levels can become disastrously low.

Dermal thrombosis is not infrequently seen in severe skin disease; in bacteremias, these thrombi may be septic; infarction is a possible consequence. Dermal edema also is a common accompaniment, and if longstanding, there may be dermal fibroplasia leading to fibrosis, usually of the stratum spongiosum (Fig. 3.18). Scale resorption with osteoclasis is sometimes encountered in these situations, but the significance or pathogenesis of the response is unknown: possibly there is interference with the vascular supply to the scale pocket. Regeneration of scales usually occurs and is heralded by the proliferation of osteoblastlike cells.

A good example of a chronic-active dermal or subdermal lesion is that associated with *Renibacterium salmoninarum*. The re-

A

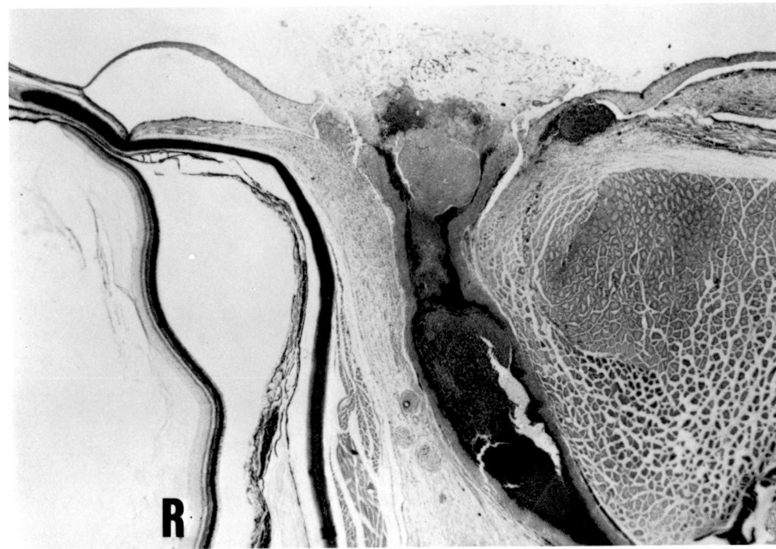

B

Fig. 3.17. A and **B.** Multifocal ulcerative dermatitis over the head of this cichlid, due to *Mycobacterium*. Tuft of secondary fungi over the eye marks the location for the histological section shown in **B.** (*R* denotes retina)

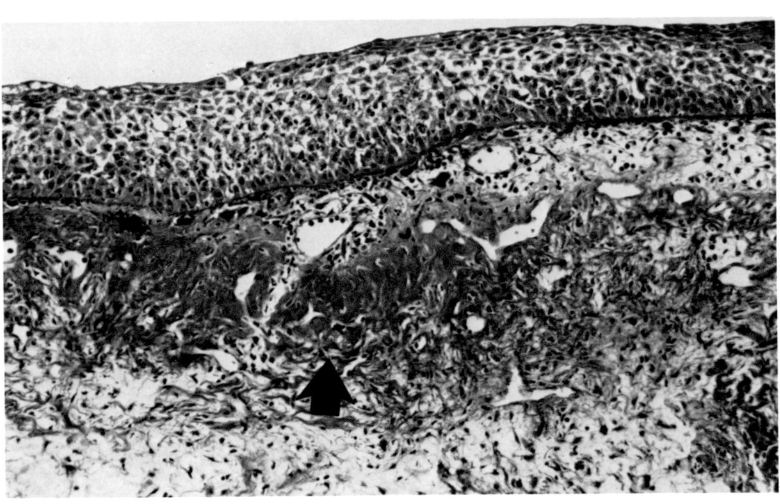

Fig. 3.18. Fibroplasia of superficial dermis (*arrow*). Epidermis also shows spongiform degeneration.

sponse, mainly in the muscle, may be so severe as to dissect the whole dermis away from the subcutis to produce large blisters (Fig. 3.19). Alternatively, the lesion may be a largely dermal one, sometimes centering around scale pockets to produce multifocal chronic-active dermatitis or so-called spawning rash (Fig. 3.20). These lesions, which may occur in the absence of detectable lesions elsewhere in the body, including the kidney, resolve after spawning. The cellular response involves mainly neutrophils and macrophages; as seen in special stains, the latter are stuffed with bacteria.

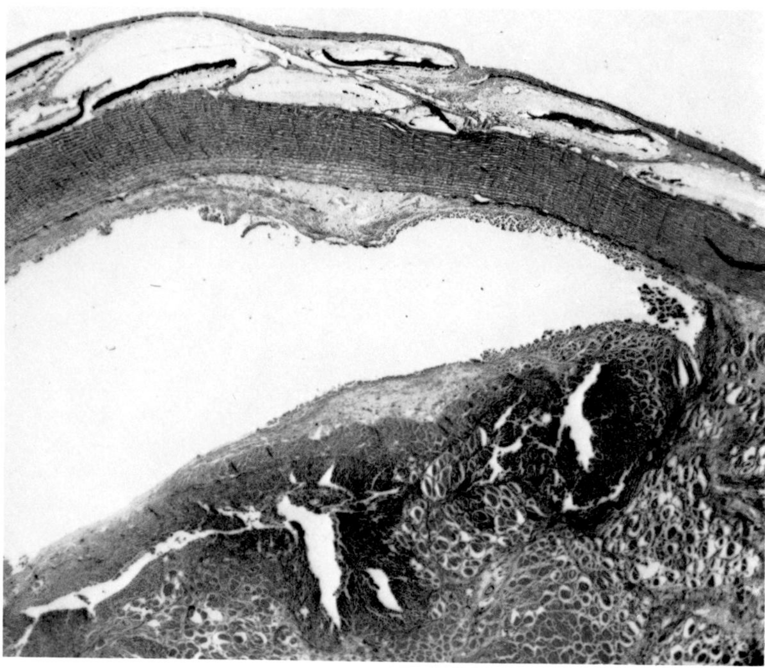

Fig. 3.19. Bacterial kidney disease in mature rainbow trout showing subdermal cavitation and some dermal edema causing lifting of scales. The cavity in this case was filled with a cloudy serous fluid.

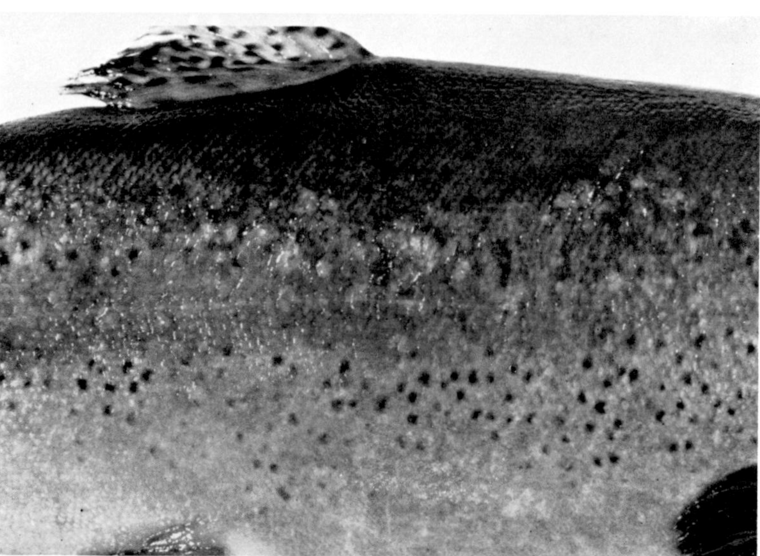

Fig. 3.20. Spawning rash in broodstock rainbow trout. This dermal variant of bacterial kidney disease, which may affect no other organ, causes multifocal granulomas that disappear after spawning. (*Micrograph courtesy of R. D. Moccia*)

Fig. 3.21. Strawberry disease in rainbow trout showing massive lichenoid infiltration of dermis with lymphocytes and macrophages. The etiology of this lesion, which may ulcerate, is unknown, but it is antibiotic sensitive. (*Material courtesy of D. P. Olson*)

A chronic lichenoid-type dermatitis, with macrophages and lymphocytes arranged in a layer between dermis and epidermis, is seen in the so-called strawberry disease of rainbow trout (Fig. 3.21). This antibiotic-responsive condition, which leads to ulceration, is of unknown etiology (Olson et al. 1985).

Cytomegaly of dermal fibroblasts is encountered in the iridovirus disease lymphocystis. This is seen in many species, but especially in higher teleosts. Clusters of individual cells enlarge greatly to produce grossly visible nodular aggregates (Fig. 3.22). Infected cells have a distinctive thick hyaline "capsule" to give them in section the appearance of eggs. Viral inclusion material within the cytoplasm may be lacy (plaice-type) or large and cordlike with blebs (flounder-type). Healing is associated with the influx of mainly lymphoid cells. Mortality is low in this disease, although fish may become reinfected. Cytomegaly of epidermal malpighian cells (gill epithelium also) with mild accompanying inflammation is also seen in herpesvirus infections of cod, turbot (*Scophthalmus maximus*), and probably other species.

Myxosporeans and microsporidians frequently have cysts in a dermal or subdermal location (Fig. 3.23), and these may subsequently rupture to the outside liberating spores; they are, however, easily recognized and usually present no diagnostic problem. By contrast, prespore myxosporeans, which may be found in the same location, are much more difficult to diagnose. Although dermis is an unusual location for the lesions (kidney and spleen are the main targets), an example of this is found in proliferative kidney disease; the characteristic granulomatous response surrounding the "mother and daughter" stages of the parasites creates a space-occupying lesion that may eventually ulcerate.

Metacercariae of digeneans are frequently found in the skin, often just beneath the epidermis. Whether or not the host mounts a significant response probably depends on the presence or absence of a parasitic capsule: its presence tends to correlate with little response. Several species, however, do stimulate melanocytes. Their investment of the parasite is grossly visible and gives rise to the popular name of black spot disease (Fig. 3.24). *Cryptocotyle lingua* is a digenetic metacercarial infection of marine fish with zoonotic potential. Parasitic numbers may occasionally be high enough to produce a diffuse and generalized melanosis

A

B

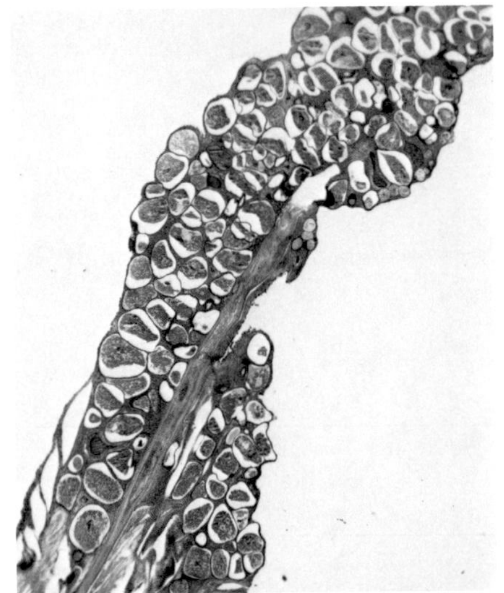
C

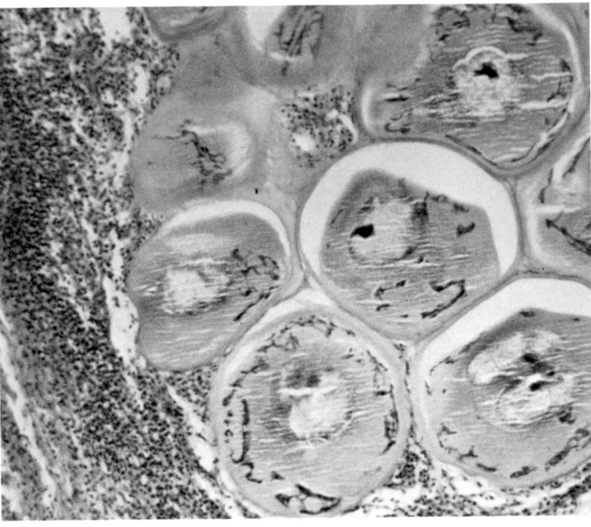

Fig. 3.22.
A. Lymphocystis disease showing common distribution of lesions on fins. **B.** Marine tropical with several lesions on body, in addition to those on fins. (*Courtesy of R. D. Moccia*) **C** and **D.** Low and high power of **A** showing cytomegalic fibroblasts with strands of inclusion material in cytoplasm and an accompanying largely lymphoid infiltrate.

Fig. 3.23. Shiner with multiple subdermal myxosporean cysts. A single black spot (encysted digenetic metacercaria) is also present on the tail.

Fig. 3.24. Black spot disease in roach due to melanin accumulation around encysted digenetic metacercariae.

of the dermis, so that large areas of skin, or indeed sometimes the entire fish, appear black.

Such changes need to be differentiated from the inherited *pigmentary anomalies* that form the basis for selective breeding programs in aquarium species. Albinism is well documented in marine flatfish culture and this can be either partial, giving a piebald appearance to the fish, or complete. It is suspected that higher-than-normal light levels may in part be responsible for the sometimes 50% incidence, although nutritional factors probably also play a major role. Generalized systemic infections often result in darkening of fish due to a relaxation of nervous control over melanophores, as well as endocrine effects. Locally severe or space-occupying lesions may lead to regionalized color changes due to compression of peripheral nerves. Eye disease such as cataract normally results in dark fish although the converse does also occur. Pigmentary lesions such as these are uncommon in wild fish, presumably at least in part because they fall easy victims to predators.

Dermal granulomas are quite common and may be found with or without the presence of epithelioid cells and multinucleated giant cells (Fig. 3.25); the latter are usually foreign-body-type, and although fish are capable of forming the classical Langhans-type, they have usually been seen only in experimental situations. Splendore-Hoeppli reactions (club colonies) as seen in botryomycosis in mammals are rare in fish.

Metazoan parasites are frequently incriminated in dermal disease, an example being various species of the copepod *Lepeophtheirus.* As reviewed by Kabata (1984) *Lepeophtheirus salmonis* does not invariably cause pronounced damage as it browses over the skin, but under intensive farming conditions of Atlantic salmon, the parasites become concentrated and severe ulcerative dermatitis may result, with loss of substantial areas of skin. On top of the head, the ulceration may be deep enough to involve the cranium. Some copepods may be deeply embedded within the skin and yet elicit surprisingly little host response, mainly localized mild dermal fibrosis and epidermal hyperplasia. By contrast, others such as *Lepeophtheirus pectoralis,* which are only superficially attached to the skin and effect slight penetration, elicit massive dermal fibroplasia with hemorrhage in severe cases and a chronic inflammatory infiltrate.

Some crustacea may elicit a host response that proliferates to such an extent

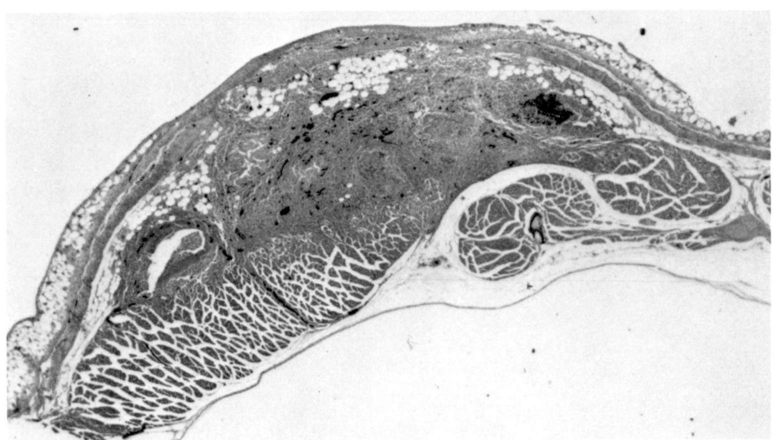
A

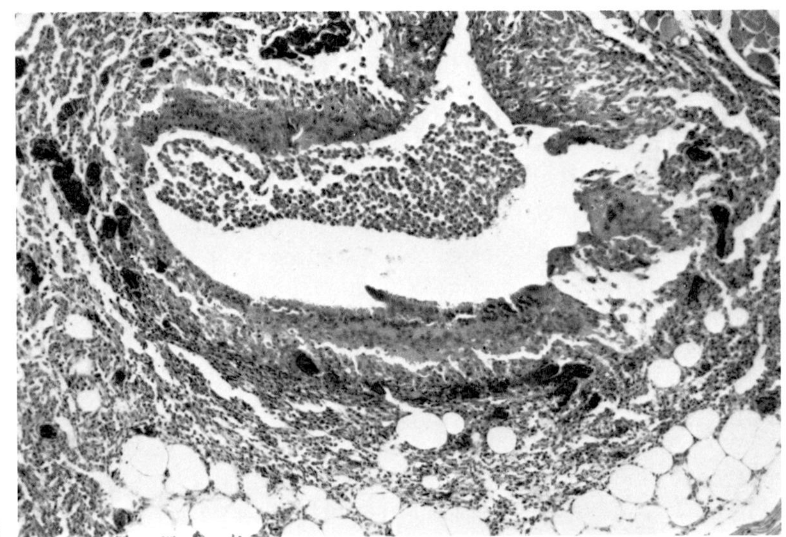
B

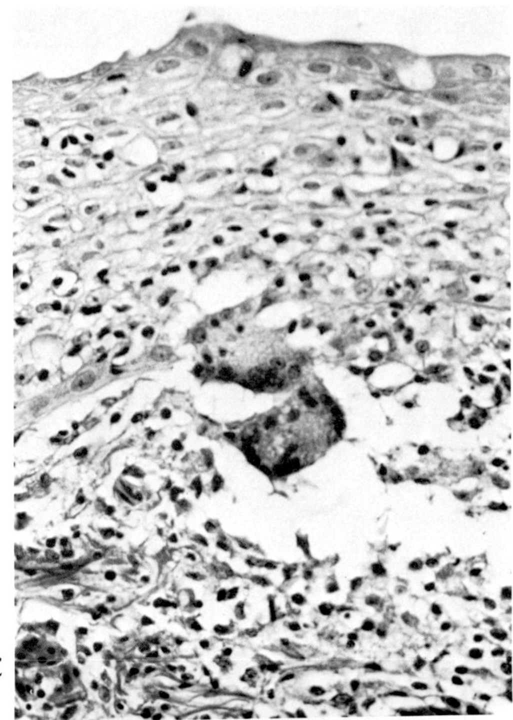
C

Fig. 3.25. A and **B.** Dermal granuloma in brown bullhead. The syncytial giant-cell-lined cavity suggests a penetrating foreign body as the cause of the reaction. **C.** Detail of response showing giant cells at dermo-epidermal junction.

that the point of attachment of the parasite becomes encased. The externally visible masses of tumorlike fibrous tissue may contain large numbers of melanocytes so as to appear grossly black. Some isopod crustacea in the family Cymothoidae form pouches (a zoocaecidium) in the flank of the host, with marked thinning of the overlying epidermis. The well-vascularized wall of the pouch acts as a food source.

By contrast with these space-occupying and local inflammatory responses, the branchiuran *Argulus* has a preoral stylet that not only causes local mechanical injury, but also releases digestive enzymes, giving rise to systemic as well as local effects. These then are just a few of the reactions caused by such parasites, which are not only seriously damaging per se but in addition pave the way for secondary attack by microorganisms.

Trauma from nets or rough handling, self-inflicted injury due to dermal irritation, or deep puncture-type wounds from birds' beaks may all cause severe erosion or dermal ulceration (Fig. 3.26).

Not surprisingly, wound healing is a temperature-dependent process, but it appears that the epidermis is less susceptible to its influence than is the dermis, the repair of which is greatly slowed by low water temperatures (Roberts 1975). Nevertheless, fish skin shows a remarkable capacity for repair, as do so many other tissues, and despite sometimes enormous lesions, grossly visible scars are not an inevitable consequence. As in mammals, vitamin C deficiency impairs collagen formation and hence wound healing.

References

Ahmed, A. T. A., and S. Egusa. 1980. Dermal fibrosarcoma in goldfish *Carassius auratus* (L.). *J. Fish Dis.* 3:249–54.

Ajmal, M., and B. C. Hobbs. 1967. Causes and effects of Columnaris-type diseases in fish. *Nature* 215:141–42.

Amin, O. M. 1979. Lymphocystis disease in Wisconsin fishes. *J. Fish Dis.* 2:207–17.

Anders, K., and H. Moller. 1985. Spawning papillomatosis of smelt, *Osmerus eperlanus* L., from the Elbe estuary. *J. Fish Dis.* 8:233–35.

Appy, R. G., and D. K. Cone. 1982. Attachment of *Myzobdella lugubris* (Hirudinea: Piscicolidae) to logperch, *Percina caprodes,* and brown bullhead, *Ictalurus nebulosus. Trans. Am. Microsc. Soc.* 101:135–41.

Arthur, J. R., and L. Margolis. 1984. *Trichodina truttae* Mueller, 1937 (Ciliophora: Peritrichida), a common pathogenic ectoparasite

Fig. 3.26. Chinook cage-reared salmon with severe multifocal dermal abrasions, probably net-induced. (*Courtesy of J. Gill*)

of cultured juvenile salmonid fishes in British Columbia: redescription and examination by scanning electron microscopy. *Can. J. Zool.* 62:1842–48.

Barrow, G. I., and M. Hewitt. 1971. Skin infection with *Mycobacterium marinum* from a tropical fish tank. *Br. Med. J.* 2:505–6.

Becker, C. D., and M. P. Fujihara. 1978. The bacterial pathogen *Flexibacter columnaris* and its epizootiology among Columbia River fish. A review and synthesis. In American Fisheries Society Monograph no. 2. 1–92. Washington, D.C.

Blackstock, N., and A. D. Pickering. 1980. Acidophilic granular cells in the epidermis of the brown trout, *Salmo trutta* L. *Cell Tissue Res.* 210:359–69.

Bullock, A. M., and R. J. Roberts. 1980. Inhibition of epidermal migration in the skin of rainbow trout *Salmo gairdneri* Richardson in the presence of achromogenic *Aeromonas salmonicida. J. Fish Dis.* 3:517–24.

______. 1981. Sunburn lesions in salmonid fry: a clinical and histological report. *J. Fish Dis.* 4:271–75.

Bullock, A. M., and D. A. Robertson. 1982. A note on the occurrence of *Ichtyobodo necator* (Henneguy, 1883) in a wild population of juvenile plaice, *Pleuronectes platessa* L. *J. Fish Dis.* 5:531–33.

Bullock, A. M., R. J. Roberts, P. Waddington, and W. D. A. Bookless. 1983. Sunburn lesions in koi carp. *Vet. Rec.* 112:551.

Burton, D. 1979. Differential chromatic activity of melanophores in integumentary patterns of winter flounder (*Pseudopleuronectes americanus* Walbaum). *Can. J. Zool.* 57:650–57.

Bylund, G., E. T. Valtonen, and E. Niemela. 1980. Observations on epidermal papillomata in wild and cultured Atlantic salmon *Salmo salar* L. in Finland. *J. Fish Dis.* 3:525–28.

Callinan, R. B. 1988. Diseases of native Australian fishes. In *Fish Diseases—Refresher Course for Veterinarians.* Proc. 106, Post-Graduate Committee in Veterinary Science, University of Sydney. 459–71.

Campbell, A. C., and J. A. Buswell. 1982. An investigation into the bacterial aetiology of "black patch necrosis" in Dover sole, *Solea solea* L. *J. Fish Dis.* 5:495–508.

Carlisle, J. C., and R. J. Roberts. 1977. An epidermal papilloma of the Atlantic salmon I. Epizootiology, pathology and immunology. *J. Wildl. Dis.* 13:230–34.

Cipriano, R. C., and Heartwell, C. M., III. 1986. Susceptibility of salmonids to furunculosis: differences between serum and mucus responses against *Aeromonas salmonicida. Trans. Am. Fish. Soc.* 115:83–88.

Cone, D. K., and P. H. Odense. 1984. Pathology of five species of *Gyrodactylus* Nordmann, 1832 (Monogenea). *Can. J. Zool.* 62:1084–88.

Cone, D. K., J. D. Miller, and W. K. Austin. 1980. The pathology of "saddleback" disease of underyearling Atlantic salmon (*Salmo salar*). *Can. J. Zool.* 58:1283–87.

Copland, J. W., and L. G. Willoughby. 1982. The pathology of *Saprolegnia* infections of *Anguilla anguilla* L. elvers. *J. Fish Dis.* 5:421–28.

Daniels, S. B., R. L. Herman, and C. N. Burke. 1976. Fine structure of an unidentified protozoon in the epithelium of rainbow trout exposed to water with *Myxosoma cerebralis. J. Protozool.* 23:402–10.

Delves-Broughton, J., J. K. Fawell, and D. Woods. 1980. The first occurrence of "cauliflower disease" of eels *Anguilla anguilla* L. in the British Isles. *J. Fish Dis.* 3:255–56.

Desser, S. S., K. Molnar, and I. Weller. 1983. Ultrastructure of sporogenesis of *Thelohanellus nikolskii* Akhmerov, 1955 (Myxozoa: Myxosporea) from the common carp, *Cyprinus carpio. J. Parasitol.* 69:504–18.

Edwards, C. J. 1978. Algal infections of fish tissue: a recent record and review. *J. Fish Dis.* 1:175–79.

Ellis, A. E., and R. Wootten. 1978. Costiasis of Atlantic salmon, *Salmo salar* L. smolts in seawater. *J. Fish. Dis.* 1:389–93.

Ellis, A. E., G. Dear, and D. J. Stewart. 1983. Histopathology of "Sekitenbyo" caused by *Pseudomonas anguilliseptica* in the European eel, *Anguilla anguilla* L., in Scotland. *J. Fish Dis.* 6:77–79.

Emerson, C. J., J. F. Payne, and A. K. Bal. 1985. Evidence for the presence of a viral non-lymphocystis type disease in winter flounder, *Pseudopleuronectes americanus* (Walbaum), from the northwest Atlantic. *J. Fish Dis.* 8:91–102.

Ferguson, H. W. 1977. Columnaris disease in

rainbow trout (*Salmo gairdneri*) in Northern Ireland. *Vet. Rec.* 101:55–56.

Ferguson, H. W., and D. H. McCarthy. 1978. Histopathology of furunculosis in brown trout *Salmo trutta* L. *J. Fish Dis.* 1:165–74.

Foissner, W., G. L. Hoffman, and A. J. Mitchell. 1985. *Heteropolaria colisarum* Foissner & Schubert, 1977 (Protozoa:Epistylididae) of North American freshwater fishes. *J. Fish Dis.* 8:145–60.

Frerichs, G. N., S. D. Millar, and R. J. Roberts. 1986. Ulcerative rhabdovirus in fish in Southeast Asia. *Nature* 322:216.

Grizzle, J. M., T. E. Schwedler, and A. L. Scott. 1981. Papillomas of black bullheads, *Ictalurus melas* (Rafinesque), living in a chlorinated sewage pond. *J. Fish Dis.* 4:345–51.

Hackett, J. L., W. H. Lynch, W. D. Paterson, and D. H. Coombs. 1984. Extracellular protease, extracellular haemolysin, and virulence in *Aeromonas salmonicida*. *Can. J. Fish. Aquat. Sci.* 41:1354–60.

Hastein, T., S. Jakob Saltveit, and R. J. Roberts. 1978. Mass mortality among minnows *Phoxinus phoxinus* (L.) in Lake Tveitevatn, Norway, due to an aberrant strain of *Aeromonas salmonicida*. *J. Fish Dis.* 1:241–49.

Hatton, J. P. 1976. The dermatology of teleost fishes—a bibliography. Scotland: University of Stirling.

Hazen, T. C., G. W. Esch, R. V. Dimock, Jr., and A. Mansfield. 1982. Chemotaxis of *Aeromonas hydrophila* to the surface mucus of fish. *Curr. Microbiol.* 7:371–75.

Hazen, T. C., M. L. Raker, G. W. Esch, and C. B. Fliermans. 1978. Ultrastructure of redsore lesions on largemouth bass (*Micropterus salmoides*): association of the ciliate *Epistylis* sp. and the bacterium *Aeromonas hydrophila*. *J. Protozool.* 25:351–55.

Huizinga, H. W., G. W. Esch, and T. C. Hazen. 1979. Histopathology of redsore disease (*Aeromonas hydrophila*) in naturally and experimentally infected largemouth bass *Micropterus salmoides* (Lacepede). *J. Fish Dis.* 2:263–77.

Hussein, S. A., and D. H. Mills. 1982. The prevalence of "cauliflower disease" of the eel, *Anguilla anguilla* L., in tributaries of the River Tweed, Scotland. *J. Fish Dis.* 5:161–65.

Jensen, N. J., and J. L. Larsen. 1979. The ulcus-syndrome in cod (*Gadus morhua*). I. A pathological and histopathological study. *Nord. Vet. Med.* 31:222–28.

Jensen, N. J., B. Bloch, and J. L. Larsen. 1979. The ulcus-syndrome in cod (*Gadus morhua*). III. A preliminary virological report. *Nord. Vet. Med.* 31:436–42.

Johansson, N., K. M. Svensson, and G. Fridberg. 1982. Studies on the pathology of ulcerative dermal necrosis (UDN) in Swedish salmon, *Salmo salar* L., and sea trout, *Salmo trutta* L., populations. *J. Fish Dis.* 5:293–308.

Kabata, Z. 1984. Diseases caused by metazoans: Crustaceans. In *Diseases of marine animals,* vol. 4, pt. 1, ed. O. Kinne, 321–97. Hamburg: Biologische Anstalt Helgoland.

Kelly, R. K., O. Nielsen, S. C. Mitchell, and T. Yamamoto. 1983. Characterization of *Herpesvirus vitreum* isolated from hyperplastic epidermal tissue of walleye, *Stizostedion vitreum vitreum* (Mitchill). *J. Fish Dis.* 6:249–60.

Kranz, H., N. Peters, G. Bresching, and H. F. Stich. 1980. On cell kinetics in skin tumours of the Pacific English sole *Parophrys vetulus* Girard. *J. Fish Dis.* 3:125–32.

Larsen, J. L., and N. J. Jensen. 1979. The ulcus-syndrome in cod (*Gadus morhua*). II. A bacteriological investigation. *Nord. Vet. Med.* 31:289–96.

Lobb, C. J. 1987. Secretory immunity induced in catfish, *Ictalurus punctatus* following bath immunization. *Dev. Comp. Immunol.* 11:727–38.

Lock, R. A. C., and A. P. van Overbeeke. 1981. Effects of mercuric chloride and methylmercuric chloride on mucus secretion in rainbow trout, *Salmo gairdneri* Richardson. *Comp. Biochem. Physiol.* 69:67–73.

Lom, J. 1981. Fish invading dinoflagellates: A synopsis of existing and newly proposed genera. *Folia Parasitol.* (Praha) 28:3–11.

Lopez-Vidriero, M. T., R. Jones, L. Reid, and T. C. Fletcher. 1980. Analysis of skin mucus of plaice *Pleuronectes platessa* L. *J. Comp. Pathol.* 90:415–20.

Lounatmaa, K., and J. Janatuinen. 1978. Electron microscopy of an ulcerative dermal necrosis (UDN)-like salmon disease in Finland. *J. Fish Dis.* 1:369–75.

McCarthy, D. H., and R. J. Roberts. 1980. Furunculosis of fish—the present state of our knowledge. *Adv. Aquat. Microbiol.* 2:293–341.

McCarthy, D. H., D. F. Amend, K. A. Johnson, and J. V. Bloom. 1983. *Aeromonas salmonicida*: determination of an antigen associated with protective immunity and evaluation of an experimental bacterin. *J. Fish Dis.* 6:155–74.

McCraw, B. M. 1952. Furunculosis of fish. In Special Scientific Report: Fisheries no. 84. U.S. Dept. of the Interior, Fish and Wildlife Service, 1–87. Washington, D.C.

McKenzie, R. A., and W. T. K. Hall. 1976. Dermal ulceration of mullet (*Mugil cephalus*). *Aust. Vet. J.* 52:230–31.

McVicar, A. H., and P. G. White. 1979. Fin and skin necrosis of cultivated Dover soie *Solea solea* (L.). *J. Fish Dis.* 2:557–62.

Miyazaki, T., and J. A. Plumb. 1985. Histopathology of *Edwardsiella ictaluri* in channel catfish, *Ictalurus punctatus* (Rafinesque). *J. Fish Dis.* 8:389–92.

Molnar, K. 1980. Cutaneous sphaerosporosis of the common carp fry. *Acta Vet. Acad. Sci. Hung.* 28:371–74.

Morrison, C., J. Cornick, G. Shum, and B. Zwicker. 1981. Microbiology and histopathology of "saddleback" disease of underyearling Atlantic salmon, *Salmo salar* L. *J. Fish Dis.* 4:243–58.

Noga, E. J., and M. J. Dykstra. 1986. Oomycete fungi associated with ulcerative mycosis in menhaden, *Brevoortia tyrannus* (Latrobe). *J. Fish Dis.* 9:47–53.

Obradovic, J., B. Maran, and R. Sabocanec. 1983. Papillomatosis in the sheat-fish, *Silurus glanis* L. *J. Fish Dis.* 6:83–84.

Olson, D. P., M. H. Beleau, R. A. Busch, S. Roberts, and R. I. Krieger. 1985. Strawberry disease in rainbow trout, *Salmo gairdneri* Richardson. *J. Fish Dis.* 8:103–11.

Olson, O. P., and N. Watabe. 1980. Studies on formation and resorption of fish scales. IV. Ultrastructure of developing scales in newly hatched fry of the sheepshead minnow, *Cyprinodon variegatus* (Atheriniformes:Cyprinodontidae). *Cell Tissue Res.* 211:303–16.

Oris, J. T., and J. P. Giesy, Jr. 1985. The photoenhanced toxicity of anthracene to juvenile sunfish (*Lepomis* spp.). *Aquat. Toxicol.* 6:133–46.

Paperna, I., I. Sabnai, and A. Colorni. 1982. An outbreak of lymphocystis in *Sparus aurata* L. in the Gulf of Aquaba, Red Sea. *J. Fish Dis.* 5:433–37.

Parisot, T. J. 1958. Tuberculosis of fish. *Bacteriol. Rev.* 22:240–45.

Paterson, W. D., D. Douey, and D. Desautels. 1980. Isolation and identification of an atypical *Aeromonas salmonicida* strain causing epizootic losses among Atlantic salmon (*Salmo salar*) reared in a Nova Scotian hatchery. *Can. J. Fish Aquat. Sci.* 37:2236–41.

Peleteiro, M. C., and R. H. Richards. 1985. Identification of lymphocytes in the epidermis of the rainbow trout, *Salmo gairdneri* Richardson. *J. Fish Dis.* 8:161–72.

Peters, N., W. Schmidt, H. Kranz, and H. F. Stich. 1983. Nuclear inclusions in the X-cells of skin papillomas of Pacific flatfish. *J. Fish Dis.* 6:533–36.

Phromsuthirak, P. 1977. Electron microscopy of wound healing in the skin of *Gasterosteus aculeatus. J. Fish Biol.* 11:193–206.

Pickering, A. D., and R. H. Richards. 1980. Factors influencing the structure, function and biota of the salmonid epidermis. *Proc. Royal Soc. Edinburgh* 79:93–104.

Poston, H. A., and M. J. Wolfe. 1985. Niacin requirement for optimum growth, feed conversion and protection of rainbow trout, *Salmo gairdneri* Richardson, from ultraviolet-B irradiation. *J. Fish Dis.* 8:451–60.

Pottinger, T. G., A. D. Pickering, and N. Blackstock. 1984. Ectoparasite induced changes in epidermal mucification of the brown trout, *Salmo trutta* L. *J. Fish Biol.* 25:123–28.

Pulsford, A., and R. A. Matthews. 1982. An ultrastructural study of *Myxobolus exiguus* Thelohan, 1895 (Myxosporea) from grey mullet, *Crenimugil labrosus* (Risso). *J. Fish Dis.* 5:509–26.

Roberts, R. J. 1972. Ulcerative dermal necrosis (UDN) of salmon (*Salmo salar* L.). Symp. Zool. Soc. Lond. 30:53–81.

———. 1975. The effect of temperature on diseases and their histopathological manifestations in fish. In *The Pathology of Fishes,* ed. W. E. Ribelin and G. Migaki, 399–428. Madison: University of Wisconsin Press.

Roberts, R. J., and A. M. Bullock. 1976. The dermatology of marine teleost fish. II. Dermatopathology of the integument. *Oceanogr. Mar. Biol. Ann. Rev.* 14:227–46.

Roberts, R. J., and B. J. Hill. 1976. Studies on ulcerative dermal necrosis of salmonids. V. The histopathology of the condition in brown trout (Salmo trutta L.). J. Fish Biol. 8:89–92.

Roberts, R. J., H. Young, and J. A. Milne. 1971. Studies on the skin of plaice (Pleuronectes platessa L.) 1. The structure and ultrastructure of normal plaice skin. J. Fish Biol. 4:87–98.

Robertson, D. A. 1985. A review of Ichthyobodo necator (Henneguy, 1883), an important and damaging fish parasite. In Recent Advances in Aquaculture, vol. 2, ed. J. F. Muir and R. J. Roberts. London and Sydney: Croom Helm.

Robertson, D. A., R. J. Roberts, and A. M. Bullock. 1981. Pathogenesis and autoradiographic studies of the epidermis of salmonids infested with Ichthyobodo necator (Henneguy, 1883). J. Fish Dis. 4:113–25.

Russell, P. H. 1974. Lymphocystis in wild plaice Pleuronectes platessa (L.), and flounder, Platichthys flesus (L.), in British coastal waters: a histopathological and serological study. J. Fish Biol. 6:771–78.

St. Louis-Cormier, E. A., C. K. Osterland, and P. D. Anderson. 1984. Evidence for a cutaneous secretory immune system in rainbow trout (Salmo gairdneri). Dev. Comp. Immunol. 8:71–80.

Sano, T., H. Fukuda, M. Furukawa, H. Hosoya, and Y. Moriya. 1985. A herpesvirus isolated from carp papilloma in Japan. In Fish and Shellfish Pathology, ed. A. E. Ellis. 307–11. London: Academic Press.

Sato, N., N. Yamane, and T. Kawamura. 1982. Systemic *Citrobacter freundii* infection among sunfish *Mola mola* in Matsushima Aquarium. *Bull. Jpn. Soc. Sci. Fish.* 48:1551–57.

Shotts, E. B., Jr., F. D. Talkington, D. G. Elliott, and D. H. McCarthy. 1980. Aetiology of an ulcerative disease in goldfish, *Carassius auratus* (L.): characterization of the causative agent. *J. Fish Dis.* 3:181–86.

Sohnle, P. G., and M. J. Chusid. 1983. Defense against infection with filamentous fungi in rainbow trout. *Comp. Biochem. Physiol.* 74:71–76.

Tatner, M. F., C. M. Johnson, and M. T. Horne. 1984. The tissue localization of *Aeromonas salmonicida* in rainbow trout, *Salmo gairdneri* Richardson, following three methods of administration. *J. Fish Biol.* 25:95–108.

Trust, T. J., E. E. Ishiguro, and H. M. Atkinson. 1980. Relationship between *Haemophilus piscium* and *Aeromonas salmonicida* revealed by *Aeromonas hydrophila* bacteriophage. *FEMS Microbiol. Lett.* 9:199–201.

Uhazy, L. S. 1978. Lesions associated with *Philometroides huronensis* (Nematoda:Philometridae) in the white sucker (*Catostomus commersoni*). *J. Wildl. Dis.* 14:401–8.

Valtonen, T. E., and A. L. Keranen. 1981. Ichthyophthiriasis of Atlantic salmon, *Salmo salar* L., at the Montta hatchery in northern Finland in 1978–1979. *J. Fish Dis.* 4:405–11.

Ventura, M. T., and I. Paperna. 1985. Histopathology of *Ichthyophthirius multifiliis* infections in fishes. *J. Fish Biol.* 27:185–203.

Watermann, B. 1982. An unidentified cell type associated with an inflammatory condition of the subcutaneous connective tissue in dab, *Limanda limanda* L. *J. Fish Dis.* 5:257–61.

Whitear, M., A. K. Mittal, and E. B. Lane. 1980. Endothelial layers in fish skin. *J. Fish Biol.* 17:43–65.

Willoughby, L. G. 1978. Saprolegnias of salmonid fish in Windermere: a critical analysis. *J. Fish Dis.* 1:51–67.

Woodhead, A. D. 1982. Skin lesions in the tail of the spiny dogfish, *Squalus acanthias* L. *J. Fish Dis.* 5:71–74.

Yamada, J., and N. Watabe. 1979. Studies on fish scale formation and resorption. I. Fine structure and calcification of the scales in *Fundulus heteroclitus* (Atheriniformes:Cyprinodontidae). *J. Morphol.* 159:49–65.

Yamamoto, T., R. K. Kelly, and O. Nielsen. 1983. Epidermal hyperplasias of Northern Pike (*Esox lucius*) associated with herpesvirus and C-type particles. *Arch. Virol.* 79:255–72.

4 Kidney

General Considerations

Fish in a freshwater environment are constantly having to fight against the osmotic influx of water that occurs across the gills during respiration. The kidney plays the major role in this fight, producing large quantities of very dilute urine. Loss of ions, which occurs through urine and passively across the gills, is compensated by active extraction from the water by the chloride cells in the gills.

In seawater by contrast, the situation is reversed and the fish has to fight against becoming dehydrated due to passive loss of water across the gills into the surrounding water, which has a much higher ionic strength than the fish's body fluids. Ingestion of water compensates for this but also introduces a large salt load that must then be excreted. This task falls to the chloride cells in the gills, which excrete mainly the monovalent ions Na^+ and Cl^-, and also to the kidneys, which excrete mainly the divalent ions. The kidney cannot produce concentrated urine, and at best it is isosmotic with blood. It is produced in low volumes, however, and there may be some further water absorption by the urinary bladder.

The kidney of teleost fish is usually a fused organ lying in a retroperitoneal location just ventral to the spinal column and often intermeshed with its processes, making its removal somewhat difficult. In addition to hemopoietic and lymphoid tissue, the anterior kidney contains endocrine elements, the chromaffin cells and interrenal tissue, which may be located around major blood vessels and which represent the equivalent of the mammalian adrenal medulla and cortex, respectively (see Fig. 11.5). The posterior kidney contains the nephrons with variable quantities of hemopoietic and lymphoid tissue in the interstitium. At the junction of anterior and posterior kidney may be found the corpuscles of Stannius (see Fig. 11.7). These are small white nodules of endocrine tissue and should not be mistaken for parasites or granulomas! Ureters fuse and may form a urinary bladder prior to ducting urine to the outside, posterior to the anus. The mammalian organization of a distinct renal cortex and medulla is not present.

The renal arteries supply blood to the glomeruli via afferent arterioles. The efferent arterioles then contribute blood to the peritubular capillaries. In many species, the other major source of blood to these capillaries is from the renal portal system, which receives blood and lymph draining from the tail region of the fish. Lymph from this area receives propulsion from muscular contraction and from a "lymph heart." Thus both anterior and posterior regions of the kidney possess a sinusoidal system of blood vessels that are lined by highly active phagocytes.

The kidney therefore has two distinct

blood supplies, arterial blood via the renal artery, afferent and efferent glomerular arterioles, and venous blood via the renal portal system. From a pathological perspective, a major consequence of this is a relative independence of glomerular from tubular lesions, at least those due to anoxia. In mammals, major interference with glomerular blood flow inevitably leads to tubular lesions due to anoxia, as they do not have an ancillary renal portal supply to protect tubules from such changes. Thus, even in situations where there is apparently complete obliteration of a teleost's glomerular blood flow, tubular lesions may be minimal. On the negative side, pathogens such as bacteria have yet another portal of entry into this nutrient-rich area. They can ascend through the urinary tract and renal portal system, or they can descend through the glomerular arterioles as embolic infection. Ascending infections may most frequently restrict themselves to the posterior kidney.

The kidney is a target organ in many diseases, particularly the economically important ones of cultured fish. A reason for this may be the affinity of the organ for circulating particulate antigens. Experimentally it has been shown in rainbow trout that more than 70% of bacteria inoculated into the dorsal aorta are initially trapped by the kidney, both anterior and posterior, although much of this redistributes to the spleen at a later stage. The trapping is mainly performed by the large number of phagocytes lining the renal sinusoids and peritubular capillaries of anterior and posterior portions (Fig. 4.1).

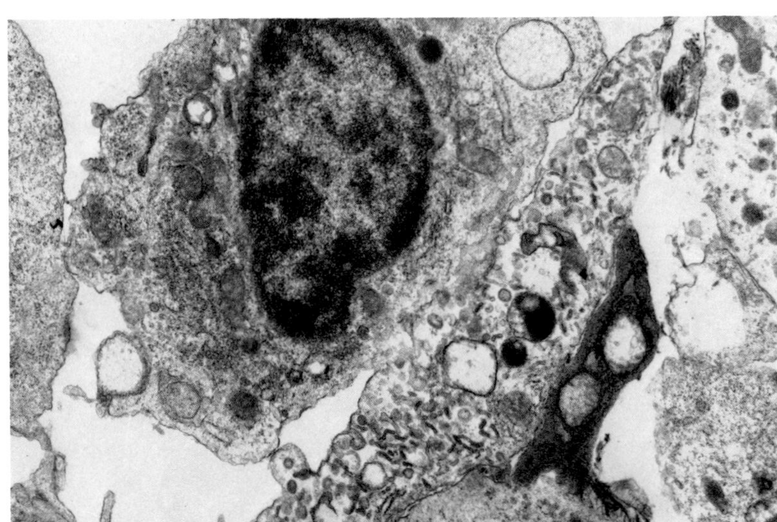

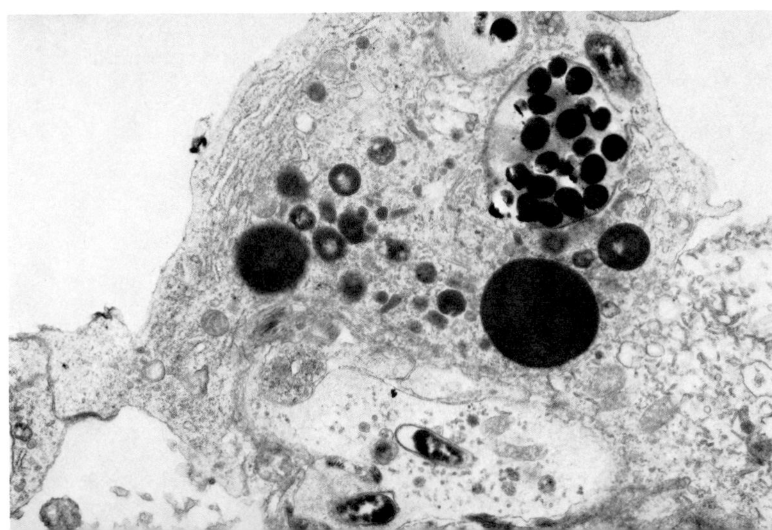

Fig. 4.1. A and **B.** Renal portal system showing macrophages resting on endothelium. Note extensive pinocytotic activity within cytoplasm of endothelial cells. Numerous lysosomes, melanin, and bacteria may be seen within macrophages.

The presence of hemopoietic tissue in the interstitium, often large amounts in the lower teleosts, can complicate the interpretation of inflammatory changes, especially in the early stages, unless there is a shift in the normal ratio of hemopoietic and lymphoid elements so that the picture becomes dominated by a particular cell type or there are accompanying degenerative changes (Fig. 4.2). Chronic nephritis is very common and is frequently severe enough to cause grossly visible changes such as bulging and distortion of the kidney or occasionally even rupture of the overlying capsule (Fig. 4.3). Other commonly observed gross lesions are color changes due to granulomatous inflammation or to mineral deposits within the parenchyma or dilating the ureters.

Melanomacrophage centers (MMC) are present in the interstitium of most teleosts except salmonids in which the aggregations are not discrete; an inexperienced observer may confuse these normal structures with granulomas. Of course, antigens and other particulates are deposited within these centers (Fig. 4.4), and as a consequence, MMCs may become the focus for granulomata formation, especially in diseases such as mycobacteriosis where the antigen persists. The number, size, and histological appearance of these structures change with age, season, state of nutrition, and antigenic exposure, and their examination should never be overlooked.

It is important to appreciate that many of the excretory and osmoregulatory func-

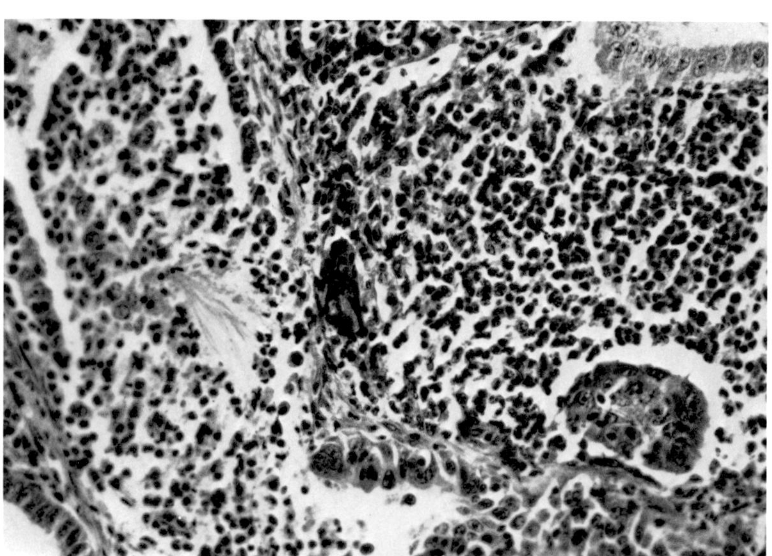

Fig. 4.2. Suppurative pyelonephritis in rainbow trout. A large number of neutrophils are present in interstitium and may be seen pouring through a degenerating epithelium into the tubular lumen.

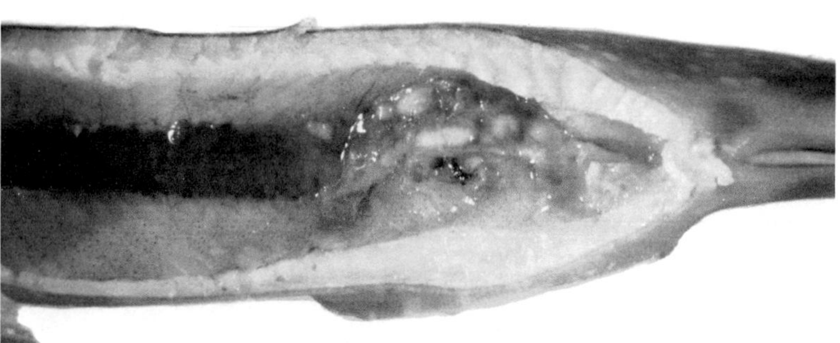

Fig. 4.3. Nephrocalcinosis in brook trout with marked distortion of posterior kidney, capsular rupture, and involvement of underlying muscle.

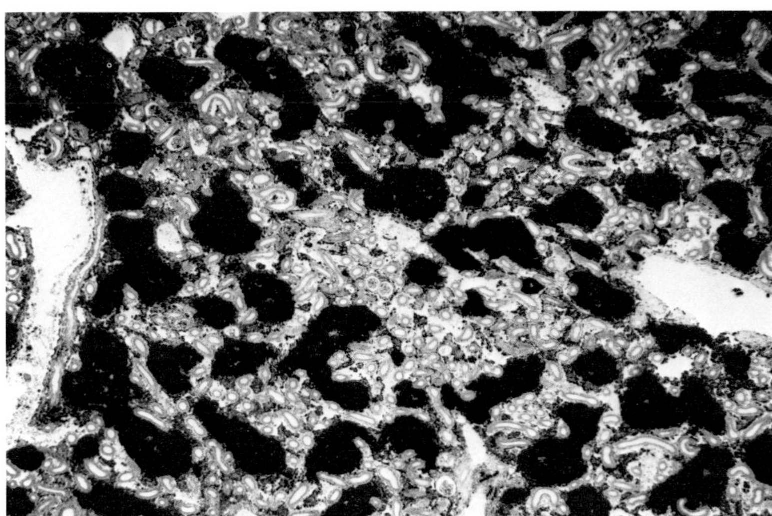

Fig. 4.4. Turbot (*Scophthalmus maximus*) injected intravenously with carbon and killed 24 hours later. Black foci represent carbon within melanomacrophage centers.

tions of the kidney in higher vertebrates are performed in teleosts by other organs, notably the gills and intestine. As a result, even severe renal disease may be accommodated by the fish, apparently with far fewer systemic consequences than would be seen with a comparable lesion in a mammal.

Involvement of the renin-angiotensin system in renal disease in fish is as yet unknown, although in rainbow trout a decrease in renal perfusion pressure does result in increased renin release, probably from the juxtaglomerular complex. Decreased angiotensin production (derived from the action of renin on angiotensinogen) as a consequence of severe renal disease could lead to impaired control over cardiac output, blood volume, and hence osmoregulation.

The Nephron

Marine species tend to have lower numbers of smaller glomeruli with fewer capillary loops than do freshwater species. Indeed some marine species are aglomerular. Adaptation from a freshwater to a seawater environment sees a reduction in glomerular filtration rate. This is accompanied by a reduced number of perfused and filtering glomeruli, by a decrease in endothelial fenestration, and by an increase in mesangial matrix.

Glomerular lesions are not uncommon, but they are not seen as frequently as in higher vertebrates. Once again, interpretation is complicated by the morphological differences between species living in a variety of environments and by the alterations that can occur when some species adapt to water of a different salinity, no doubt partly a reflection of the large physiological adaptations required. Glomerular involvement is seen in viremias, bacteremias, and parasitemias (Fig. 4.5), but acute glomerulonephritis is relatively uncommon compared to interstitial disease. In general, a full spectrum of pathological changes is seen with synechial attachment of visceral to parietal epithelium, followed by proliferation of epithelial and mesangial components, and eventually sclerosis in chronic disease (Fig. 4.6), which usually also involves Bowman's capsule. Proliferative glomerulonephritis is seen in eels associated with a virus similar to that of infectious pancreatic necrosis (IPN); hyaline droplets are also found in tubular epithelial cells, probably a consequence of altered glomerular filtration. Other associated lesions include multifocal necrosis in renal interstitium, liver, and spleen, as well as proliferative branchitis.

Diffuse or segmental thickening of the

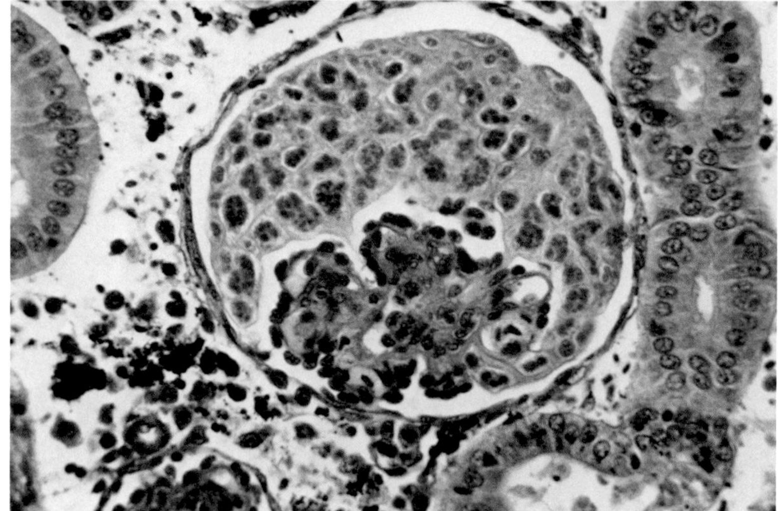

Fig. 4.5. Various stages of the myxosporean *Chloromyxum* filling Bowman's space in rainbow trout. (*Material courtesy of C. Smith*)

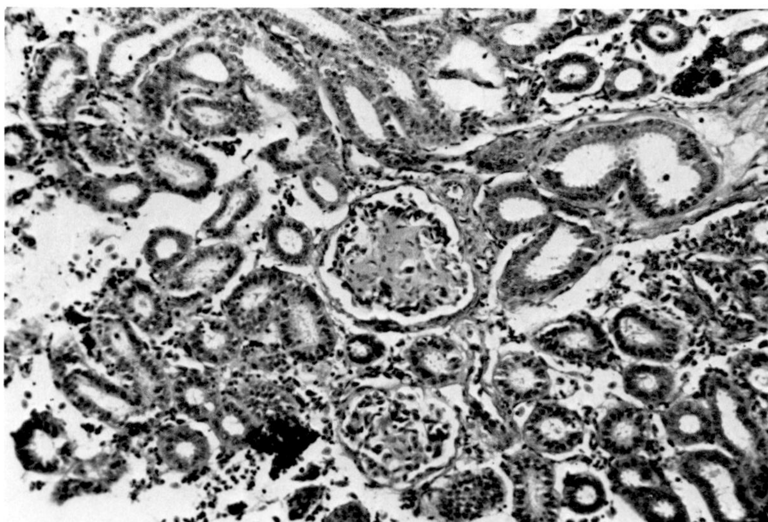

A

Fig. 4.6. A and **B.** Chronic glomerulonephritis represented in **A** by mesangial proliferation and in **B** by fibrosis of Bowman's capsule with fibrinoid deposits (*arrow*) plus numerous synechiae of sclerosing tuft.

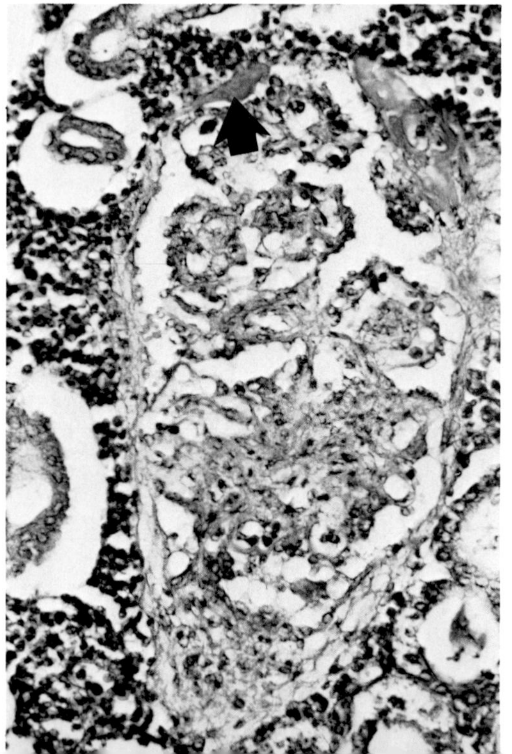

B

basement membranes is not uncommon in a variety of infectious and noninfectious conditions including chronic viral hemorrhagic septicemia (VHS) and Sekoke disease, a diabetes mellituslike syndrome in carp. Although subendothelial electron-dense deposits have been reported in several species (Fig. 4.7) with or without accompanying inflammatory cells, these have not been conclusively shown to be immunoglobulin in origin. Hence, immune-complex-mediated glomerulonephritis remains a largely unknown entity in fish.

Although there is great interspecies variation, *tubules* may be divided into neck seg-

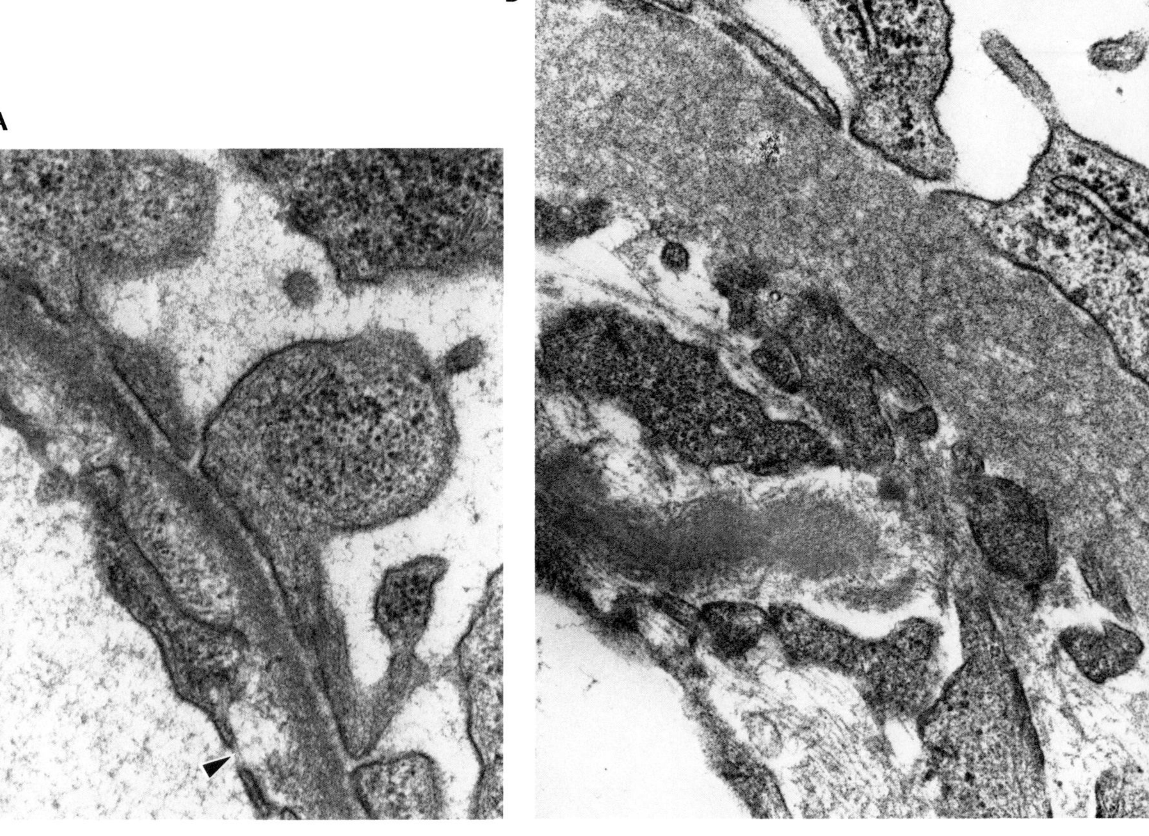

Fig. 4.7. A and **B.** Glomerulonephritis in white perch showing, in **A,** a normal capillary endothelium (fenestration, *arrow*) with podocytes sitting outside. Compare with **B** where basement membrane is several times thicker, and dense deposits lie beneath endothelium. Whether these represent immune complexes or merely fibrin is not known. The podocyte processes do not seem thicker than normal, nor are they fused. These fish also had marked fibrinoid necrosis of afferent and efferent glomerular arterioles as well as some splenic vessels, suggesting a possible Shwartzman-like reaction (see Fig. 6.16).

ments, proximal segments I and II, intermediate and distal segments (the latter is not usually found in marine teleosts), and collecting duct. The proximal tubular epithelium of columnar cells has a prominent brush border and abundant rough endoplasmic reticulum and mitochondria. The distal segment is most common in freshwater species and is considered to be the major site for dilution of tubular filtrate. Glomerular disease can result in protein leakage into the filtrate, recognized histologically as homogeneous eosinophilic deposits within tubular lumina and presumably leading to proteinuria, although protein is reabsorbed from the lumen. Such deposits need to be differentiated from myoglobin or hemoglobin. Although toxic nephrosis from these endogenous pigments is anticipated in mammals, this is not a widely recognized response in fish. Many species of apparently otherwise normal fish will often have brightly eosinophilic droplets within the cytoplasm of proximal tubular epithelial cells

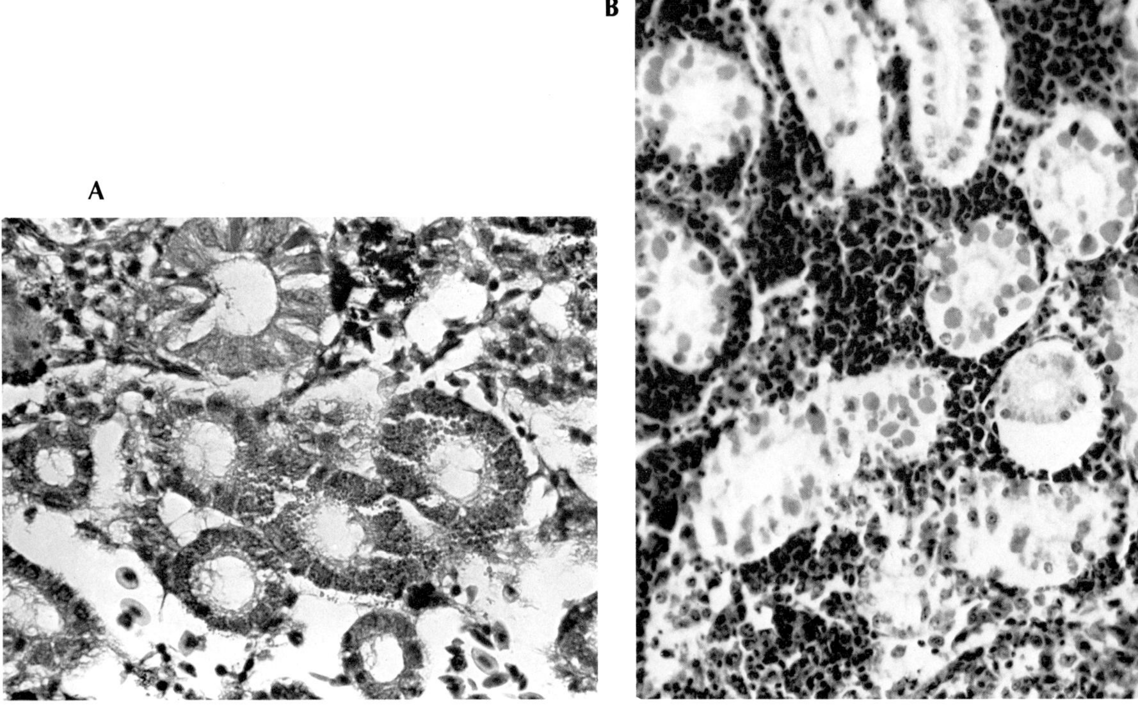

Fig. 4.8. A. Rainbow trout kidney with numerous fine granular droplets within tubular epithelial cells. **B.** Siamese fighting fish with larger eosinophilic droplets.

(Fig. 4.8). These may be fine and numerous or large and few in number. Thought to be lipoproteinaceous material and possibly reabsorbed from the tubular lumen, their significance is unknown, although several authors have noted in salmonids an association with toxicants and high ammonia levels. Great care must be exercised when interpreting tubular changes, particularly degeneration, because artifact and normal inter- and intraspecies variations are all common.

While tubular necrosis or fibrosis may occur as a result of extension from a pathological process taking place in the adjacent hemopoietic tissue or renal portal vasculature, tubules develop pathological changes that are of significance per se. In addition to the reabsorption of salt and glucose, the renal tubular epithelium has a major function in the excretion of divalent ions and of organic anions and cations. As such, it is a major route for the excretion of foreign chemicals. Thus pollution with heavy metals such as mercury or cadmium is highly likely to affect these cells. Indeed, the kidney is a target organ for many heavy metals and organic xenobiotics studied to date. The whole subject of teleost renal toxicology is superbly reviewed by Pritchard and Renfro (1984). Trump et al. (1975), using isolated flounder tubules, and Hawkins et al. (1980), in entire spots (*Leiostomus xanthurus*), described in these teleosts the pathology associated with mercury and cadmium toxicity, respectively. Proximal tubular epithelium was mainly affected, and the sequence of changes mimicked those seen in mammals under similar circumstances. In the early stages, mitochondria were dense and contracted although the cell was swollen. Irreversible damage was characterized by calcium deposition within mitochondrial cristae. These investigations showed that the heavy metal binding to the exposed sulfhydryl groups caused an increase in the permeability of the cell membrane. Subsequent entry of the metal into the cell caused further damage to mitochondrial ATP production, with impaired cellular fluid regulation and eventual cell death as consequences. In addition to accidental exposure, drug-induced tubular cell degeneration or necrosis is not uncommon, especially in fish treated with high levels for extended periods of time (Fig. 4.9). As with other vertebrates, it is probable that following necrosis, regeneration of tubular epithelium depends on the basement membrane remaining intact; otherwise repair is by fibrosis.

Mineral deposits are frequently encountered within tubules, collecting ducts, and ureters of both farmed and wild fish. The condition is known as nephrocalcinosis. It is especially common in intensively reared rainbow trout and brook trout in which the grossly visible chalky white caseous material can distort the normal appearance of the kidney (Fig. 4.10). Mortality is usually low, and although food conversion efficiency is probably impaired, the major concern about the condition centers on a reduction in carcass quality at slaughter. In severe cases, the muscle dorsal to the kidney may also be affected (Fig. 4.3). Histologically, mineral deposits within the parenchyma elicit a pronounced granulomatous response involving both tubules and interstitium and resulting in fibrosis with ballooning of tubules and

A

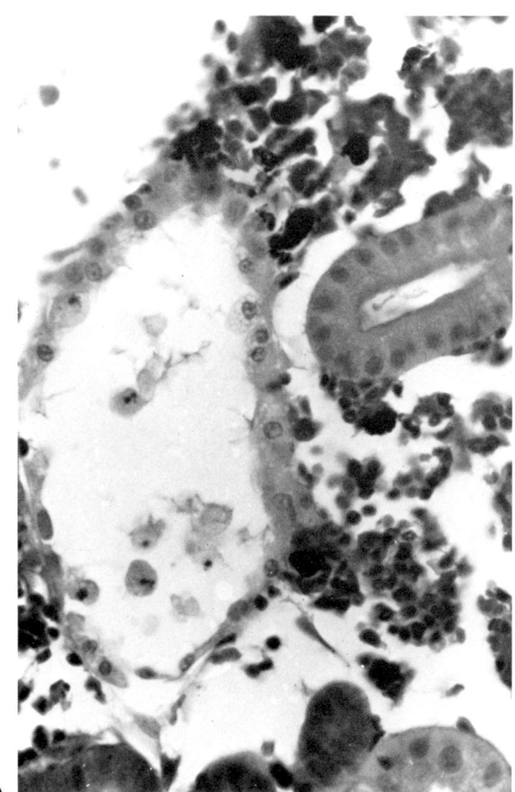

B

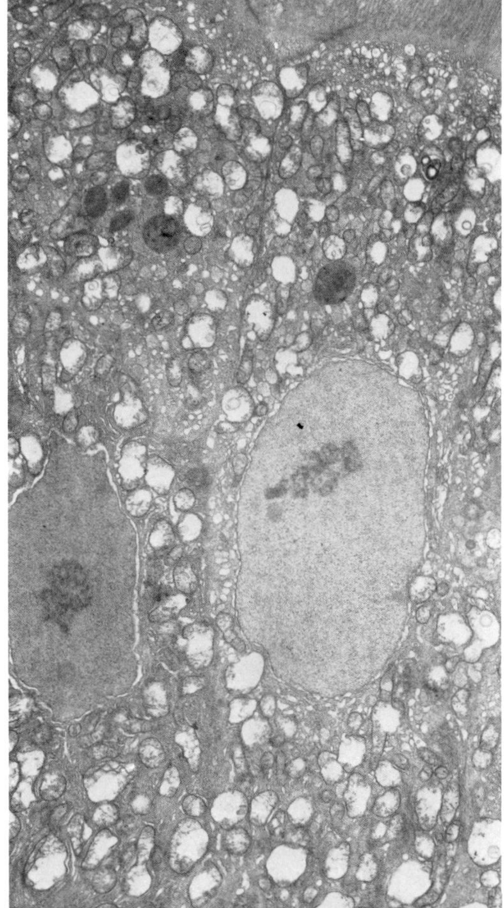

Fig. 4.9. A. Kidney from rainbow trout following oral administration of nitrofurantoin at dose rate of 3.5 g per 100 lb fish for 32 days. There is extensive necrosis of tubular epithelium and exfoliation into the lumen. **B.** Second proximal segment of nephron from rainbow trout treated with formalin (22 ppm) for 16 days, showing high amplitude swelling of mitochondria containing few cristae. (*Courtesy of B. Hicks*)

A

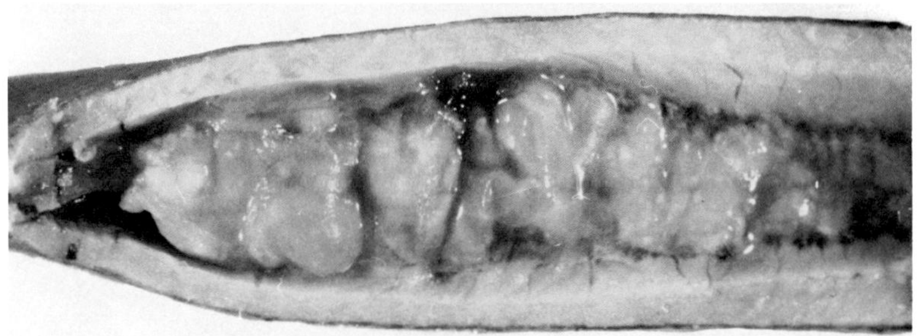
B

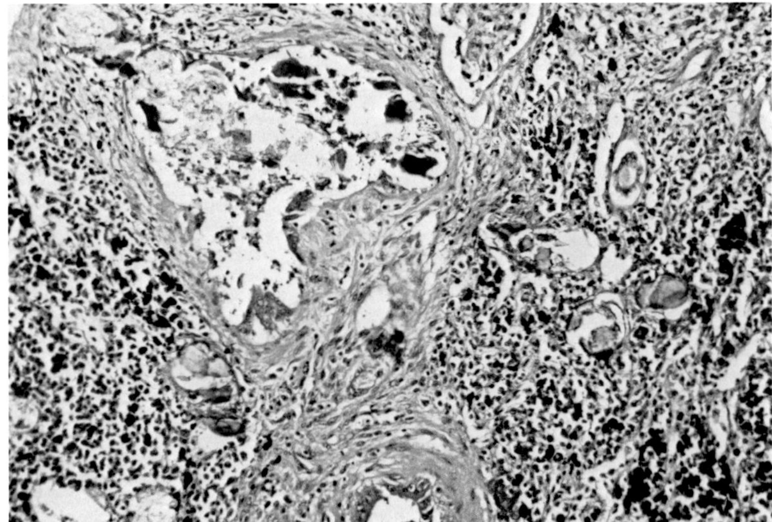
C

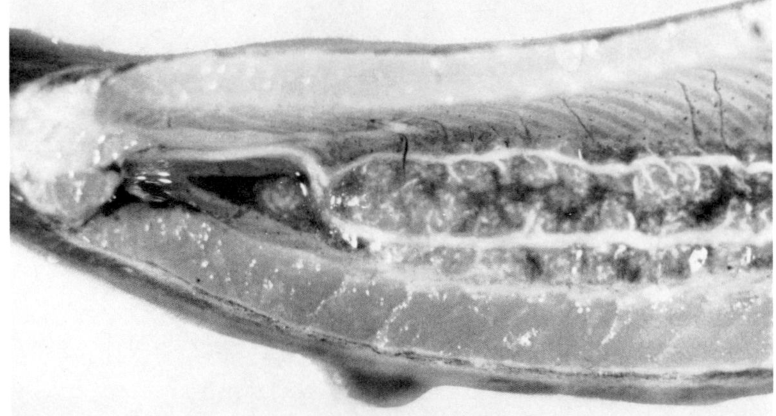
D

Fig. 4.10. A and **B.** Intensively cultured rainbow trout with nephrocalcinosis. The gray masses were gritty when sectioned and frequently cavitated. **C.** Histology of **A** showing large quantities of mineral within a greatly dilated and fibrotic tubule. Some mineral is also present within the interstitium. **D.** Experimentally produced selenium toxicity causing nephrocalcinosis. Collecting ducts and ureters are filled with mineral.

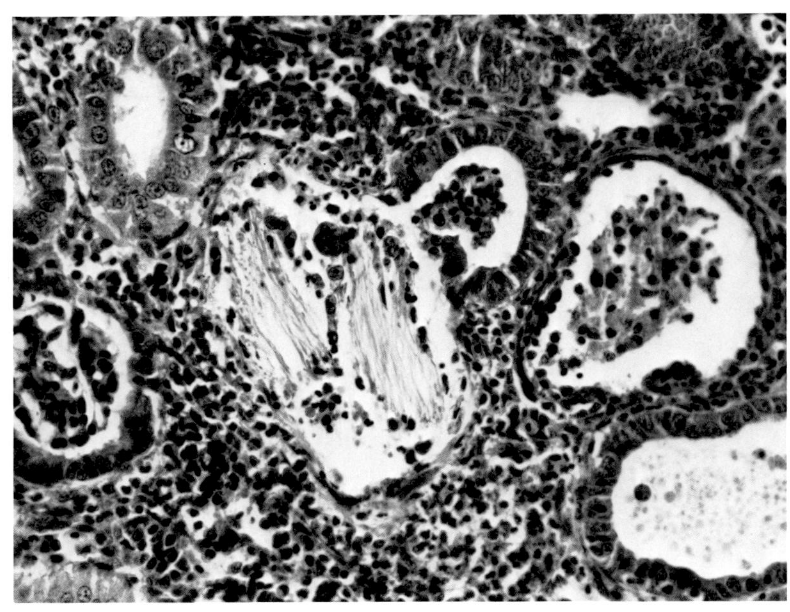

Fig. 4.11. Rainbow trout with acute degeneration and necrosis of tubular epithelium. Large numbers of sloughed and dead epithelial cells are present in the lumen, together with inflammatory cells, mainly neutrophils, which are also present in large numbers in the interstitium. One tubule contains prominent crystalline deposits. Whether these are the cause or the result of the tubulonephritis is unknown, and whether this is merely an early stage of the nephrocalcinosis syndrome, common among farmed salmonids, is also unknown. Fish with lesions like these, however, often have large numbers of gram-negative bacilli in their kidneys; these are encountered on impression smears only, and their signficance is unknown.

collecting ducts. Associated lesions include gastric submucosal granulomas. The cause (or causes) of the condition is unknown, although high CO_2 levels are implicated and have experimentally reproduced the lesions, as indeed have magnesium deficiency, selenium toxicity (Fig. 4.10D), and a diet low in minerals. The disease frequently presents in an advanced or essentially end-stage form, and it is likely that several very different etiologies could end up producing similar lesions.

Similar chronic renal changes are seen in "visceral granuloma," a condition of trout reported in North America and in gilthead bream (*Sparus aurata*) in Israel and differentiated from nephrocalcinosis by widespread involvement of other visceral organs and mesentery. The cause is unknown, but it is thought to have a dietary association. In the gilthead bream, the renal tubules are packed with tyrosine crystals, which, however, disappear as the condition advances, suggesting they may not be the primary problem.

Heavy metals or feed binders have been considered the cause of comparable granulomatous conditions in farmed turbot (*Scoph-*

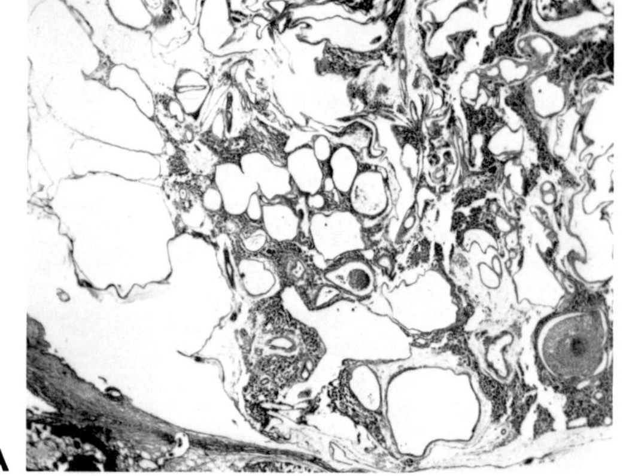

A

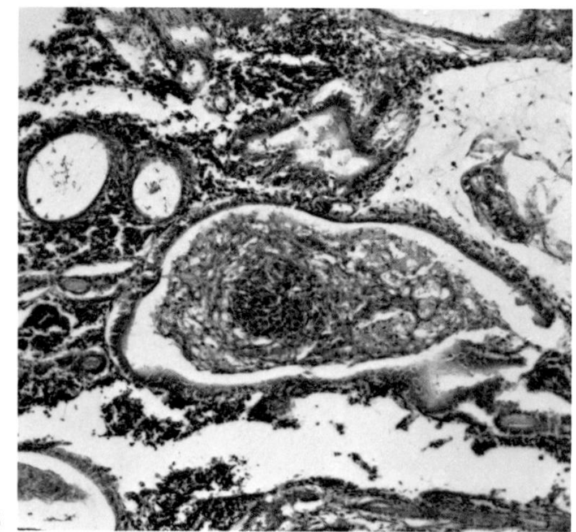

B

Fig. 4.12. A and **B.** Polycystic dilation of tubules in desert pupfish kept in captivity far beyond its normal lifespan. Cause of the dilation is unknown but may be associated with intratubular granulomas (**B**) seen in several locations. The cause of these is also unknown.

thalmus maximus). Oxalate crystals are also occasionally seen within the tubules (and renal but not splenic MMC) of a variety of species, particularly wild and aquarium fish; their significance is usually unknown, although they have been linked with pyridoxine deficiency in trout (Fig. 4.11) and vitamin A deficiency in clown fish. Polycystic dilatation of tubules is occasionally encountered, and although the cause and significance are rarely determined, the severity of the lesions may suggest renal failure (Fig. 4.12).

Degenerative changes within tubular epithelium, and cast formation comprised of cellular and proteinaceous debris, both accompany renal tubular parasitism. An important example of this is myxosporidiosis associated with the genera *Mitrospora* and *Sphaerospora.* While most of the kidney-infecting species of *Sphaerospora* so far described are not conspicuously pathogenic (Fig. 4.13), some do produce nephrosis or other lesions. Their importance, however, lies in the exciting recent discoveries concerning the extrarenal prespore stages of these parasites, which in the case of *S. renicola* have been shown to be a cause of swimbladder inflammation of carp fry. It has also been suggested that proliferative kidney disease (PKD) of salmonids is caused

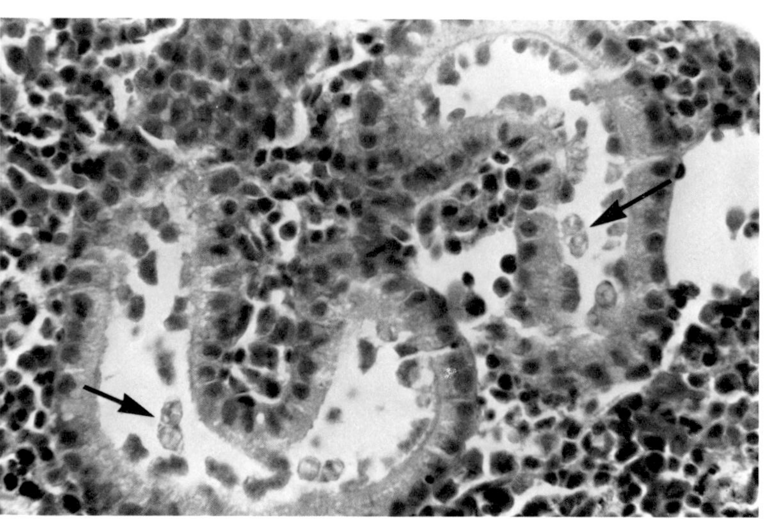

Fig. 4.13. *Sphaerospora*-like myxosporeans (*arrows*) associated with brush border of tubules in brown trout. The fish was taken from a river supplying water to a farm that suffers annual heavy mortality from proliferative kidney disease.

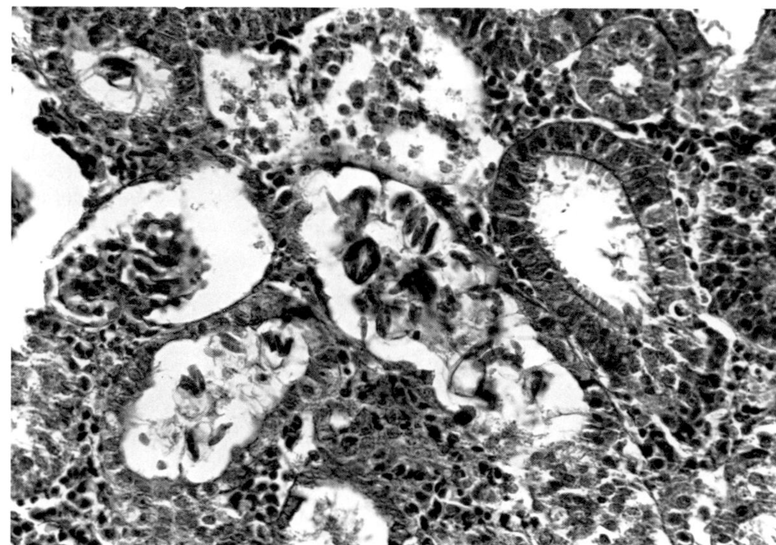

Fig. 4.14. Coccidia within kidney of common shiner causing a flattening of tubular epithelium. The majority of the renal parenchyma, however, is normal.

by a similar parasite, since tubular stages have recently been observed in the kidneys of North American fish. Tubular coccidiosis is occasionally seen in minnows, but the significance of the infection to the fish remains unknown (Fig. 4.14).

Inflammation of the urethra and urinary bladder is especially common in young salmonids with acute bacteremias and may be pronounced enough for the whole region to appear grossly red. This suggests either a descending infection following renal colonization or possibly an ascending one (Fig. 4.15). Although normally regarded as residents of skin or gills, parasites such as *Trichodina* occasionally may be encountered within the distal urinary tract (Fig. 4.15); their presence is usually regarded as incidental. Unilateral aplasia of a ureter, or chronic inflammation, may result in cystic dilation of that portion proximal to the obstruction. Although rare, such lesions are sometimes seen (Fig. 4.15).

Renal Interstitium

Many bacteria localize in the kidney, probably partly due to the extensive blood supply and trapping abilities of the organ. In acute bacteremias such as those associated with acute *Aeromonas, Pseudomonas, Vibrio,* or Enterobacteriaceae infection, the organisms can frequently be seen associated with the portal vasculature (Fig. 4.16). Toxin release or inflammation, leading to necrosis, can extend to tubules or hemopoietic elements.

The renal interstitium is also a favored location for granulomas due to many organisms, including mycobacteria, *Nocardia* spp., and systemic mycoses such as *Exophiala* spp., *Ochroconis* spp., and *Ichthyophonus* (Fig. 4.17). These lesions can sometimes be so extensive as to virtually obliterate the organ or cause compression of adjacent structures. In particular, there may be compression of the peritubular capillaries with resulting anoxic damage to tubules. Giant cells may be encountered as part of the host response that otherwise comprises epithelioid cells, macrophages, and lymphocytes with eventual caseous necrotic cores encased by fibrous tissue, especially in the mycobacterial granulomas. One of the most economically important diseases of salmonid culture is bacterial kidney disease (BKD).

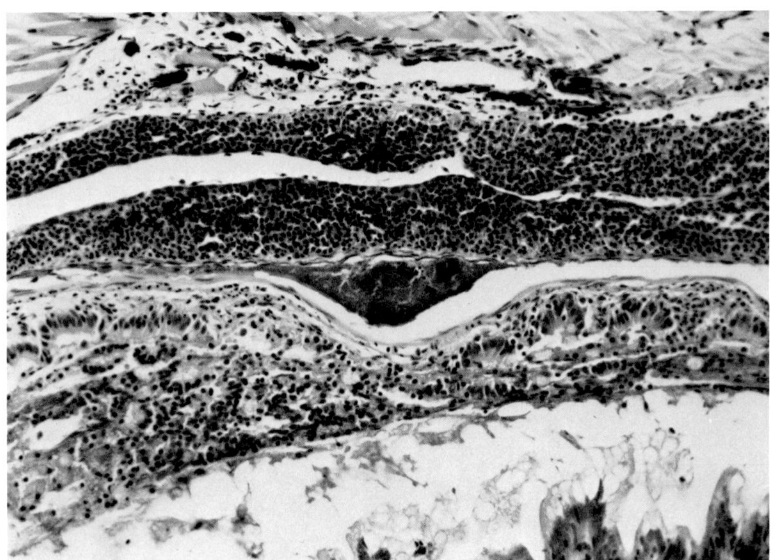
A

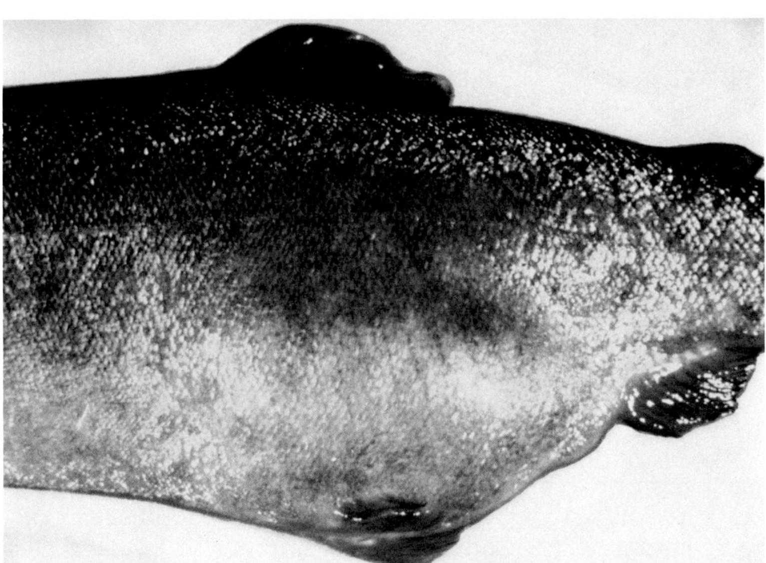
B

C

Fig. 4.15. A. Acute furunculosis in Atlantic salmon fry showing severe congestion of "urethra" and bacterial colony located beneath. **B** and **C.** Cystic dilation of ureter in trout, probably due to hypoplasia. **D.** Common shiner with several *Trichodina* present as an incidental finding within urethral lumen. Numerous rodlet cells (*arrow*) may also be seen.

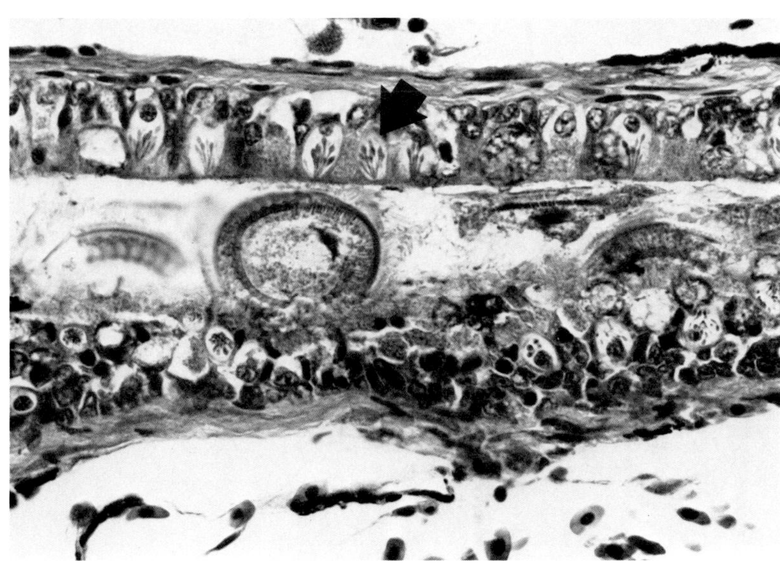

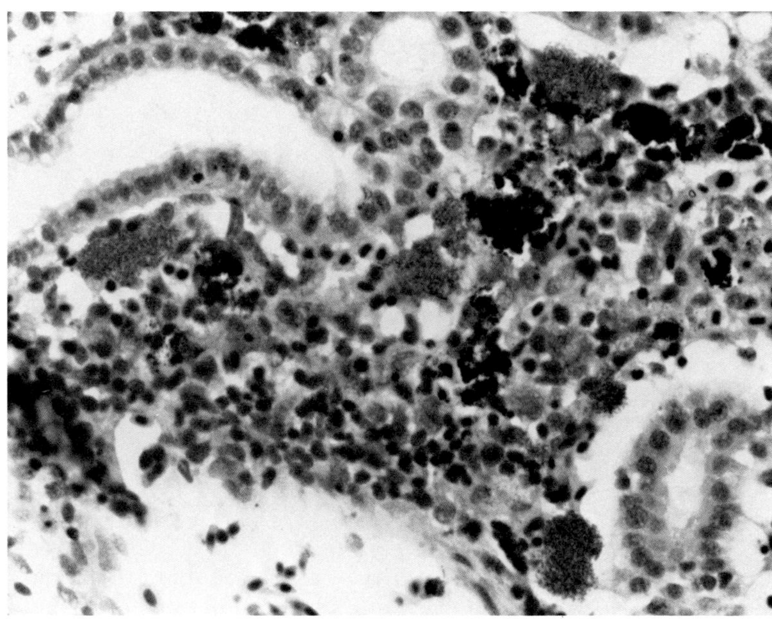

Fig. 4.16. Furunculosis in brown trout. Colonies of organisms are associated with endothelial lining of peritubular capillaries, part of the renal portal system.

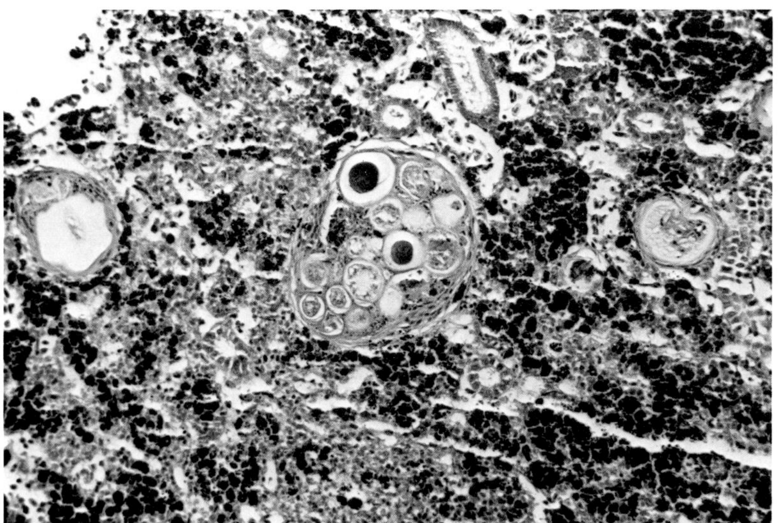

A

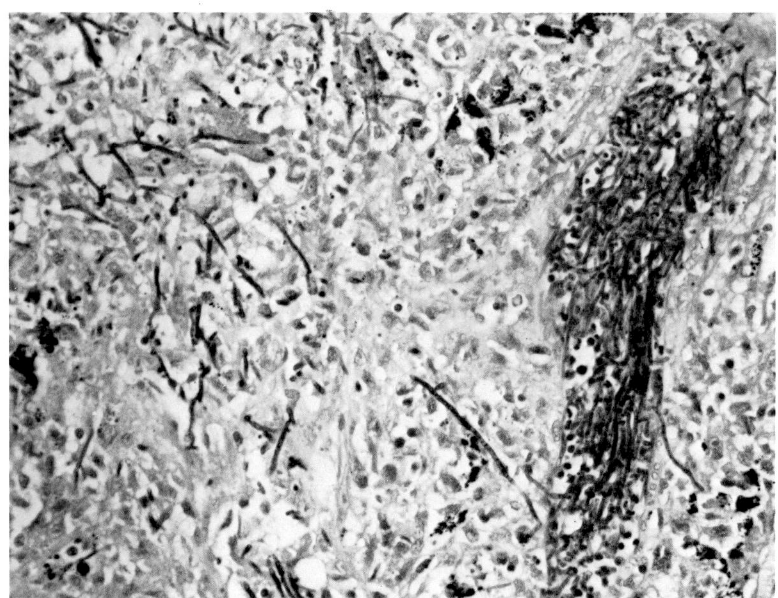

B

Fig. 4.17. A. *Ichthyophonus* in cultured rainbow trout. (*Material courtesy of J. Humphrey*) **B.** *Exophiala* sp. in Atlantic salmon. Lumen of the tubule is packed with fungal hyphae, which also extend into interstitium. (*Material courtesy of T. Poppe*)

Bacterial Kidney Disease

Renibacterium salmoninarum, a small gram-positive diplobacillus, specifically targets the salmonid kidney, although lesions are also found elsewhere (Fig. 4.18). The organism favors an intracellular location, and the host response is granulomatous or pyogranulomatous. In severe cases, the lesions may caseate and cavitate and contain a serous flocculent fluid. In addition to anterior and posterior kidney, which may be massively enlarged, nodular, and gray, another common site for lesions is the heart, which often has a diphtheritic epicarditis as well as a necrotizing and granulomatous myocarditis (see Fig. 6.12) with lesions frequently being present at the compact-spongy interface. In some cases the epicarditis can be so pronounced that there is gross distortion of the normal outline of the heart. Focal granulomas may be present in the spleen and liver and elsewhere in the viscera. Occasionally a peritonitis is seen, and this may present as a fibrinous cast over hepatic and splenic capsules. The ovary and developing eggs may be affected, allowing for vertical transmission of the disease. The eye is commonly involved, granulomas being present in the retrobulbar fat and connective tissues, choroid gland, and uveal tract—exophthalmia is frequently noted in severely affected fish and indeed may at times be the only grossly visible evidence of disease. The gills are often affected (embolic trapping?), and epithelial cells containing organisms may be seen exfoliating from the lamellar surfaces. Less commonly affected sites include lamina propria of the gut and the meninges of the brain. Subdermal blisters are not uncommon in larger fish, the reaction extending down into the underlying muscle.

A syndrome so far peculiar to southern Ontario is known colloquially as "spawning rash" (see Fig. 3.20). Affected fish have small intradermal vesicles and granulomas that are detected at spawning time but disappear at other times of the year. Lesions elsewhere in the body are not usually present. Probably a variant of BKD, the typical organisms are certainly capable of experimentally causing visceral disease in other young salmonids and cyprinids.

Histologically the response is dominated by macrophages, which are often stuffed with organisms, and by neutrophils, which occasionally contain one or two, thereby emphasizing the very different roles these cells play in teleosts by comparison with mammals. Of interest, the bacteria are often difficult to see in a routine H and E–stained section despite almost obliterating cells by sheer numbers; a gram- or PAS-stained section or impression smear reveals the extent of tissue involvement. Fibrin exudation is seen in some locations, especially when there are vascular lesions, which are quite common and indeed may be a major feature of the naturally occurring disease. These can take the form of necrosis of the media or endothelium, and in severe cases limited hemorrhage or thrombosis may also be evident. These lesions may be seen in most tissues although grossly the epicardium is a prime target (see Fig. 6.12A).

Young and Chapman (1978) provide an excellent ultrastructural description of the developing lesions in the nephrons of experimentally and naturally infected brook trout. In the glomerulus they showed subendothelial electron-dense deposits as the earliest change; with time, deposits were also noted in the mesangium, which proliferated. Fusion of foot processes, thickening of basement membrane, and blebbing of endothelial cells were noted later. Tubular damage was also seen; changes observed included swollen rough endoplasmic reticulum (RER) and mitochondria and, eventually, evidence of irreversible damage such as ruptured mitochondria. Infiltration of neutrophils and macrophages was seen, with some fibrosis at a later stage. In our experience, BKD is largely an interstitial reaction with secondary involvement of the nephron, probably due to interference with blood supply by compression. This, however, is probably determined by route of infection, the renal portal macrophages acting as an early nidus, with later spread to endothelium and underlying hemopoietic elements.

Proliferative Kidney Disease

One of the most serious diseases of farmed salmonids, this condition is presumed to be caused by a protozoan parasite, probably a prespore myxosporean. As the name suggests, the kidney is again the target organ, and it may become so massively enlarged as to be visible externally along the lateral line. Exophthalmia, ascites, and anemia are other gross features. Both anterior and posterior portions of the kidney may be thrown into gray bulbous ridges or discrete nodules, depending on the severity of in-

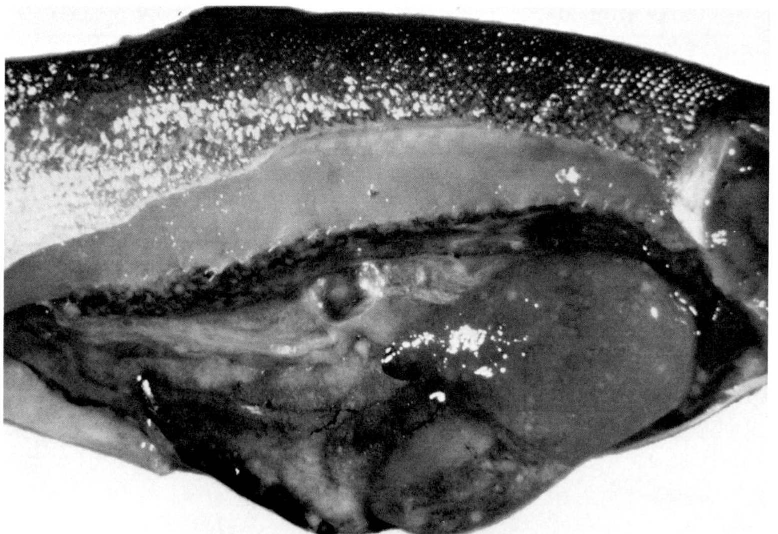
A

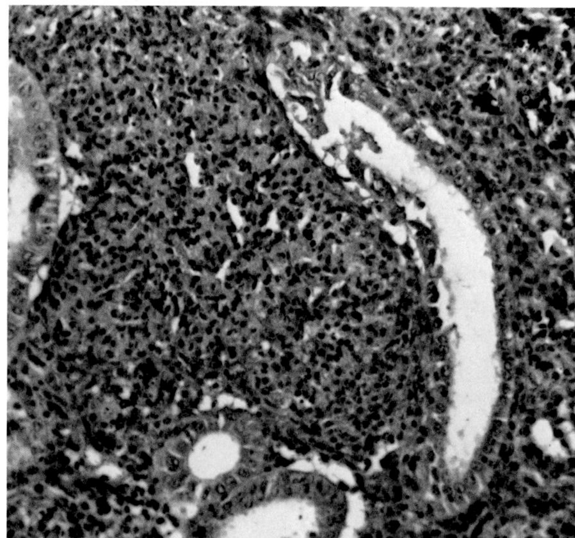
B

Fig. 4.18. A. Landlocked Atlantic salmon with bacterial kidney disease showing multifocal granulomas throughout viscera and kidney. **B.** Focal interstitial granulomas in bacterial kidney disease may cause compression of portal vasculature and tubules, some of which may subsequently drop out. **C.** Coho salmon with bacterial kidney disease showing large number of bacteria within renal portal macrophage, often within phagolysosomes. A single organism may be seen within the cytoplasm of the underlying endothelial cell (*arrow*). **D.** Neutrophil within renal interstitium of same fish as in **C.** Despite ample opportunity, only a single organism has been phagocytosed. Although this particular cell may be immature, comparing Figs. **C** and **D** nevertheless gives a clear illustration of the very different roles played by these two cell types.

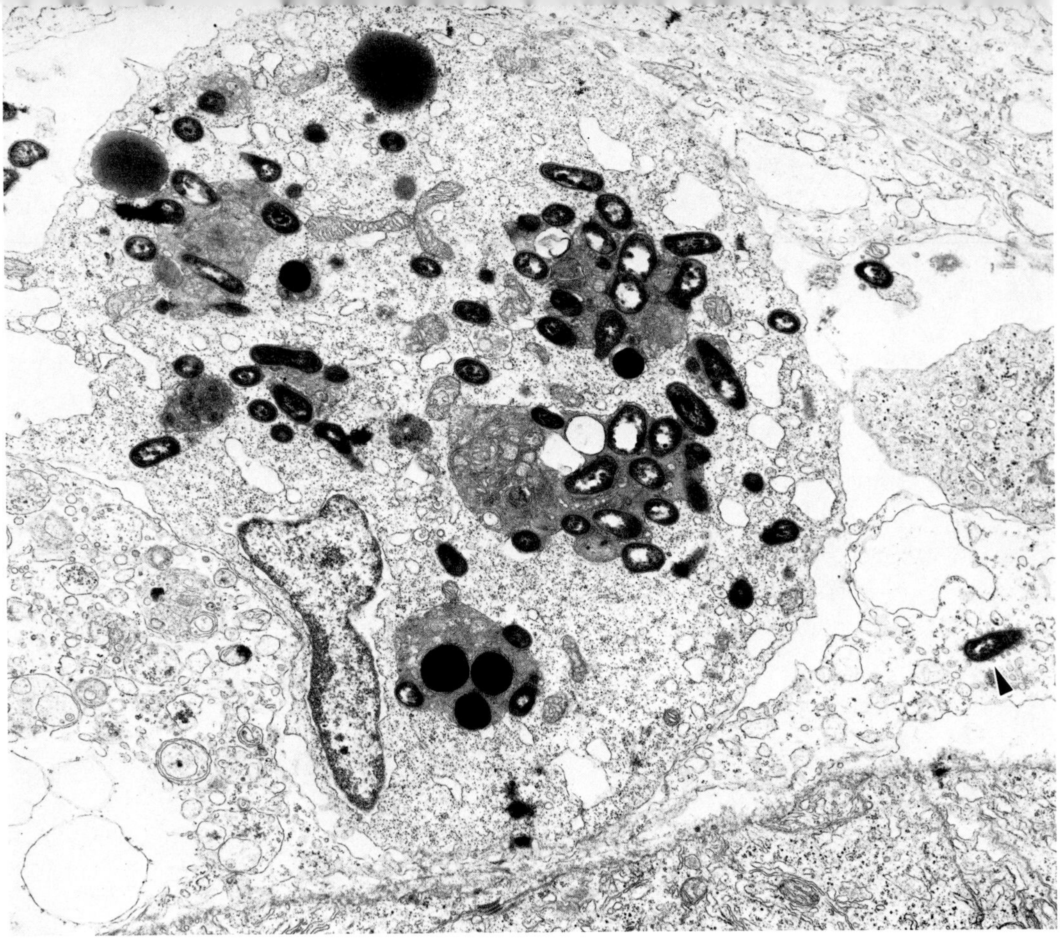

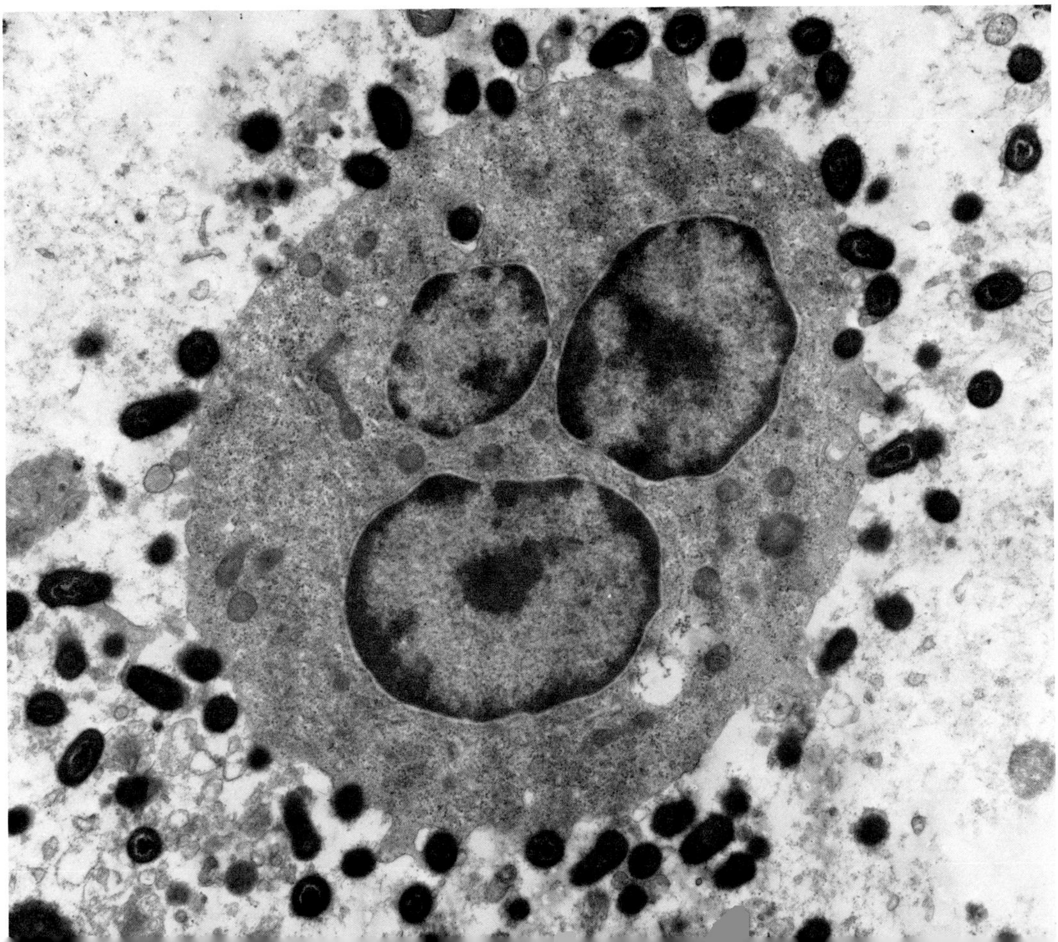

volvement (Fig. 4.19). Splenomegaly is usually present and indeed may be one of the earliest indications of infection. Histology reveals a multifocal granulomatous inflammation, widely disseminated in many organs in severely affected fish but invariably present in kidney and possibly spleen even in mild infections. Early lesions center on the interstitium, and as a feature accompanying this and many other granulomatous renal lesions, there is an apparent reduction in melanin. The reason for this is unknown although it may be related to macrophage activation. The response in all organs comprises mainly macrophages and lymphocytes, although in the kidney there is also

Fig. 4.19.

A. Proliferative kidney disease (PKD) in Arctic char showing marked hypertrophy of kidney, which is largely gray and thrown into bulbous ridges.
B. Typical appearance of PKD within renal interstitum of rainbow trout. Note lack of melanomacrophages and presence of many large "mother" or "primary" cells, many of which have an accompanying cuff of macrophages and lymphocytes (*arrow*).
C. Early PKD in rainbow trout showing the mitotic activity (in this case two neutrophils are in mitosis), which accompanies the parasites (*arrow*).

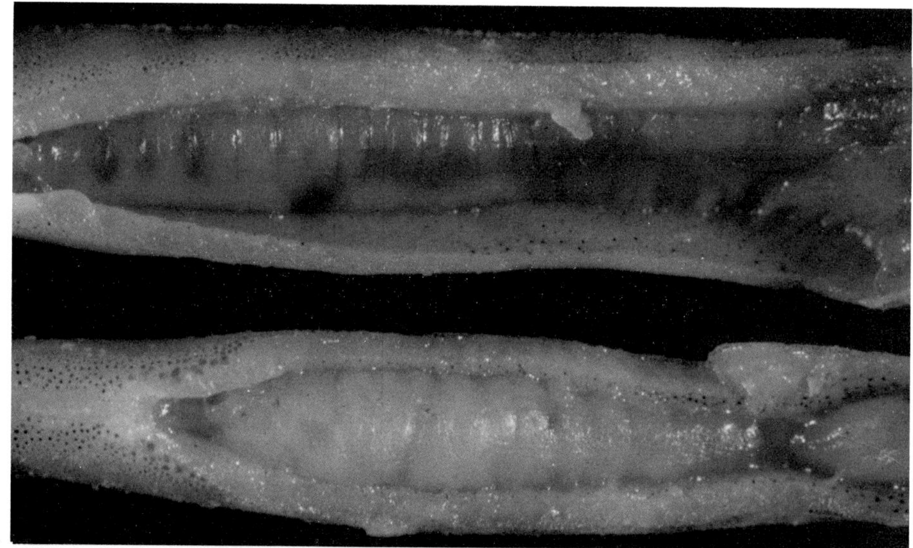
A

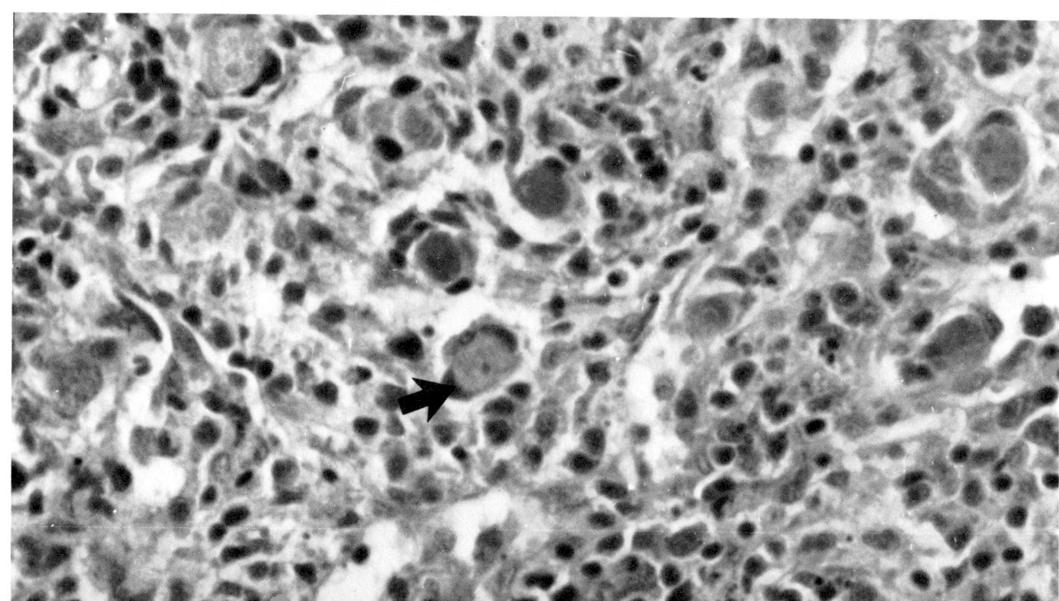
B

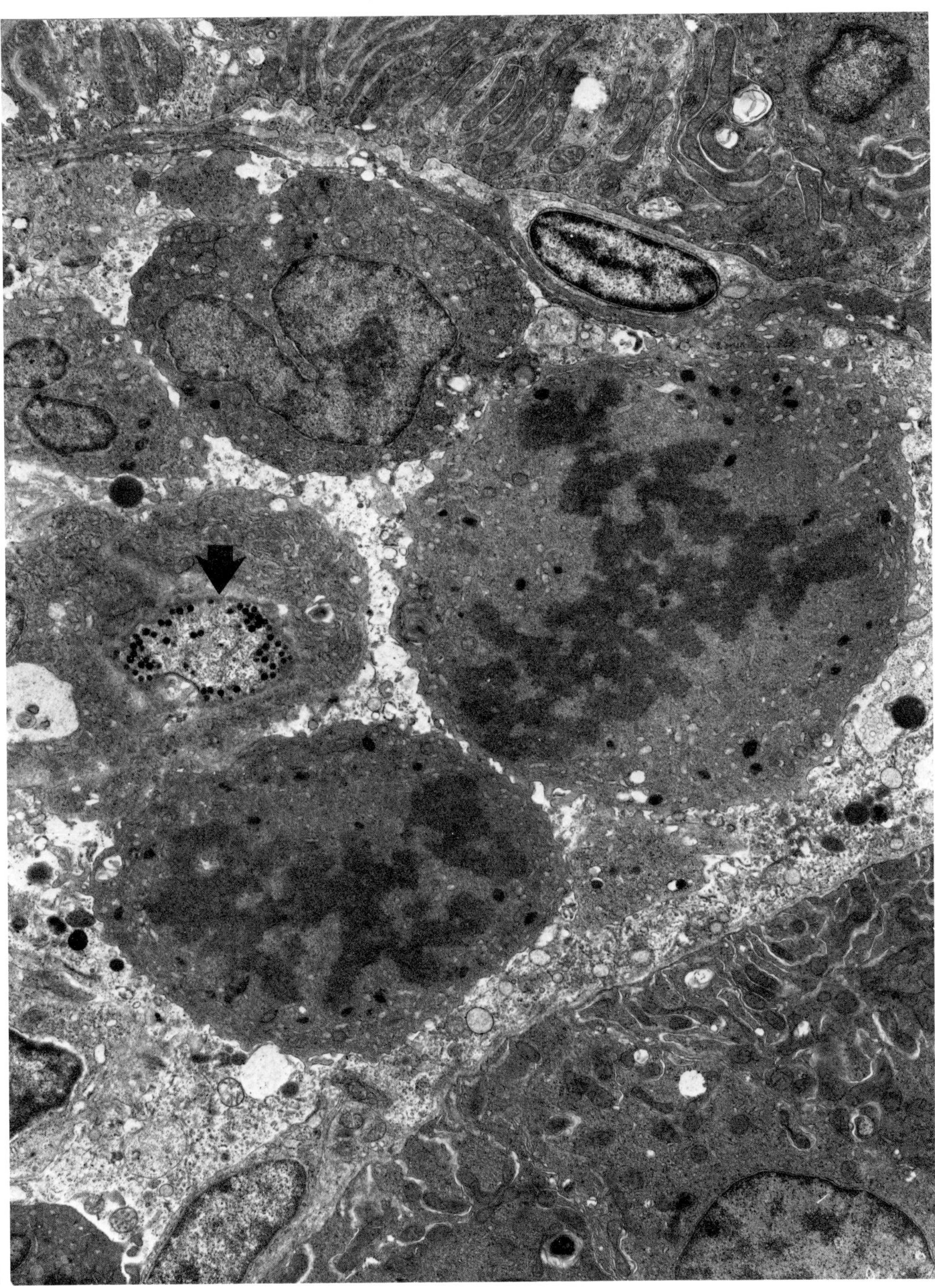

D. PKD in rainbow trout showing a "mother" parasite within the lumen of a vessel, surrounded by host macrophages. A single lymphocyte is also present—note its close apposition to the plasmalemma of the "mother" parasite.
E. PKD in rainbow trout showing secondary cell within a primary. Note presence of numerous dense spherical bodies within cytoplasm of primary; these are absent from secondary cell.
F. High-power micrograph of dense bodies. Note that their electron-lucent bar is perpendicular to the plasmalemma when they touch it. Two also have "tails," possibly a remnant of the Golgi from which they may derive.

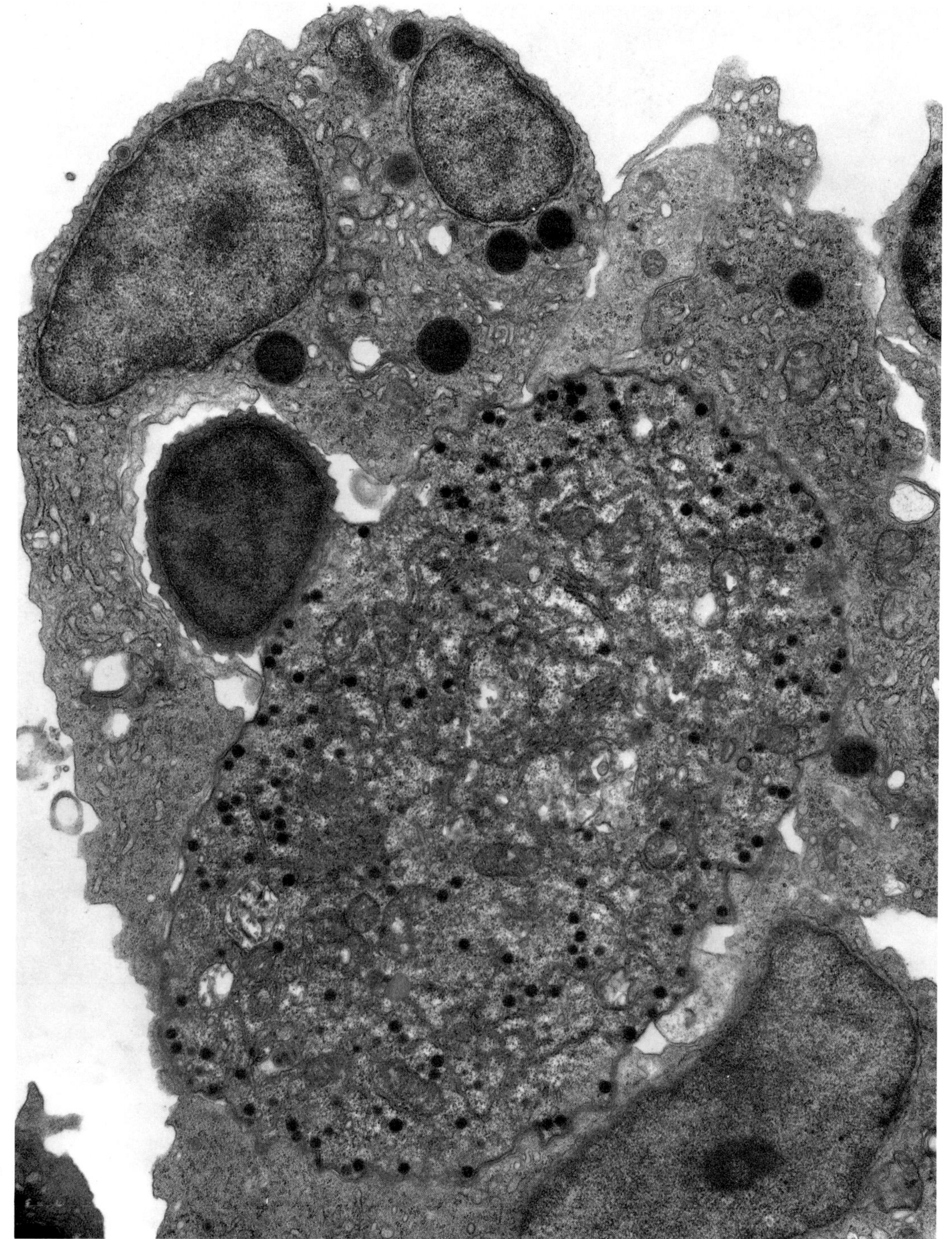

D

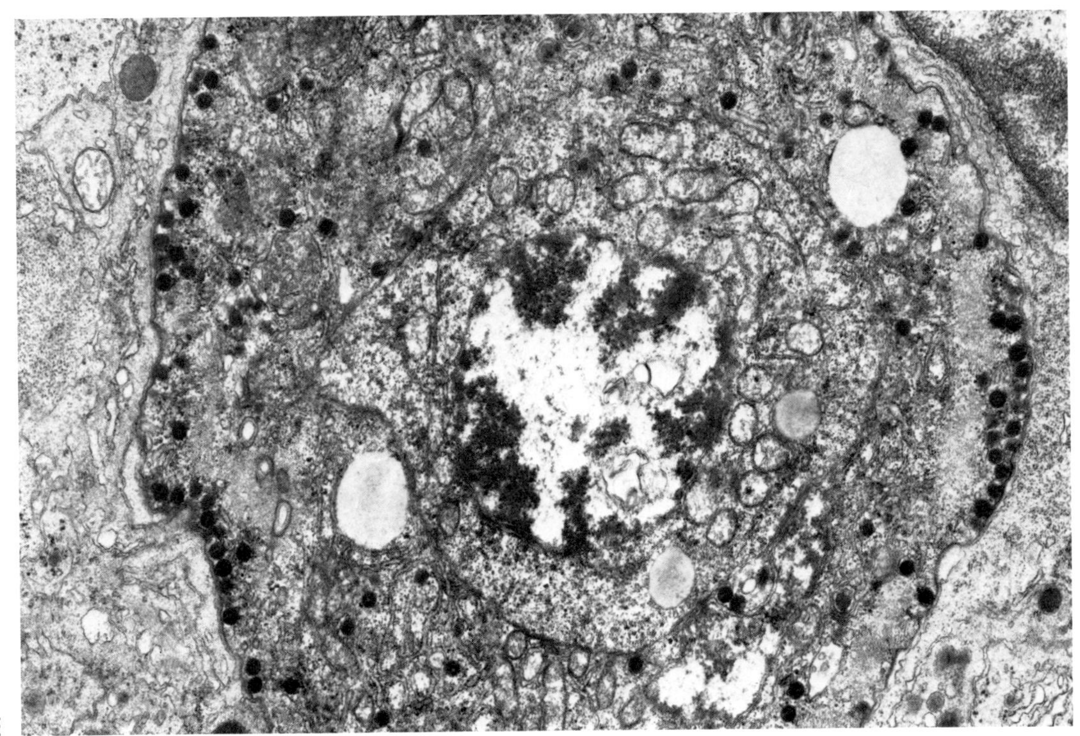

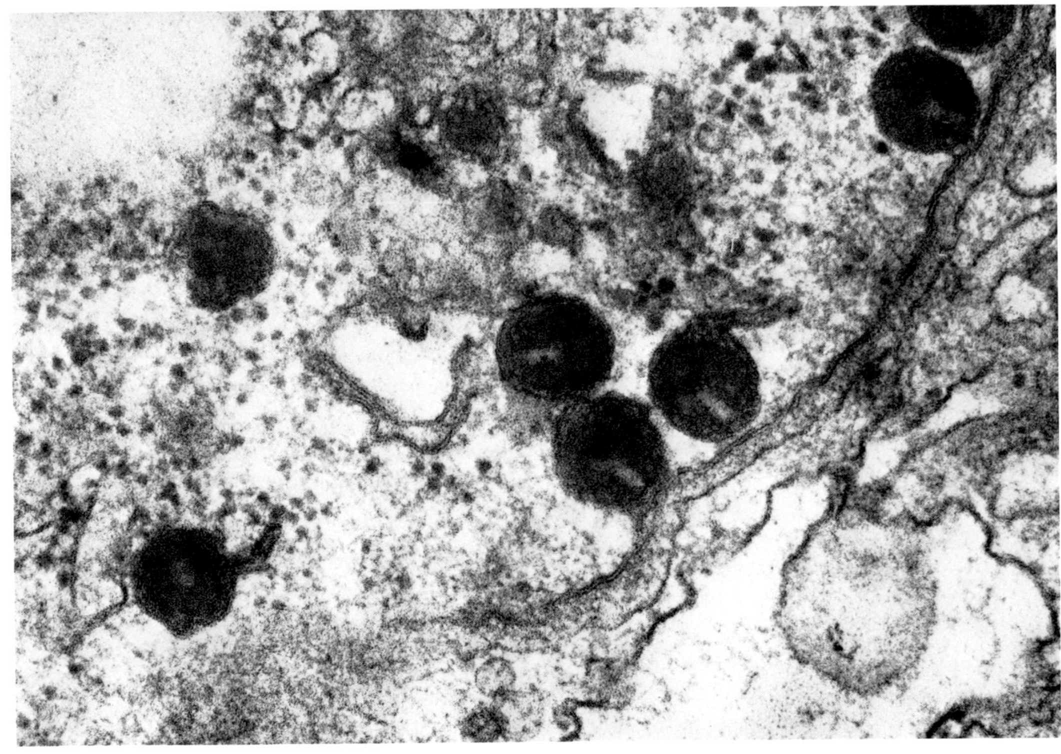

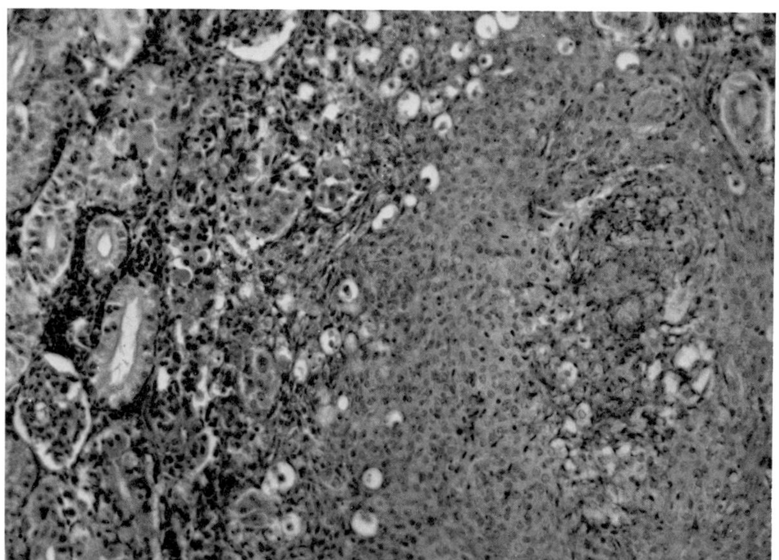

Fig. 4.20. Hexamitiasis in aquarium fish. The protozoa are present within vacuoles at periphery of a large granuloma with a caseonecrotic core.

hyperplasia of the neutrophil series. The reactions center around the characteristic large eosinophilic primary (mother) and secondary (daughter) cells. The endothelium of the portal vessels is a frequent location for the parasites (due to phagocytosis by the resident phagocytes?), and often large aggregates of cells may be seen to virtually obliterate the lumen of the vessel.

The ultrastructural appearance of the parasites is very characteristic, largely due to the presence, within the primary cells alone, of small spherical electron-dense bodies, which are closely associated with the plasmalemma and may be Golgi derived. As the disease progresses, there may be destruction of nephrons, possibly by compression, although it is probably the concurrent anemia rather than renal failure that accounts for the often heavy mortality. Recovered fish may have mild fibrosis, but in most cases it is very difficult to see any evidence of previous damage, illustrating yet again the great recuperative powers of teleosts. In North America, recovering fish may have tubular stages of *Sphaerospora* sp., and these astute observations suggested to Kent and Hedrick (1985) that the cause of PKD may be the prespore stages of this myxosporean.

Other Interstitial Diseases

Ceratomyxa shasta is another myxosporean of salmonids that induces granulomatous lesions throughout the viscera, including the kidney. Its economic impact is lessened, however, because it is restricted geographically to the west coast of North America. Other protozoa are also sometimes encountered in the interstitium when systemic disease is present (Fig. 4.20).

The renal hematopoietic tissue is a target for many acute virus diseases, especially the rhabdoviruses such as infectious hematopoietic necrosis (IHN) (Fig. 4.21) and viral hemorrhagic septicemia (VHS) of salmonids or the recently described hirame rhabdovirus (HRV). There is marked necrosis of all elements including endothelium of the renal portal system and in later stages, the tubules too. The herpesviruses affecting salmonids and channel catfish also produce hematopoietic and tubular necrosis. Edema and hemorrhage may accompany any virus disease that affects vascular endothelium.

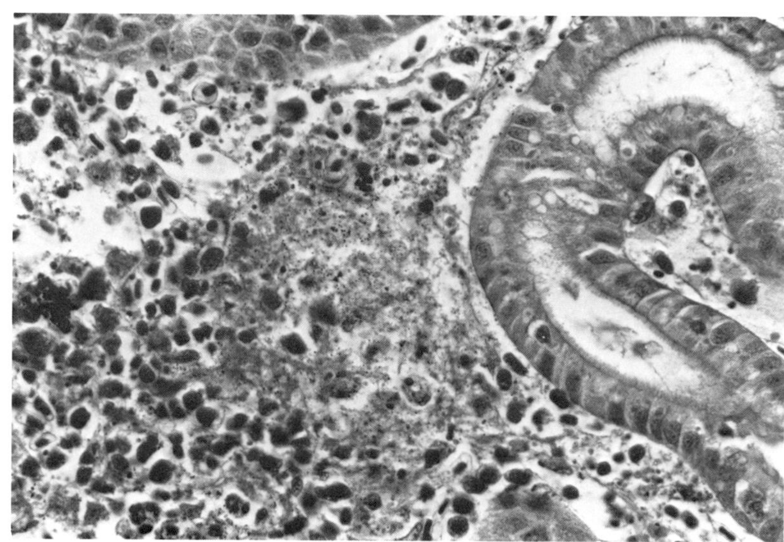

Fig. 4.21. Experimentally produced infectious hematopoietic necrosis in rainbow trout. There is marked necrosis of all interstitial elements, including melanomacrophages; pyknotic debris and melanin granules are scattered throughout. Adjacent tubule is for the most part normal.

References

Aldrin, J.-F., M. Mevel, J.-Y. Robert, M. Vigneulle, and F. Baudin-Laurencin. 1978. Incidences metaboliques de la corynebacteriose experimentale chez le saumon coho. *Bull. Soc. Sci. Vet. et Med. comparee* 80:79–90.

Anderson, C. D., R. J. Roberts, K. MacKenzie, and A. H. McVicar. 1976. The hepato-renal syndrome in cultured turbot (*Scophthalmus maximus* L.). *J. Fish Biol.* 8:331–41.

Bell, G. R., D. A. Higgs, and G. S. Traxler. 1984. The effect of dietary ascorbate, zinc, and manganese on the development of experimentally induced bacterial kidney disease in sockeye salmon (*Oncorhynchus nerka*). *Aquaculture* 36:293–311.

Bruno, D. W. 1986. Histopathology of bacterial kidney disease in laboratory infected rainbow trout, *Salmo gairdneri* Richardson, and Atlantic salmon, *Salmo salar* L, with reference to naturally infected fish. *J. Fish Dis.* 9:523–37.

Capreol, S. V., and L. E. Sutherland. 1968. Comparative morphology of juxta glomerular cells. I. Juxtaglomerular cells in fish. *Can. J. Zool.* 46:249.

Clifton-Hadley, R. S., D. Bucke, and R. H. Richards. 1984. Proliferative kidney disease of salmonid fish: a review. *J. Fish Dis.* 7:363–77.

Colville, T. P., R. H. Richards, and J. W. Dobbie. 1983. Variations in renal corpuscular morphology with adaptation to sea water in the rainbow trout, *Salmo gairdneri* Richardson. *J. Fish Biol.* 23:451–56.

Cowey, C. B., D. Knox, J. W. Adron, S. George, and B. Pirie. 1977. The production of renal calcinosis by magnesium deficiency in rainbow trout (*Salmo gairdneri*). *Br. J. Nutr.* 38:127–35.

Desser, S. S., K. Molnar, and I. Horvath. 1983. An ultrastructural study of the myxosporeans, *Sphaerospora angulata* and *Sphaerospora carassii,* in the common carp, *Cyprinus carpio* L. *J. Protozool.* 30:415–22.

Elger, M., and H. Hentschel. 1983. Glomerular disease in cultured rainbow trout, *Salmo gairdneri* Richardson, suffering from presumptive chronic viral haemorrhagic septicaemia. *J. Fish Dis.* 6:211–29.

Evelyn, T. P. T., J. E. Ketcheson, and L. Prosperi-Porta. 1984. Further evidence for the presence of *Renibacterium salmoninarum* in salmonid eggs and for the failure of povidone-iodine to reduce the intra-ovum infection rate in water-hardened eggs. *J. Fish Dis.* 7:173–82.

Ferguson, H. W. 1984. Renal portal phagocytosis of bacteria in rainbow trout (*Salmo gairdneri* Richardson): ultrastructural observations. *Can. J. Zool.* 62:2505–11.

Ferguson, H. W., and E. A. Needham. 1978. Proliferative kidney disease in rainbow trout

Salmo gairdneri Richardson. *J. Fish Dis.* 1:91–108.

Ferguson, H. W., R. D. Moccia, and N. E. Down. 1982. An epizootic of diffuse glomerulonephritis in white perch (*Roccus americanus*) from a heavily polluted industrial basin in Lake Ontario. *Vet. Pathol.* 19:638–45.

Harrison, J. G., and R. H. Richards. 1979. The pathology and histopathology of nephrocalcinosis in rainbow trout *Salmo gairdneri* Richardson in fresh water. *J. Fish Dis.* 2:1–12.

Hawkins, W. E., L. G. Tate, and T. G. Sarphie. 1980. Acute effects of cadmium on the spot *Leiostomus xanthurus* (Teleostei): tissue distribution and renal ultrastructure. *J. Toxicol. Environ. Health.* 6:283–95.

Hicks, B. D., J. W. Hilton, and H. W. Ferguson. 1984. Influence of dietary selenium on the occurrence of nephrocalcinosis in rainbow trout, *Salmo gairdneri* Richardson. *J. Fish Dis.* 7:379–89.

Hinton, D. E., R. T. Jones, and R. L. Herman. 1976. Glomerular mesangial fibrosis in hatchery-reared rainbow trout (*Salmo gairdneri*). *J. Fish. Res. Board Can.* 33:2551–59.

Hoffmann, R., and R. Lommel. 1984. Haematological studies in proliferative kidney disease of rainbow trout, *Salmo gairdneri* Richardson. *J. Fish Dis.* 7:323–26.

Kent, M. L., and R. P. Hedrick. 1985. PKX, the causative agent of proliferative kidney disease (PKD) in Pacific salmonid fishes and its affinities with the myxozoa. *J. Protozool.* 32:254–60.

______. 1986. Development of the PKX myxosporean in rainbow trout *Salmo gairdneri. Dis. Aquat. Org.* 1:169–82.

Lall, S. P., W. D. Paterson, J. A. Hines, and N. J. Adams. 1985. Control of bacterial kidney disease in Atlantic salmon, *Salmo salar* L., by dietary modification. *J. Fish Dis.* 8:113–24.

Langvad, F., O. Pedersen, and K. Engjom. 1985. A fungal disease caused by *Exophiala* sp. nova in farmed Atlantic salmon in western Norway. In *Fish and Shellfish Pathology,* ed. A. E. Ellis, 323–28. Orlando, Fla.: Academic Press.

McCarthy, D. H., T. R. Croy, and D. F. Amend. 1984. Immunization of rainbow trout, *Salmo gairdneri* Richardson, against bacterial kidney disease: preliminary efficacy evaluation. *J. Fish Dis.* 7:65–71.

Mitchum, D. L., L. E. Sherman, and G. T. Baxter. 1979. Bacterial kidney disease in feral populations of brook trout (*Salvelinus fontinalis*), brown trout (*Salmo trutta*), and rainbow trout (*Salmo gairdneri*). *J. Fish. Res. Board Can.* 36:1370–76.

Molnar, K. 1980. Renal sphaerosporosis in the common carp *Cyprinus carpio* L. *J. Fish Dis.* 3:11–19.

Otis, E. J., R. E. Wolke, and V. S. Blazer. 1985. Infection of *Exophiala salmonis* in Atlantic salmon (*Salmo salar*). *J. Wildl. Dis.* 21:61–64.

Paperna, I. H. 1982. *Kudoa* infection in the glomeruli, mesentery and peritoneum of cultured *Sparus aurata* L. *J. Fish Dis.* 5:539–43.

Paperna, I. H., J. G. Harrison, and G. W. Kissil. 1980. Pathology and histopathology of a systemic granuloma in *Sparus aurata* (L.) cultured in the Gulf of Aqaba. *J. Fish Dis.* 3:213–21.

Paterson, W. D., D. Desautels, and J. M. Weber. 1981. The immune response of Atlantic salmon, *Salmo salar* L., to the causative agent of bacterial kidney disease, *Renibacterium salmoninarum. J. Fish Dis.* 4:99–111.

Paterson, W. D., C. Gallant, D. Desautels, and L. Marshall. 1979. Detection of bacterial kidney disease in wild salmonids in the Margaree River system and adjacent waters using an indirect fluorescent antibody technique. *J. Fish. Res. Board Can.* 36:1464–68.

Pritchard, J. B., and J. L. Renfro. 1984. Interactions of xenobiotics with teleost renal function. In *Aquatic Toxicology,* vol. 2, ed. L. J. Weber, 51–106. New York: Raven Press.

Sanders, J. E., and J. L. Fryer. 1980. *Renibacterium salmoninarum* gen. nov., sp. nov., the causative agent of bacterial kidney disease in salmonid fishes. *Int. J. Syst. Bact.* 30:496–502.

Sanders, J. E., K. S. Pilcher, and J. L. Fryer. 1978. Relation of water temperature to bacterial kidney disease in coho salmon (*Oncorhynchus kisutch*), sockeye salmon (*O. nerka*), and steelhead trout (*Salmo gairdneri*). *J. Fish. Res. Board Can.* 35:8–11.

Seagrave, C. P., D. Bucke, and D. J. Alderman. 1980. Ultrastructure of a haplosporean-like organism: the possible causative agent of proliferative kidney disease in rainbow trout. *J. Fish Biol.* 16:453–59.

Smart, G. R., D. Knox, J. G. Harrison, J. A. Ralph, R. H. Richards, and C. B. Cowey. 1979. Nephrocalcinosis in rainbow trout *Salmo gairdneri* Richardson; the effect of exposure to elevated CO_2 concentrations. *J. Fish Dis.* 2:279–89.

Smith, C. E., J. K. Morrison, H. W. Ramsey, and H. W. Ferguson. 1984. Proliferative kidney disease: first reported outbreak in North America. *J. Fish Dis.* 7:207–16.

Trump, B. F., R. T. Jones, and S. Sahaphong. 1975. Cellular effects of mercury on fish kidney tubules. In *The Pathology of Fishes,* ed. W. E. Ribelin and G. Migaki, 585–612. Madison: University of Wisconsin Press.

Wood, E. M., and W. T. Yasutake. 1956. Histopathology of kidney disease in fish. *Am. J. Pathol.* 32:845–57.

Yasutake, W. T. 1978. Histopathology of yearling sockeye salmon (*Oncorhynchus nerka*) infected with infectious hematopoietic necrosis (IHN). *Fish Pathol.* 14:59–64.

Young, C. L., G. B. Chapman. 1978. Ultrastructural aspects of the causative agent and renal histopathology of bacterial kidney disease in brook trout (*Salvelinus fontinalis*). *J. Fish. Res. Board Can.* 35:1234–48.

Zapata, A. 1979. Ultrastructural study of the teleost fish kidney. *Dev. Comp. Immunol.* 3:55–65.

5

Spleen, Blood and Lymph, Thymus, and Reticulo-endothelial System

Spleen

The spleen is usually a discrete organ, although in some species there may be several nodules. In the absense of lymph nodes, the spleen in teleosts represents one of the major filters in the vascular system, removing circulating antigens and effete blood cells. As a consequence, virtually any systemic infectious disease involves the spleen to some degree. In many species, such as rainbow trout, the spleen is the major source of erythrocytes. Therefore, diseases such as infectious hematopoietic necrosis (IHN), which can destroy splenic red pulp, may impair the ability of the fish to respond to loss of erythrocytes from hemorrhage or increased rate of breakdown.

The splenic capsule is very thin, with little evidence of contractile ability, but in some species such as pleuronectids and gadoids, pancreatic tissue is often present. Splenic arterioles terminate by dividing into several thick-walled capillaries, the ellipsoids, which themselves open into the pulp spaces. The ellipsoids have an endothelial-lined axial vessel (Fig. 5.1) surrounded by a

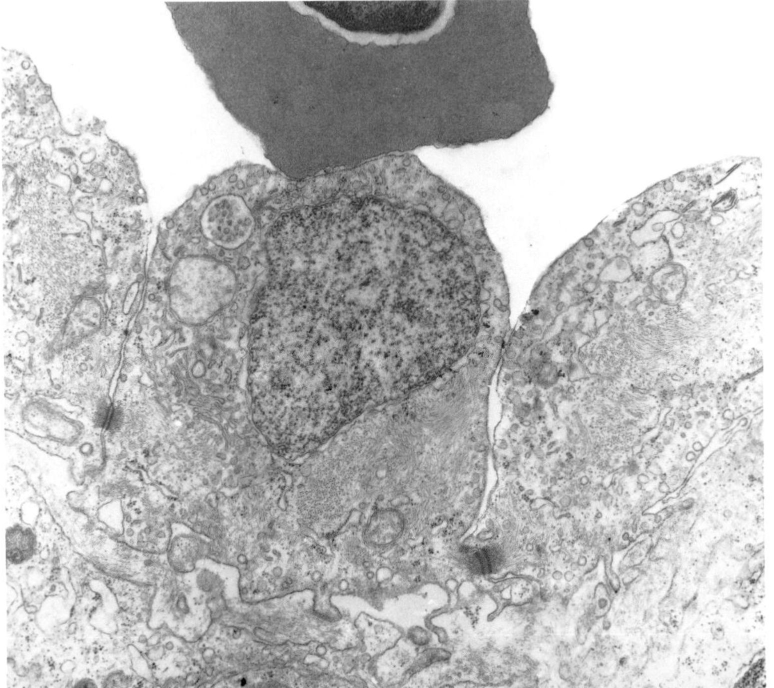

A

Fig. 5.1. A. Endothelial lining of axial vessel in splenic ellipsoid. The large number of intracytoplasmic fibrils suggests the possibility that these cells are contractile, helping to shunt blood into the sheaths of the ellipsoids. **B.** Plaice injected with carbon 1 hour before it was killed. Walls of the ellipsoids are outlined by the marker while the lumen of the axial vessels remains clear.

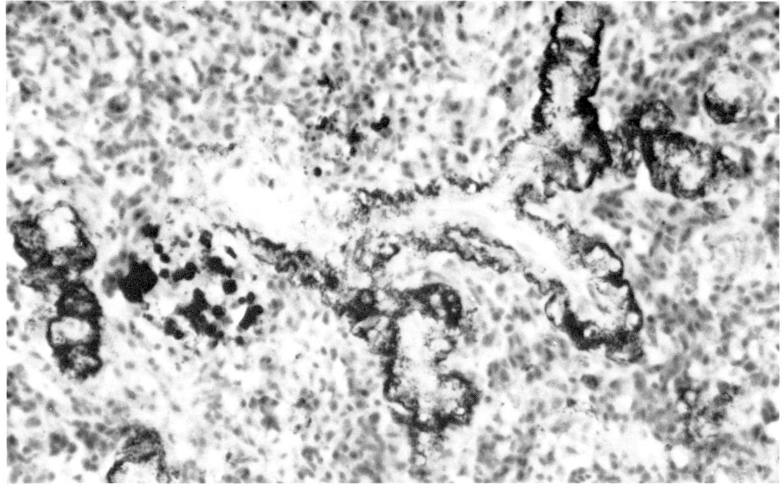

B

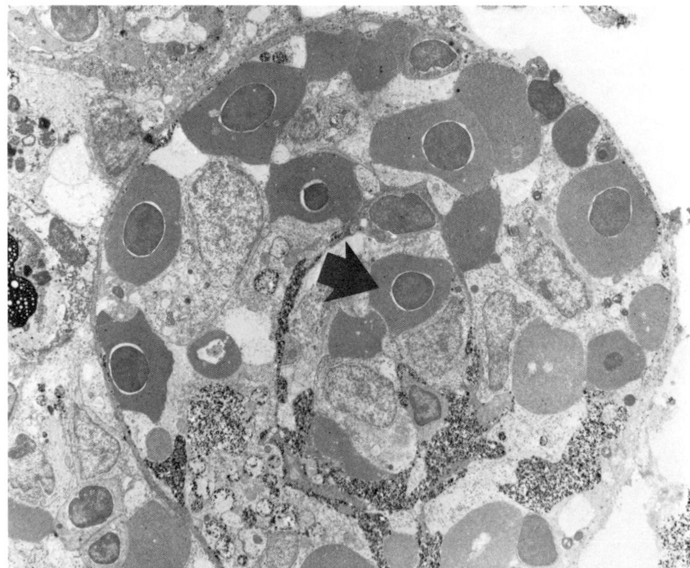

C

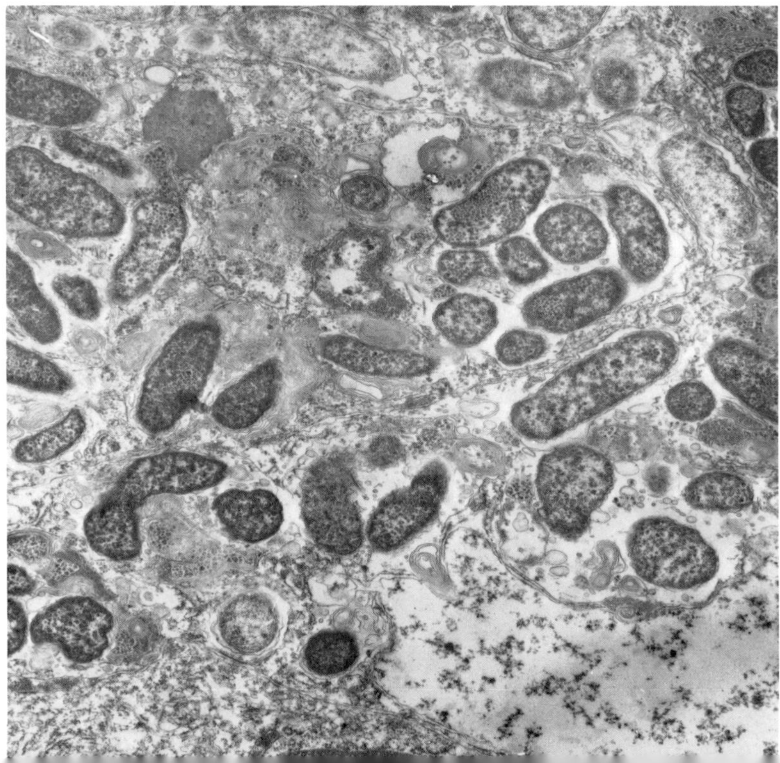

C. A single ellipsoid, shown here with carbon in reticular cells of sheath. Numerous erythrocytes are also present between reticular cells as well as within the lumen of the axial vessel (*arrow*). Movement of erythrocytes from the axial vessel into the sheath is the main method for removal of effete cells. **D.** Hitra disease of Atlantic salmon showing large number of bacteria (*Vibrio "salmonicida"*) within splenic ellipsoid. There is almost no necrosis, unlike with *Aeromonas salmonicida;* see Fig. 5.2B. (*Material courtesy of T. Hastein*)

sheath of reticular cells that are supported by reticulin fibers. The outer wall of the ellipsoid is itself surrounded by endothelial-lined sinusoidal blood spaces, including possibly lymphatics. A few species, such as dace, do not have ellipsoids.

Lymphoid tissue tends to be more diffuse in teleosts than in elasmobranchs in which it is very well marked, although it does form a fine investment to ellipsoids and to the melanomacrophage centers (MMC). The latter are discrete encapsulated structures, and they are a particular feature of the teleost spleen. They are not, however, found in salmonids, in which melanomacrophages, while prominent, do not aggregate into encapsulated centers. In addition to the spleen, MMC are found in kidney, liver, and occasionally other sites such as gonads and thyroid. Chronically sick fish, those in polluted waters, old fish, or those in which catabolism has been excessive (as in starvation) all have increased size and numbers of MMC, for they serve as "repositories" for end products of cell breakdown such as phospholipids, and also for antigens or other particulate material. The color of MMC pigment in normal fish varies from pink to golden brown (ceroid or lipofuscin), but it tends to be darker (melanin) in older or sick fish.

Particulate antigens are initially trapped by the ellipsoids and may be seen within the fine reticular meshwork or intracellularly within the phagocytic cells of the sheaths (Fig. 5.1). Toxin release may destroy these structures with a consequent loss of vascular integrity. If severe, this may cause marked enlargement of the spleen. It is a particular feature of some viremias and bacteremias such as furunculosis (Fig. 5.2). Splenomeg-

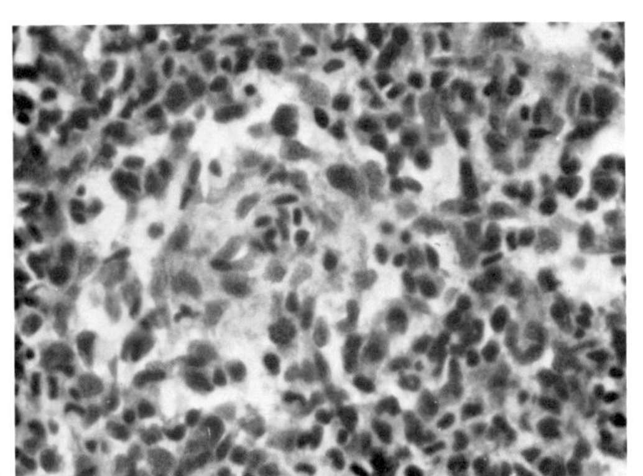
A

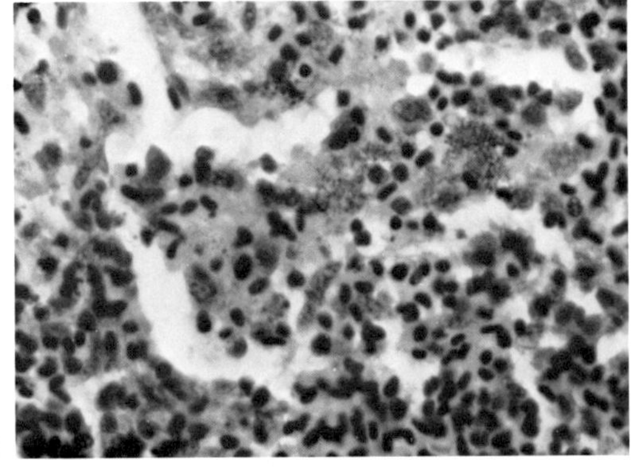
B

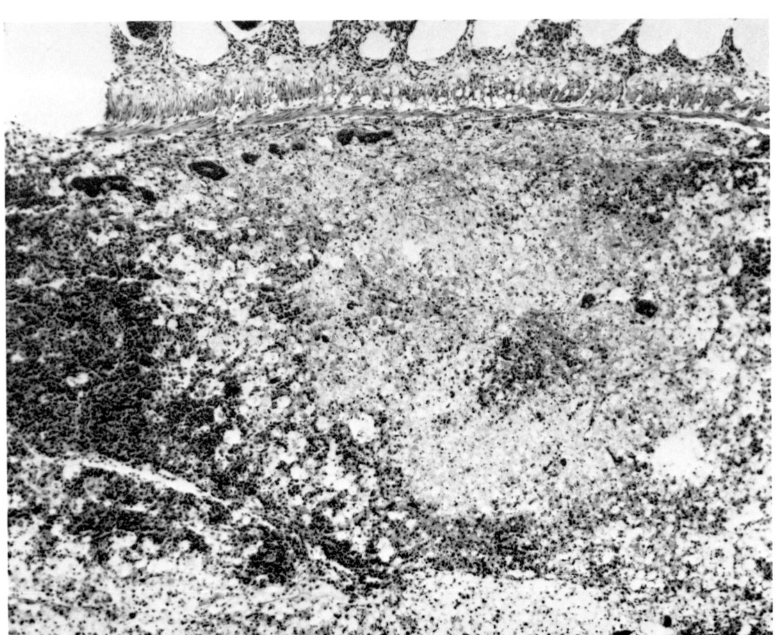
C

Fig. 5.2. A. Ellipsoid from normal brown trout. **B.** Ellipsoid from brown trout with furunculosis. Bacteria have been trapped by the ellipsoid and have caused necrosis. If a large number of vessels are affected, as is often the case, the consequential loss of vascular integrity results in splenomegaly. **C.** Splenic necrosis in goldfish due to bacteria. The whole structure of the spleen has disintegrated, and other viscera are involved.

aly is also a feature of some blood protozoan diseases, notably *Cryptobia* (*Trypanoplasma*) *salmositica* and *Haemogregarina sachai.* It may be also encountered in normal salmonids at spawning time; this is thought to represent blood storage. Drugs, asphyxia, or severe exercise may cause release of erythrocytes stored in this way, with a consequent rise in hematocrit. In acute hemolytic crises, the ellipsoidal sheaths are initially outlined by hemosiderin, which thus appear as dark brown; most of the blood pigments accumulate in the spleen rather than the kidney.

After the initial trapping, antigens and metabolic products such as iron-pigments are transferred to the MMC within macrophages (Fig. 5.3). It is interesting to note that antigens initially trapped at other sites in the body, such as the extensive network of macrophages associated with the renal portal system, may eventually relocate to the spleen, probably carried there within the macrophages. This is certainly true in rainbow trout: at one hour postinoculation with killed bacteria, the spleen contains only 1.5% of the total number of organisms

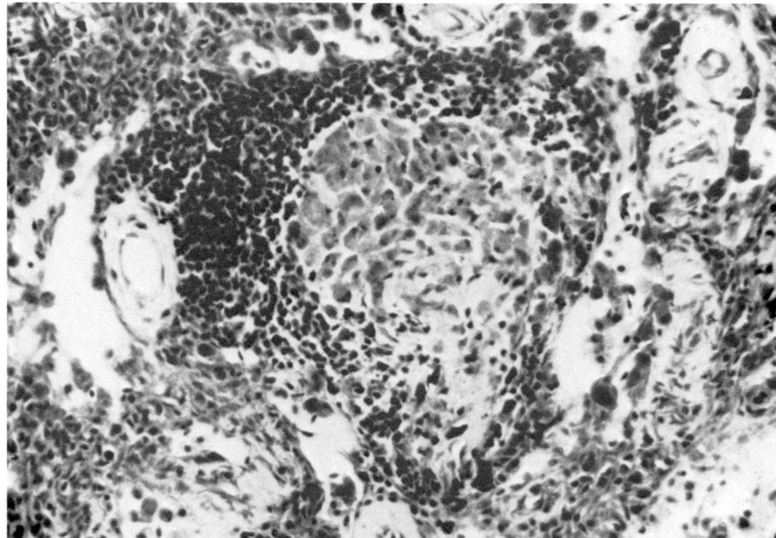

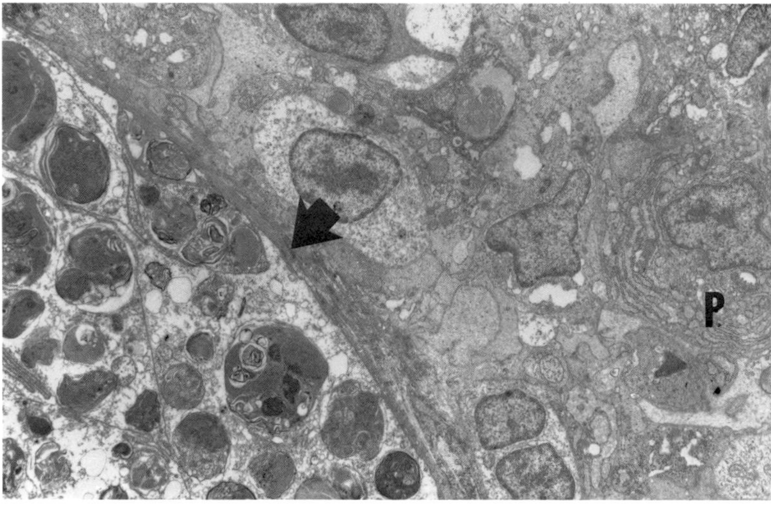

Fig. 5.3. A. Melanomacrophage center (MMC) and surrounding white pulp in spleen of turbot infected with *Haemogregarina sachai.* **B.** Same fish as in **A.** Remnants of parasites and other cellular debris are present within the MMC, which is surrounded by a distinct capsule (*arrow*). (*P* denotes plasma cell)

present in the body, whereas by 18 hours it contains 30%. In chronic disease, there may be an increase in white pulp, especially around the ellipsoids and MMC. In some cases, large numbers of plasma cells may be readily discerned.

In view of its propensity for trapping, it should be no surprise to find that the spleen is almost invariably involved in the chronic granulomatous diseases of bacterial or parasitic cause. Lesions can be so extensive as to grossly distort the thin capsule and impart a marbled appearance to the parenchyma on cut surface. Alternatively, they may be miliary and appreciated only on histology. Some of the best examples of such diseases include mycobacteriosis (Fig. 5.4), bacterial kidney disease (BKD) (Fig. 5.5), and proliferative kidney disease (PKD).

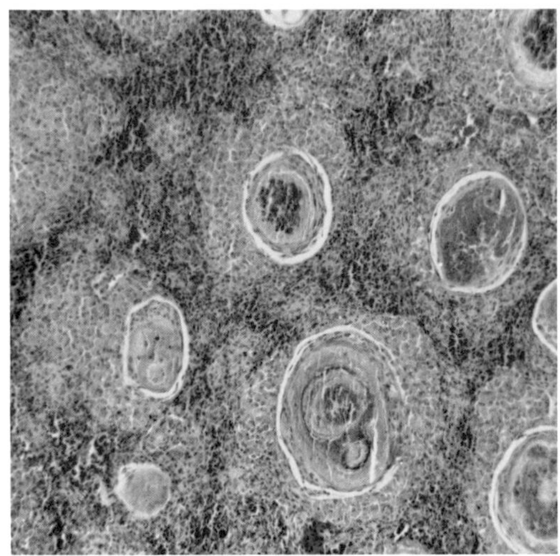

Fig. 5.4. Mycobacteriosis in cichlid showing granulomas *within* the melanomacrophage center. Bacteria have been carried there by macrophages and have continued to survive.

Blood and Lymph

Estimates of teleost blood volume vary widely but are probably in the order of 2% to 4% of body weight, compared to 5% to 8% for other vertebrate groups. Confined, nonanesthetised rainbow trout, for instance, have been shown, using ^{51}Cr-labeled erythrocytes, to have a total blood volume of 4.09%. Earlier overestimations of blood volumes were probably the result of using plasma markers such as Evans blue dye, which in teleosts tends to leak out of the blood compartment due to the greater permeability of their capillaries to proteins than is the case in mammals (Nichols 1987). By contrast with this relatively low volume, lymph is 4 to 5 times blood volume. It is tempting to speculate therefore that lymph plays a much greater role in distribution of metabolites and infectious agents than the literature suggests (Wardle 1971).

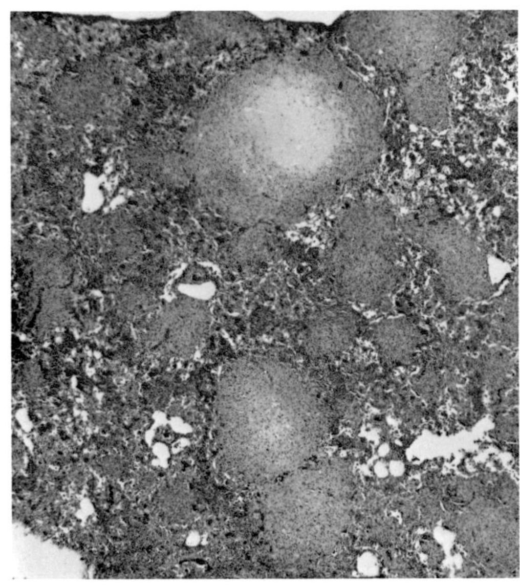

Fig. 5.5. Bacterial kidney disease in trout showing multifocal granulomas (one with a necrotic core) each centered on an ellipsoid.

Nevertheless, a substantial effort has been directed towards elucidating clinically useful indicators of disease in the blood of farmed and wild fish. Major obstacles to progress lie in the relative disproportions of blood and lymph volumes. Hence small changes in one (due to altered rates of formation or removal of lymph) lead to large changes in the other, mainly in the form of hematocrit. In addition, other factors have a major influence on the values recorded. These include age, sex, diet, species (and strains within a species), time of year, water temperature, and sampling methods such as type and duration of anesthetic or sample site (cardiac or caudal vein). When collecting blood from diseased fish, therefore, it is crucial to also take samples from clinically healthy control animals of the same age, under the same conditions, and using the same methods. As may be imagined, osmotic effects can easily dominate and obscure real changes unless this is taken into account.

Anemia. As in all other vertebrates except mammals, fish erythrocytes are nucleated flattened elliptical cells; they do, however, increase in size as they mature. Nuclear chromatin is typically clumped, and vestiges of endoplasmic reticulum and Golgi vacuoles may be seen in the cytoplasm. Nonnucleated erythrocytes are occasionally seen in some species (Wingstrand 1956). The ability of many teleosts to survive in a relative or functional absence of hemoglobin is well established. Indeed some species, such as the Antarctic icefish (*Chaenocephalus aceratus*), have no erythrocytes at all, relying entirely on plasma for gaseous transport.

Compensatory mechanisms for an experimentally produced 70% reduction in hematocrit in rainbow trout enable resting levels of oxygen uptake to be maintained. These mechanisms include increased cardiac output, which is achieved almost entirely by increasing stroke volume rather than heart rate, thereby pumping greater volumes of the reduced oxygen-holding blood through the gills (facilitated by decreasing peripheral vascular resistance). Thus even very low hematocrits may cause little in the way of observable clinical symptoms. The usual gross manifestation of anemia is branchial pallor, although care must be taken not to confuse this with pallor caused by the presence of excess mucus, severe diffuse branchial hyperplasia, or indeed even autolysis.

Anemias may be classified according to numbers, size, and hemoglobin concentration of the erythrocytes. They may be classified as responsive (regenerative) or nonresponsive. Or, they may be classified according to the pathophysiological mechanism that caused them, i.e., blood loss (hemorrhage), increased rate of breakdown (hemolytic), or reduced rate of production (hypoplastic). From a diagnostic standpoint, the latter classification is the most useful.

Hemorrhagic anemias are not common, although they are occasionally seen following trauma, especially to the gills, or following minor surgical exercises, especially cardiac puncture for the purposes of blood sampling. In those species with a compact outer layer to the myocardium, and hence coronary vessels, an unfortunately placed hypodermic needle may cause acute pericardial hemorrhage and sometimes cardiac tamponade. An alternative sequel is bleeding into the peritoneal cavity, especially if there have been multiple attempts at sampling. Abdominal hemorrhage is also encountered as a poststripping event in artificial spawning, especially if the handlers are inexperienced enough to be rough or the fish themselves are not ripe. Water temperatures at these times may dictate the severity of the consequences (low temperatures necessitate lower oxygen requirements and hence reduced dependence on hemoglobin) and the speed of recovery. External parasites such as leeches or lampreys can cause excessive blood loss, as can some bacterial or viral diseases where vascular endothelium is destroyed. Examples of the latter include viral hemorrhagic septicemia (VHS) of trout and channel catfish virus disease. In mammals, diseases causing severe vascular injury may compound the hemorrhagic anemia by fragmentation hemolysis. It is unclear if the blood pressure seen in fish is high enough to cause such damage. One exception may be the afferent vessels of the gills.

The dermal ulcers associated with *Aeromonas salmonicida* may also cause hemorrhage. Granulomatous or neoplastic diseases can lead to chronic bleeding, especially where there is dermal ulceration; sometimes this is compounded by self-induced trauma if the lesions project. Acute hemorrhage of the gills accompanies the mass exodus of *Sanguinicola klamathensis* miracidia. In these situations, there will be increased numbers of circulating polychromatocytes or immature red blood cells (a responsive anemia). Fish polychromatocytes are smaller than mature erythrocytes and increase in size as they age (the converse of the mammalian situation). It is important to bear this in mind whenever classifying a fish anemia as microcytic: it means one thing in mammals but quite another in fish. Vitamin K deficiency in trout leads to visceral hemorrhage. Since vitamin K is required for prothrombin and thromboplastin synthesis, a deficiency impairs coagulation.

Acute abdominal hemorrhage is sometimes seen in fish with excessively fatty livers. While it is likely that this is often traumatic in origin due to extreme friability of the organ, the possibility of impaired production of clotting factors should also be borne in mind.

Hemolytic anemias are characterized by hemosiderin accumulation in the spleen. If pronounced, the ellipsoidal sheaths become demarcated with pigment, followed by heavy accumulation within the melanomacrophage centers. Pigment may also be seen in renal tubular epithelium, although there is usually little indication for nephrosis, as might be expected, assuming that hemoglobin is nephrotoxic, as it is in mammals. Unless there is concurrent destruction or suppression of hemopoietic tissue, the anemia will be responsive, and the blood picture will see an increasing number of polychromatocytes. In severe cases, there may be activation of ancillary foci of hemopoietic tissue. Such foci may be found in the liver around portal tracts, and in the pericardium.

Common causes include bacterial hemolysins, notably *Vibrio* spp., and hematozoa such as *Cryptobia* (*Trypanoplasma*) spp., *Trypanosoma* spp., and *Haemogregarina* spp. Inclusions within the cytoplasm of erythrocytes are seen in many species. Originally thought to be protozoan, these were shown in the cod, and other species, to be viral, and the disease was termed piscine (or viral) erythrocytic necrosis (PEN). Although the virus (provisionally classified as an iridovirus) has not been cultured in vitro, the disease has been experimentally reproduced in cod, *Gadus morhua,* inclusions initially appearing in immature erythrocytes. Epizootics have also been reported in Pacific herring.

Toxic causes of hemolytic anemia are protean, but commonly include chlorine exposure. Not only is there destruction of erythrocytes, but in the presence of nitrogenous wastes, chloramines are also formed. These lead to methemoglobin formation, thereby reducing the oxygen-carrying capacity of the blood. Nitrite, aniline, and nitrobenzene toxicity also result in methemoglobin formation.

Anemia is a feature of lipoid liver disease in salmonids, and probably other species being fed diets containing highly oxidized (rancid) oils or diets in which vitamin E is deficient. Not only is there an increased rate of breakdown of erythrocytes due to increased fragility, but there is an apparent failure of maturation of polychromatocytes. These findings are consistent with the concept of the need for vitamin E to maintain cell membrane integrity.

Anemia is also a feature of other diseases, such as PKD of salmonids. Indeed, in this condition anemia is probably the cause of death. The precise pathogenesis is not completely understood, but it is probably mainly hemolytic, with large numbers of polychromatocytes appearing in the peripheral circulation. Thus, despite often massive replacement of hemopoietic tissue by the granulomatous inflammation, myelophthisis would not seem to be a dominating event. Immune-mediated erythrolysis has not been demonstrated in this disease, but the possibility is worthy of investigation.

Hypoplastic anemias can be divided

into those in which there is deficient cell production and those in which there is a deficiency of hemoglobin synthesis. Folic acid and vitamin B_{12} have complementary roles and a deficiency of either will cause anemia. Salmon and trout with inadequate B_{12} develop a hypochromic anemia, whereas with folic acid, erythrocytes are normochromic but macrocytic. Other nutrient deficiencies leading to anemia include niacin, pyridoxine, inositol, riboflavin, and vitamin C. Iron-deficiency anemia of fish is probably only an experimental disease and even then, only when the water is very low in minerals; adequate levels will otherwise probably be absorbed mainly via the gills.

Aplasia of erythropoietic tissue may affect all cell types causing a pancytopenia. Possible causes include radiation damage or radiomimetic drugs. The clinical presentation in these cases may be compounded by hemorrhage due to thrombocytopenia. Chloramphenicol can cause an irreversible myeloid aplasia in humans. Although commonly used in aquarium species, there is little information on its effect in fish. Certainly, in our hands, short-term use would seem to cause little problem. Outright destruction of hemopoietic tissue or erythropoietin-forming tissues by viruses, such as IHN (see Fig. 4.21) or the recently described hirame rhabdovirus (HRV), may in the long term lead to anemia if the tissue is unable to recover before erythrocytes become effete.

Myelophthisic anemia due to fibrosis of hemopoietic tissue is a distinct possibility in the chronic granulomatous diseases such as PKD, mycobacteriosis, or bacterial kidney disease (BKD). In addition, toxic depression of erythropoiesis by nonexcreted metabolites or depressed erythropoietin production should also be considered, especially if, as is almost invariably the case in these diseases, there is involvement of nephrons. In teleosts, the gills are mainly responsible for nitrogenous excretion, but it is not known whether chronic gill disease or elevated ammonia levels could suppress hemopoietic tissue before the animal succumbs to oxygen lack or to the toxic action of ammonia on the central nervous system. Tumors involving hemopoietic tissue could also cause myelophthisic anemia.

Leukemia is the neoplastic proliferation of the leukocyte-forming tissues and usually results in the presence of neoplastic leukocytes in the peripheral blood. The best example of this myeloproliferative disease in fish is the retrovirus-associated lymphosarcoma of northern pike and muskellunge in which the tumorous creamy white nodules seen in a variety of tissues (see Fig. 13.10) reflect a leukemic increase in circulating lymphocytes. The pathogenesis of the anemia that accompanies this disease is uncertain, but in another retroviral lymphosarcoma seen in cats, the anemia is thought to be largely the result of aplasia. Thymomas are also encountered in salmonids (see Fig. 13.11), and these too may be leukemic with involvement of other tissues.

It is important to differentiate purposeless proliferation of one or more cell lines (neoplasia) from directed physiological proliferation in response to one of a variety of stimuli (leukocytosis). Such differentiation is not always easy, especially if a cause is not readily apparent. Protozoan diseases in particular may cause one or more cell types to proliferate, leading to either an absolute or a relative increase in their numbers in the peripheral blood. Such an example is seen in *Haemogregarina sachai* infection of cultured turbot in which nodules of lymphoreticular tissue are found throughout the body, together with anemia and an increase in circulating monocytes and neutrophils (Fig. 5.6). At first sight, therefore, the disease mimics lymphosarcoma until close inspection reveals the apicomplexan protozoa within parasitized leukocytes.

Thymus

These lymphoid structures are superficially situated on the dorsal aspect of the branchial cavity (see Fig. 13.10). They are

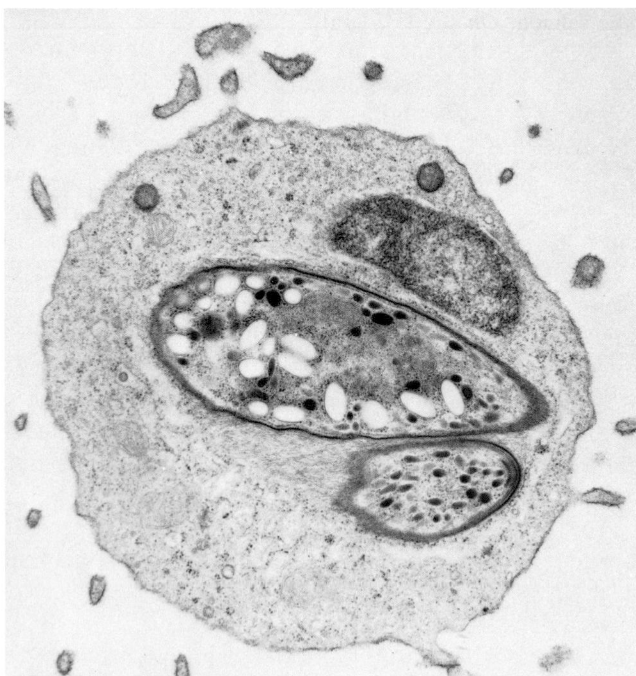

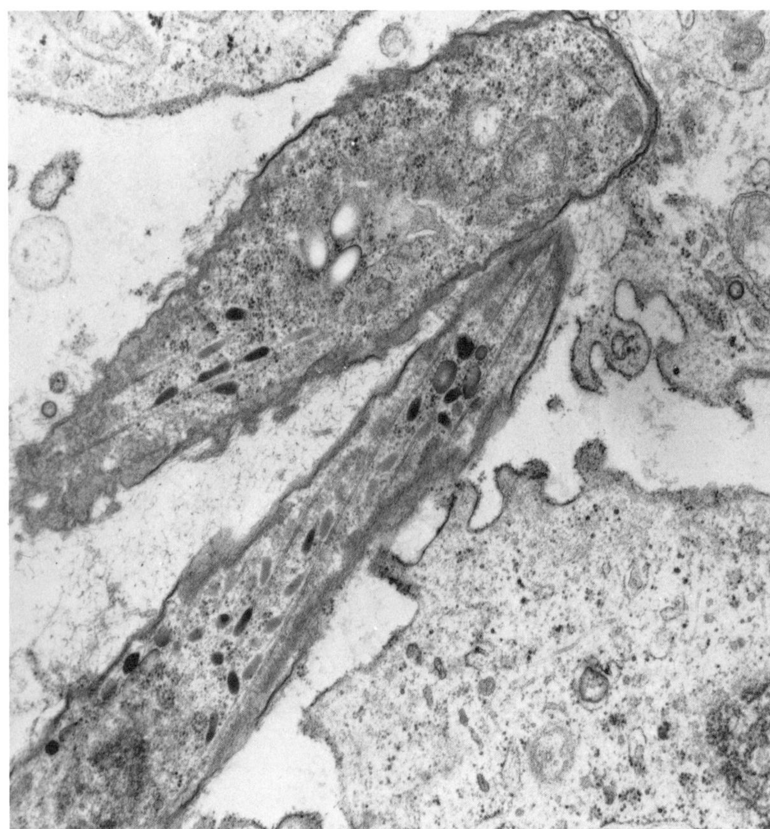

Fig. 5.6. *Haemogregarina sachai* infection of cultured turbot (*Scophthalmus maximus*) showing parasites within a circulating leukocyte (**A**) and within one of the granulomas they elicit (**B**).

covered by a very thin epithelium and have a fine connective tissue stroma. The major cellular component is the thymocyte, but epithelioid or macrophage-like cells and granular eosinophilic cells may also be present. In some species they may be arranged to form an apparent cortex and medulla; Hassall's corpuscles may also be seen. The thymus can be detected in some species a few days before hatching, whereas in others like the plaice, it is seen only after metamorphosis of the larvae into flatfish. As in mammals, involution of the organ is seen in older fish, often at spawning.

The superficial location of the thymus suggests a certain vulnerability, and indeed in young fish with severe bacterial and mycotic infections of the branchial cavity, necrosis of the epithelium overlying the thymus, with invasion and destruction of thymic parenchyma, are not infrequent observations. The significance of these observations and the implications for the possible induction of immune tolerance or suppression have yet to be properly assessed. Experimentally, thymectomy of rainbow trout has been shown to depress some components of immunity, to increase survival to a challenge administered after the fish had been vaccinated, or to decrease survival to challenge in unvaccinated controls (Tatner 1987).

Reticuloendothelial System

This comprises the highly phagocytic cells distributed in various tissues throughout the body. Their precise location has been dealt with under the appropriate system, but to summarize, they are found in the sinusoids and peri-tubular capillaries of the kidney, in the sheaths of the splenic ellipsoids, in the peritoneal cavity as free cells, and as the cells forming the endothelial lining (endocardium) for the myocardial trabeculae in the atrium and to a lesser extent the ventricle also. Comparative pathologists will no doubt be surprised to see the liver missing from this list. In fact, Kupffer cells have not been found in the species of teleosts so far examined (elasmobranchs may be different), although endothelial cells have sometimes been mistakenly identified as such.

While there is very little information on the subject, it is likely that circulating monocytes, which are themselves modestly phagocytic, are the source for histiocytes under normal situations and for macrophages in inflammatory lesions. Whether monocytes are responsible for populating the fixed macrophages of larval tissues such as the kidney, and whether these cells become self-sustaining by in situ replication or still require replacement by circulating cells are all unknown.

Other cells with limited phagocytic abilities include thrombocytes, pillar cells and lamellar epithelial cells in the gills, fat-storing cells (cells of Ito) in the liver, and possibly the ependymal lining of the ventricles in the brain.

References

Alexander, N., R. M. Laurs, A. McIntosh, and S. W. Russell. 1980. Haematological characteristics of albacore, *Thunnus alalunga* (Bonnaterre), and skipjack, *Katsuwonus pelamis* (Linnaeus). *J. Fish Biol.* 16:383–95.

Bamford, O. S. 1974. Oxygen reception in the rainbow trout (*Salmo gairdneri*). *Comp. Biochem. Physiol.* 48:69–76.

Barber, D. L., and J. E. Mills Westermann. 1975. Morphological and histochemical studies on a PAS-positive granular leukocyte in blood and connective tissues of *Catostomus commersonnii* Lacepede (Teleostei:Pisces). *Am. J. Anat.* 142:205–20.

Blaxhall, P. C. 1972. The haematological assessment of the health of freshwater fish: a review of selected literature. *J. Fish Biol.* 4:593–604.

______. 1983. Electron microscope studies of fish lymphocytes and thrombocytes. *J. Fish Biol.* 22:223–29.

Bower, S. M., and L. Margolis. 1985. Effects of temperature and salinity on the course of infection with the haemoflagellate *Cryptobia salmositica* in juvenile Pacific salmon, *Oncorhynchus* spp. *J. Fish Dis.* 8:25–33.

Bruno, D. W. 1986. Changes in serum parameters of rainbow trout, *Salmo gairdneri* Richardson and Atlantic salmon, *Salmo salar* L., infected with *Renibacterium salmoninarum. J. Fish Dis.* 9:205–12.

Bruno, D. W., and A. L. S. Munro. 1986. Haematological assessment of rainbow trout, *Salmo gairdneri* Richardson and Atlantic salmon, *Salmo salar* L., infected with *Renibacterium salmoninarum. J. Fish Dis.* 9:195–204.

Buckley, J. A. 1976. Heinz body hemolytic anemia in coho salmon (*Oncorhynchus kisutch*) exposed to chlorinated wastewater. *J. Fish. Res. Board Can.* 34:215–24.

Burggren, W. W., and J. N. Cameron. 1980. Anaerobic metabolism, gas exchange, and acid-base balance during hypoxic exposure in the channel catfish, *Ictalurus punctatus. J. Exp. Zool.* 213:405–16.

Cameron, J. N. 1984. Acid-base status of fish at different temperatures. *Am. J. Physiol.* 246:R452–R459.

Cannon, M. S., H. H. Mollenhauer, T. E. Eurell, D. H. Lewis, A. M. Cannon, and C. Tompkins. 1980. An ultrastructural study of the leukocytes of the channel catfish, *Ictalurus punctatus. J. Morph.* 164:1–23.

Casillas, E., M. S. Myers, L. D. Rhodes, and B. B. McCain. 1985. Serum chemistry of diseased English sole, *Parophrys vetulus* Girard, from polluted areas of Puget Sound, Washington. *J. Fish Dis.* 8:437–49.

Conroy, D. A. 1972. Studies on the haematology of the Atlantic salmon (*Salmo salar* L.). *Symp. Zool. Soc. Lond.* 30:101–27.

Dalwani, R., J. M. Dave, and K. Datta. 1985. Alterations in hepatic heme metabolism in fish exposed to sublethal cadmium levels. *Biochem. Int.* 10:33–42.

Doolittle, R. F., and D. M. Surgenor. 1962. Blood coagulation in fish. *Am. J. Physiol.* 203:964–70.

Dykova, I., and J. Lom. 1979. Histopathological changes in *Trypanosoma danilewskyi* Laveran & Mesnil, 1904 and *Trypanoplasma borelli* Laveran & Mesnil, 1902 infections of goldfish, *Carassius auratus* (L.). *J. Fish Dis.* 2:381–390.

Eddy, F. B. 1982. Osmotic and ionic regulation in captive fish with particular reference in salmonids. Comp. Biochem. Physiol. 73:125–41.

Eddy, F. B., J. P. Lomholt, R. E. Weber, and K. Johansen. 1977. Blood respiratory properties of rainbow trout (*Salmo gairdneri*) kept in water of high CO_2 tension. *J. Exp. Biol.* 67:37–47.

Evans, W. A. 1974. Growth, mortality, and hematology of cutthroat trout experimentally infected with the blood fluke *Sanguinicola klamathensis. J. Wildl. Dis.* 10:341–46.

Fang, L-S. 1987. Study on the heme catabolism of fish. *Comp. Biochem. Physiol.* 88B:667–73.

Fange, R., and S. Nilsson. 1985. The fish spleen: structure and function. *Experientia* 41:152–58.

Ferguson, H. W. 1976. The relationship between ellipsoids and melanomacrophage centres in the spleen of turbot (*Scophthalmus maximus*). *J. Comp. Pathol.* 86:377–80.

———. 1976. The ultrastructure of plaice (*Pleuronectes platessa*) leucocytes. *J. Fish Biol.* 8:139–42.

Ferguson, H. W., and R. J. Roberts. 1975. Myeloid leucosis associated with sporozoan infection in cultured turbot (*Scophthalmus maximus* L.). *J. Comp. Pathol.* 85:317–26.

Fletcher, G. L., and J. C. Smith. 1980. Evidence for permanent population differences in the annual cycle of plasma "antifreeze" levels of winter flounder. *Can. J. Zool.* 58:507–12.

Fletcher, P. E., and G. L. Fletcher. 1980. Zinc- and copper-binding proteins in the plasma of winter flounder (*Pseudopleuronectes americanus*). *Can. J. Zool.* 58:609–13.

Foda, A. 1973. Changes in hematocrit and hemoglobin in Atlantic salmon (*Salmo salar*) as a result of furunculosis disease. *J. Fish. Res. Board Can.* 30:467–68.

Froehly, M. F., and P. A. Deschaux. 1986. Presence of tonofilaments and thymic serum factor (FTS) in thymic epithelial cells of a freshwater fish (Carp:Cyprinus carpio) and a seawater fish (Bass:Dicentrarchus labrax). *Thymus* 8:235–44.

Gingerich, W. H., R. A. Pityer, and J. J. Rach. 1987. Estimates of plasma, packed cell and total blood volume in tissues of the rainbow trout (*Salmo gairdneri*). *Comp. Biochem. Physiol.* 87A:251–56.

Graham, M. S., R. L. Haedrich, and G. L. Fletcher. 1985. Hematology of three deep-sea fishes: a reflection of low metabolic rates. *Comp. Biochem. Physiol.* 80:79–84.

Grammeltvedt, A. F. 1974. A method of obtaining chromosome preparations from rainbow trout (*Salmo gairdneri*) by leucocyte culture. *Norw. J. Zool.* 22:129–34.

Heckman, J. R., F. W. Allendorf, and J. E. Wright. 1971. Trout leukocytes: growth in oxygenated cultures. *Science* 173:246–47.

Herraez, M. P., and A. G. Zapata. 1986. Structure and function of the melanomacrophage centres of the goldfish *Carassius auratus*. *Vet. Immunol. Immunopathol.* 12:117–26.

Hew, C. L., D. Slaughter, G. L. Fletcher, and S. B. Joshi. 1981. Antifreeze glycoproteins in the plasma of Newfoundland Atlantic cod (*Gadus morhua*). *Can. J. Zool.* 59:2186–92.

Hille, S. 1982. A literature review of the blood chemistry of rainbow trout, *Salmo gairdneri* Rich. *J. Fish Biol.* 20:535–69.

Houston, A. H., K. M. Mearow, and J. S. Smeda. 1976. Further observations upon the hemoglobin systems of thermally-acclimated freshwater teleosts: pumpkinseed (*Lepomis gibbosus*), white sucker (*Catostomus commersoni*), carp (*Cyprinus carpio*), goldfish (*Carassius auratus*) and carp–goldfish hybrids. *Comp. Biochem. Physiol.* 54:267–73.

Kawatsu, H. 1966. Studies on the anemia of fish. 1. Anemia of rainbow trout caused by starvation. *Bull. Freshwater Fish. Res. Lab.* 15:167–73.

______. 1976. Studies on the anemia of fish. VII. Folic acid anemia in brook trout. *Bull. Freshwater Fish. Lab.* 25:21–28.

Kelenyi, G. 1972. Phylogenesis of the azurophil leucocyte granules in vertebrates. *Sep. Exp.* 28:1094–96.

Kiceniuk, J. W., and D. R. Jones. 1977. The oxygen transport system in trout (*Salmo gairdneri*) during sustained exercise. *J. Exp. Biol.* 69:247–60.

Klontz, G. W. 1972. Haematological techniques and the immune response in rainbow trout. *Symp. Zool. Soc. Lond.* 30:89–99.

Kobayashi, K. A., and C. M. Wood. 1980. The response of the kidney of the freshwater rainbow trout to true metabolic acidosis. *J. Exp. Biol.* 84:227–44.

Lane, H. C. 1980. The response of the haemoglobin system of fed and starved rainbow trout, *Salmo gairdneri* Richardson, to bleeding. *J. Fish Biol.* 16:405–11.

Lester, R. J. G., and J. Budd. 1979. Some changes in the blood cells of diseased coho salmon. *Can. J. Zool.* 57:1458–64.

Lester, R. J. G., and S. S. Desser. 1975. Ultrastructural observations on the granulocytic leucocytes of the teleost *Catostomus commersoni*. *Can. J. Zool.* 53:1648–57.

Lom, J., I. Dykova, and M. Pavlaskova. 1983. "Unidentified" mobile protozoans from the blood of carp and some unsolved problems of myxosporean life cycles. *J. Protozool.* 30:497–508.

McCarthy, D. H., J. P. Stevenson, and M. S. Roberts. 1973. Some blood parameters of the rainbow trout (*Salmo gairdneri* Richardson). *J. Fish Biol.* 5:1–8.

McDonald, D. G., H. Hobe, and C. M. Wood. 1980. The influence of calcium on the physiological responses of the rainbow trout, *Salmo gairdneri,* to low environmental pH. *J. Exp. Biol.* 88:109–31.

Mackie, I. M., and C. S. Wardle. 1971. Electrophoretic identification of lymph from muscle tissue of plaice (*Pleuronectes platessa* L.). *Int. J. Biochem.* 2:409–13.

Margiocco, C., A. Arillo, P. Mensi, and G. Schenone. 1983. Nitrite bioaccumulation in *Salmo gairdneri* Rich. and hematological consequences. *Aquat. Tox.* 3:261–70.

Mattisson, A. G. M., and R. Fange. 1977. Light- and electron-microscopic observations on the blood cells of the Atlantic hagfish, *Myxine glutinosa* (L.). *Acta Zool.* (Stockh.) 58:205–21.

Miller, W. R., III, A. C. Hendricks, and J. Cairns, Jr. 1983. Normal ranges for diagnostically important hematological and blood chemistry characteristics of rainbow trout (*Salmo gairdneri*). *Can. J. Fish. Aquat. Sci.* 40:420–25.

Mulcahy, M. F. 1967. Serum protein changes in diseased Atlantic salmon. *Nature* 215:143–44.

Munkittrick, K. R., and J. F. Leatherland. 1983. Haematocrit values in feral goldfish, *Carassius auratus* L., as indicators of the health of the population. *J. Fish Biol.* 23:153–61.

Nichols, D. J. 1987. Fluid volumes in rainbow trout, *Salmo gairdneri*: application of compartmental analysis. *Comp. Biochem. Physiol.* 87A:703–9.

Nikinmaa, M., and A. Soivio. 1979. Oxygen dissociation curves and oxygen capacities of blood of a freshwater fish, *Salmo gairdneri*. *Ann. Zool. Fennici* 16:217–21.

______. 1980. The oxygen-binding properties of erythrocyte suspensions of *Salmo gairdneri* and of haemolysates in various buffers in the physiological pH range. *Ann. Zool. Fennici* 17:43–46.

Page, M., and A. F. Rowley. 1983. A cytochemical, light and electron microscopical study of the leucocytes of the adult river lamprey, *Lampetra fluviatilis* (L. Gray). *J. Fish Biol.* 22:503–17.

Palachek, R. M., and J. R. Tomasso. 1984. Toxicity of nitrite to channel catfish (*Ictalurus punctatus*), tilapia (*Tilapia aurea*), and largemouth bass (*Micropterus salmoides*): evidence for a nitrite exclusion mechanism. *Can. J. Fish. Aquat. Sci.* 41:1739–44.

Perry, S. F., and M. G. Vermette. 1987. The effects of prolonged epinephrine infusion on the physiology of the rainbow trout, *Salmo gairdneri*. I. Blood respiratory, acid-base and ionic states. *J. Exp. Biol.* 128:235–53.

Poston, H. A., G. F. Combs, Jr., and L. Leibovitz. 1976. Vitamin E and selenium interrelations in the diet of Atlantic salmon (*Salmo salar*): gross, histological and biochemical deficiency signs. *J. Nutr.* 106:892–904.

Pottinger, T. G., and A. D. Pickering. 1987. Androgen levels and erythrocytosis in maturing brown trout, *Salmo trutta* L. *Fish Physiol. Biochem.* 3:121–26.

Quentel, C., and J. F. Aldrin. 1986. Blood changes in catheterized rainbow trout (*Salmo gairdneri*) intraperitoneally inoculated with *Yersinia ruckeri*. *Aquaculture* 53:169–86.

Railo, E., M. Nikinmaa, and A. Soivio. 1985. Effects of sampling on blood parameters in the rainbow trout, *Salmo gairdneri* Richardson. *J. Fish Biol.* 26:725–32.

Reno, P. W., K. Kleftis, S. W. Sherburne, and B. L. Nicholson. 1986. Experimental infection and pathogenesis of viral erythrocytic necrosis (VEN) in Atlantic cod, *Gadus morhua*. *Can. J. Fish. Aquat. Sci.* 43:945–51.

Reshetnikov, Y. S. 1976. The application of biochemical indices in the investigation of Salmonidae. *Ichthyologia* 8:91–99.

Richards, R. H., and A. D. Pickering. 1979. Changes in serum parameters of *Saprolegnia*-infected brown trout, *Salmo trutta* L. *J. Fish Dis.* 2:107–206.

Roch, M., and E. J. Maly. 1979. Relationship of cadmium-induced hypocalcemia with mortality in rainbow trout (*Salmo gairdneri*) and the influence of temperature on toxicity. *J. Fish. Res. Board Can.* 36:1297–1303.

Schneider, V. B., and H. Ambrosius. 1982. On the development of lymphoid organs of *Cyrpinus carpio* L. (light- and electron-microscopic investigations). *Zool. Jb. Anat.* 107:136–49.

Sekhon, S. S., and H. W. Beams. 1969. Fine structure of the developing trout erythrocytes and thrombocytes with special reference to the marginal band and the cytoplasmic organelles. *Am. J. Anat.* 125:353–74.

Shieh, H. S. 1978. Changes of blood enzymes in brook trout induced by infection with *Aeromonas salmonicida*. *J. Fish Biol.* 12:13–18.

Smail, D. A., and S. I. Egglestone. 1980. Virus infections of marine fish erythrocytes: prevalence of piscine erythrocytic necrosis in cod *Gadus morhua* L. and blenny *Blennius pholis* L. in coastal and offshore waters of the United Kingdom. *J. Fish Dis.* 3:41–46.

Smith, C. E. 1968. Hematological changes in coho salmon fed a folic acid deficient diet. *J. Fish. Res. Board Can.* 25:151–56.

Smith, C. E., and J. E. Halver. 1969. Folic acid anemia in coho salmon. *J. Fish. Res. Board Can.* 26:111–14.

Soivio, A., and A. Oikari. 1976. Haematological effects of stress on a teleost, *Esox lucius* L. *J. Fish Biol.* 8:397–411.

Soivio, A., M. Mikinmaa, and K. Westman. 1980. The blood oxygen binding properties of hypoxic *Salmo gairdneri*. *J. Comp. Physiol.* 136:83–87.

Tatner, M. F. 1987. The effect of thymectomy on the vaccine-induced protection to Yersinia ruckeri in rainbow trout, *Salmo gairdneri*. *Dev. Comp. Immunol.* 11, 427–430.

Tavassoli, M. 1986. Bone marrow in boneless fish: lessons of evolution. *Med. Hypotheses* 20:9–15.

Wardle, C. S. 1971. New observations on the lymph system of the plaice *Pleuronectes platessa* and other teleosts. *J. Mar. Biol. Ass. U.K.* 51:977–90.

Wingstrand, K. G. 1956. Non-nucleated erythrocytes in a teleostean fish *Maurolicus mulleri* (Gmelin). *Z. Zellforsch.* 45:195–200.

Wood, C. M., and F. H. Caldwell. 1978. Renal regulation of acid-base balance in a freshwater fish. *J. Exp. Zool.* 205:301–7.

Wood, C. M., and E. B. Jackson. 1980. Blood acid-base regulation during environmental hyperoxia in the rainbow trout (*Salmo gairdneri*). *Resp. Physiol.* 42:351–72.

Wood, C. M., B. R. McMahon, and D. G. McDonald. 1977. An analysis of changes in blood pH following exhausting activity in the starry flounder, *Platichthys stellatus. J. Exp. Biol.* 69:173–85.

______. 1979. Respiratory, ventilatory, and cardiovascular responses to experimental anaemia in the starry flounder, *Platichthys stellatus. J. Exp. Biol.* 82:139–62.

Wood, C. M., P. Pieprzak, and J. N. Trott. 1979. The influence of temperature and anaemia on the adrenergic and cholinergic mechanisms controlling heart rate in the rainbow trout. *Can. J. Zool.* 57:2440–47.

Woodward, J. J., E. Casillas, L. S. Smith, and B. G. D'Aoust. 1979. Rapid decompression stress accelerates fibrinolysis in fingerling salmon. *J. Fish. Res. Board Can.* 36:592–94.

Yamamoto, K., Y. Itazawa, and H. Kobayashi. 1985. Direct observation of fish spleen by an abdominal window method and its application to exercised and hypoxic yellowtail. *Jpn. J. Ichthyol.* 31:427–33.

Yoffey, J. M. 1928. A contribution to the study of the comparative histology and physiology of the spleen, with reference chiefly to its cellular constituents. *J. Anat.* 63:314–44.

6 Cardiovascular System

The Heart

The morphology and innervation of the fish heart is beautifully reviewed by Santer (1985) and the reader is directed there for detailed information. The teleost heart is located in an antero-ventral position, separated from the peritoneal cavity by a thin septum that effectively comprises the posterior part of the pericardial sac. In sequence, returning venous blood enters the thin-walled largely connective tissue sinus venosus, then through sinoatrial valves into the thin but spongy atrium; it then enters the ventricle through the atrio-ventricular funnel and thence into the white, elastic, highly distensible bulbus arteriosus through a pair

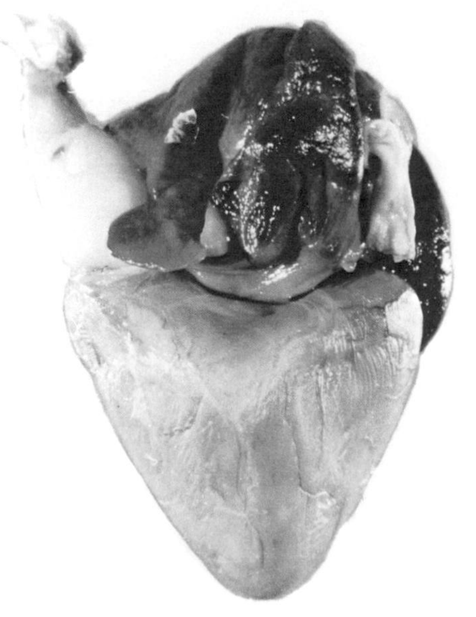
A

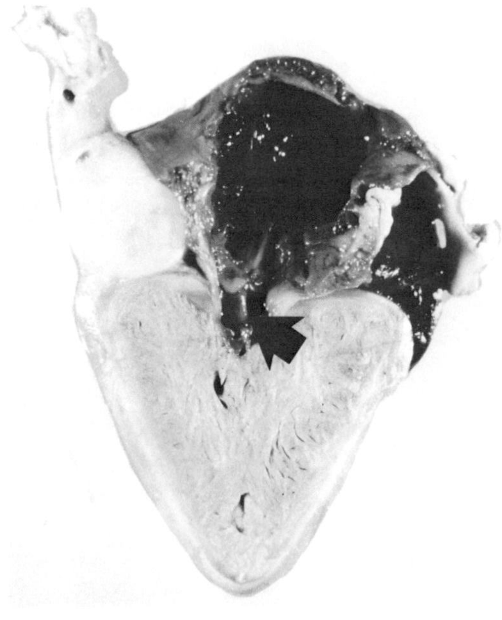
B

Fig. 6.1. A. Formalin-fixed heart from normal coho salmon showing sizes and relationships of the different chambers. The large sac-like atrium sits on top of the pyramidal ventricle, on the surface of which may be seen coronary vessels. The sinus venosus is the white collapsed structure on the right of the atrium, while the bulbus arteriosus projects from the coronary groove to the left (located ventrally in the fish). **B.** The heart has been sliced away to reveal spongy appearance of ventricular myocardium, while atrium has a lumen. Joining the two chambers is the atrioventricular funnel (*arrow*).

of semilunar valves, and out into the ventral aorta to the gills (Fig. 6.1). The distensible nature of the bulbus arteriosus ensures a continuous flow of blood in the ventral aorta throughout the cycle.

Heart rate is governed by a sinoatrial pacemaker (see Fig. 9.7E). This is usually omitted at postmortem because of the fragility of the sinus venosus. Major influences on heart rate include vagal inhibition (and in some teleosts, a weak excitatory adrenergic innervation) and temperature. Increased cardiac output is accomplished mainly by elevating stroke volume rather than heart rate, although this too does increase.

Teleost ventricles have varying levels of organization, with or without a compact outer layer in addition to an inner spongy component (Fig. 6.2). In species that have both, the two layers are separated by a thin connective tissue septum. In general, actively swimming fish tend to have a compact outer layer that may account for 16% to 73% of the total ventricle, depending on the species. There are metabolic differences between compact and spongy layers. For one thing the compact layer is supplied with oxygen from a coronary circulation derived from efferent branchial arteries (some species have other supplies), while the spongy layer derives its oxygen directly from the venous blood it is pumping. During high levels of activity, the oxygen content of the venous blood will be reduced, but some protection against anoxic damage is afforded by the myocardial myoglobin. Actively swimming fish have the highest levels of this pigment. (It is interesting to note that icefish, which survive at polar temperatures, have neither hemoglobin nor myocardial myoglobin.) Cardiac muscle from mammals prefers fatty acids to fuel its obligatory aerobic and very active energy metabolism. By comparison, teleosts are able to utilize both fatty acids and carbohydrates. Elasmobranchs, on the other hand, are unable to catabolize fatty acids (Sidell et al. 1987).

It is important to realize that variations

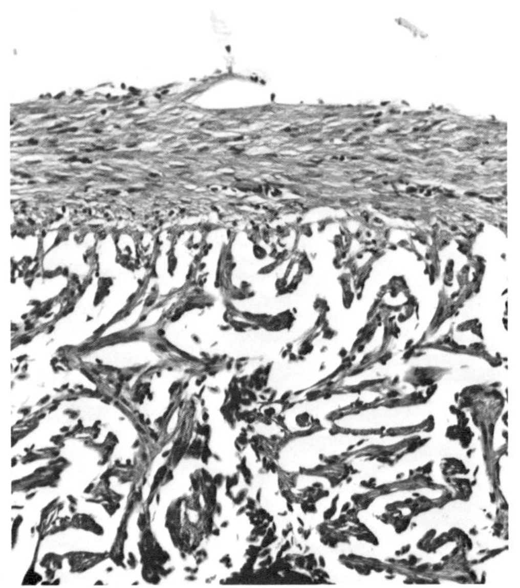

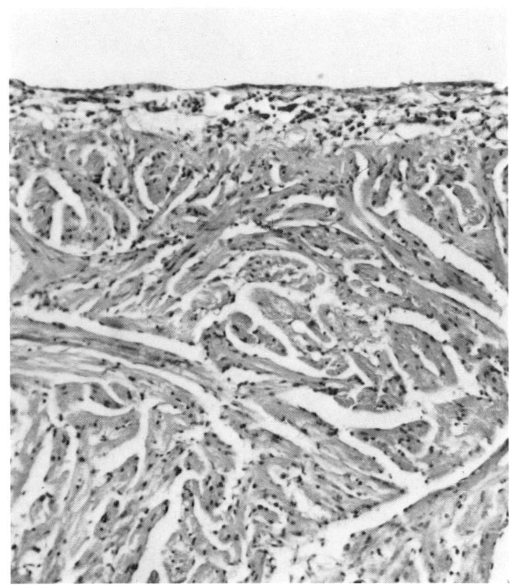

Fig. 6.2. A. Ventricle from young salmonid. Although still immature, a well-developed compact outer myocardium is obvious. This becomes thicker with age. **B.** Ventricle from marine flatfish. Note that a compact outer myocardium is not present and that the whole of the organization is a spongy meshwork of fibers. The epicardium is quite thick.

in organization and proportions of compact-versus-spongy myocardium may occur at different stages in the life cycle of a particular species.

The meshwork of mononuclear myocytes that comprise the spongy portions of myocardium is lined by a one-cell-thick endocardial (Fig. 6.3). The endocardial cells

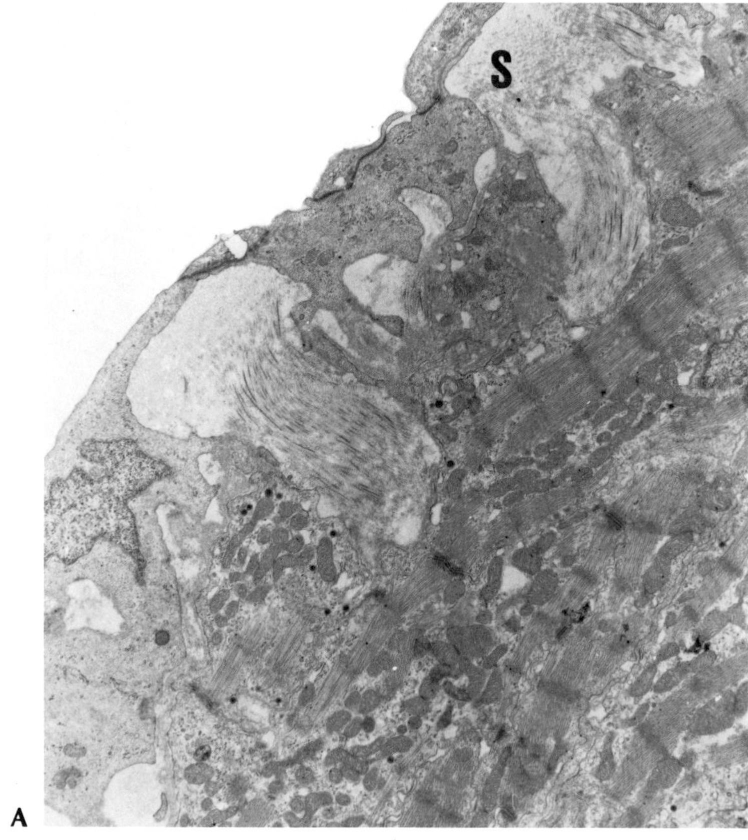

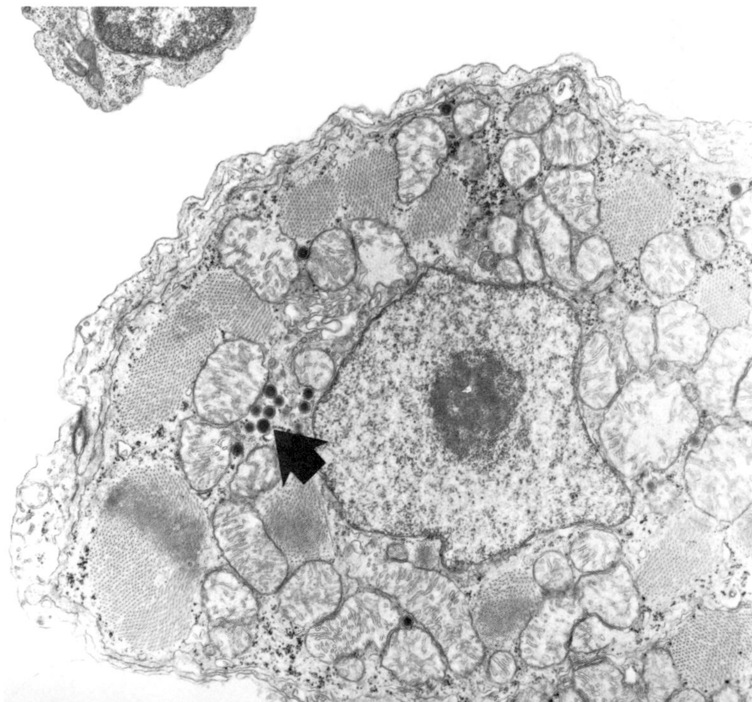

Fig. 6.3. Series of electron micrographs from ventricle of *Pleuronectes platessa,* a marine flatfish, showing some ultrastructural features of teleost myocardium. **A.** Epicardial surface showing several cells joined by numerous desmosomes and overlying the subepicardial space (*S*). Within this space is a buffy ground substance and numerous collagen fibers. **B.** Single myocardial cell showing centrally placed nucleus with prominent nucleolus and numerous large mitochondria distributed among myofibrils, most of which are cut in transverse section. A small cluster of dense-cored vesicles is also present (*arrow*). The cell is lined by a thin but distinct endocardium. **C.** High-power view of endocardium in normal fish. Note the subendocardial collagen (*arrow*). **D.** Intercalated discs. Z bands and a faint M band are present. No transverse tubule system is present, however, and the sarcoplasmic reticulum is greatly reduced, compared to mammals.

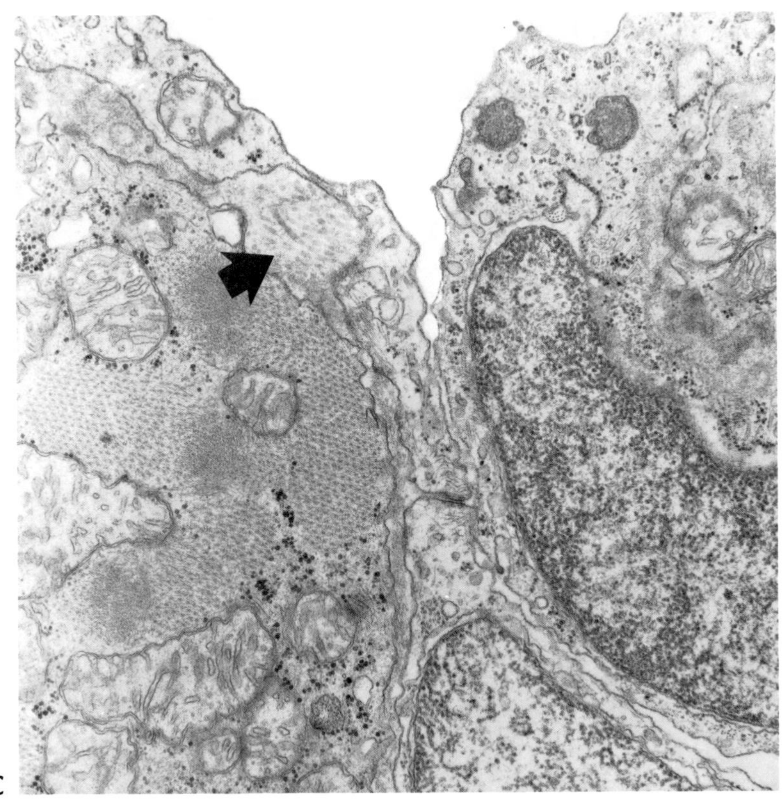

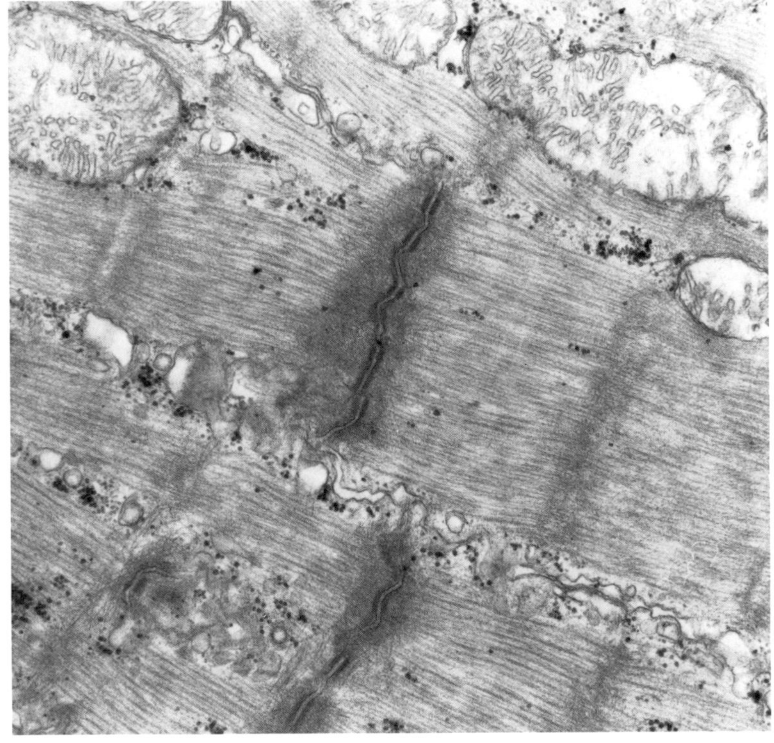

are phagocytic to varying degrees (between species) but are most avid in the atrium (Fig. 6.4). While this location is almost ideal from the standpoint of monitoring blood-borne antigens, at the same time it places the fish in a vulnerable position if the antigen encountered is highly pathogenic.

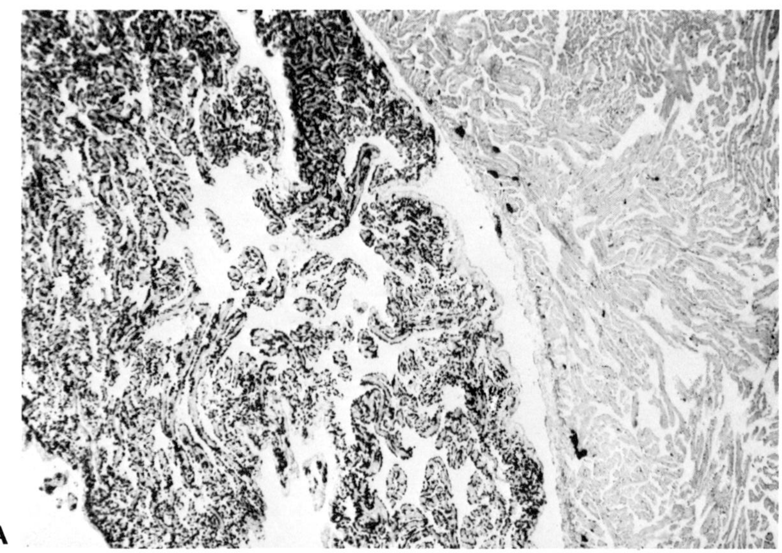

A

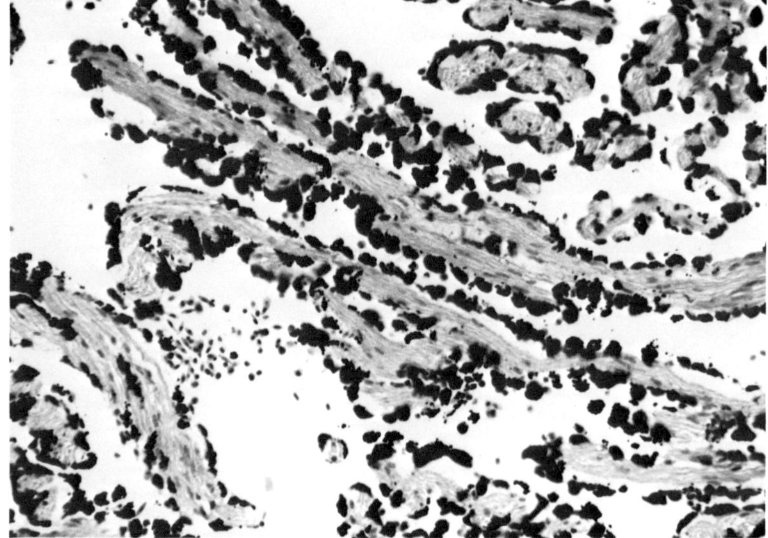

B

Fig. 6.4. Plaice injected with carbon particles (intravenously) and killed 1 hour later. **A.** Low-power picture demonstrates relative uptake of carbon particles. Ventricle (*right*) has trapped very little, whereas atrium is quite black. **B.** High-power view of atrium. Endocardial cells are prominently marked. **C.** Ultrastructure of atrium showing moderate quantities of carbon within numerous phagolysosomes of endocardial cell. A macropinocytotic vesicle adjacent to an erythrocyte was caught (by fixation) in the act of endocytosing some more carbon particles along with some buffy material, probably plasma components. **D.** Even such large particles as yeast may be endocytosed by the atrial endocardium as demonstrated in this plaice.

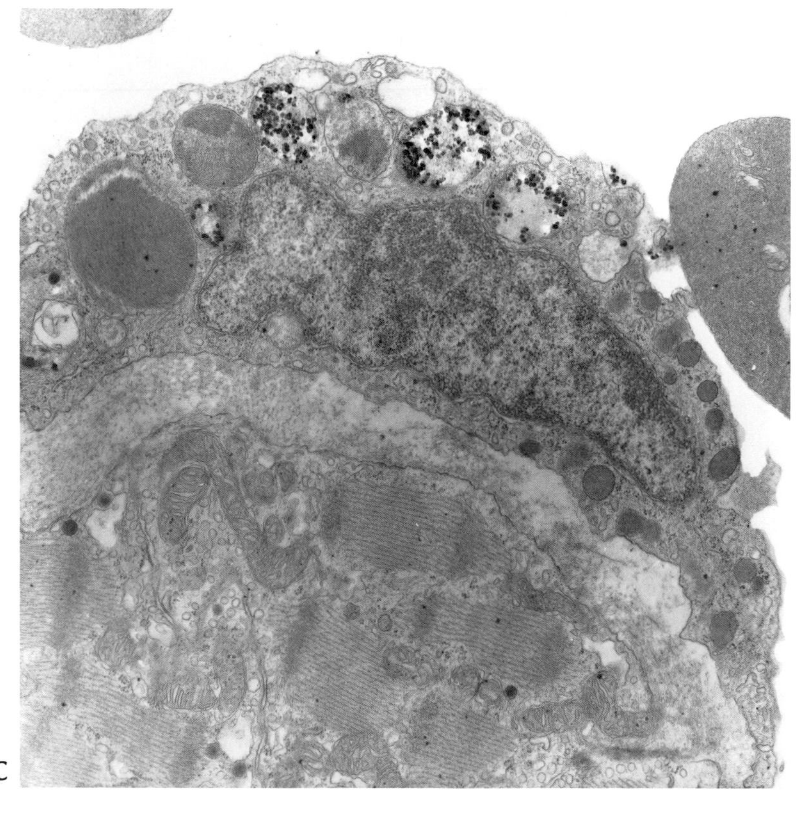

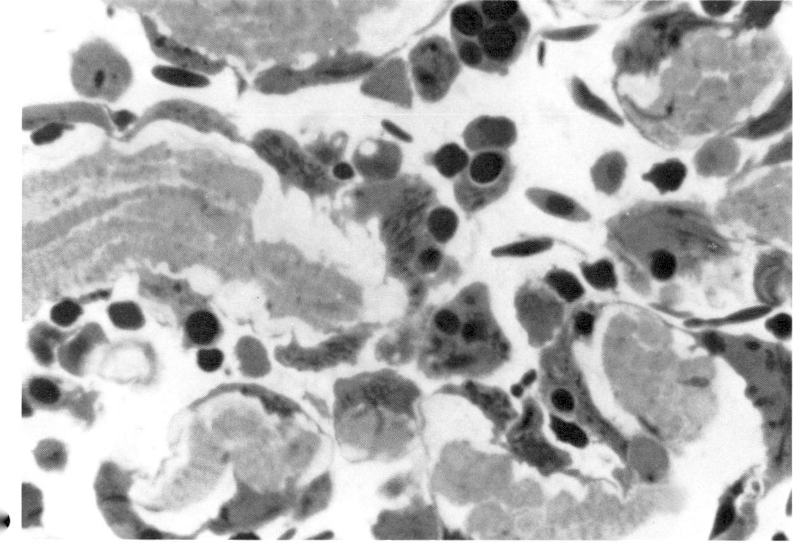

Distribution of Lesions. A common location for lesions is the extreme periphery of the spongy ventricular myocardium. Depending on the species, this may be the compact-spongy interface (Fig. 6.5), or it may be just beneath the epicardium (Fig. 6.6). The reason for this predilection is unknown, although in the case of bacteria, it may represent a relatively low-flow environment that thus facilitates attachment to, or penetration of, the endocardium. Alternatively, the site may be vulnerable for mechanical or metabolic reasons or due to inadequate oxygenation.

The phagocytic properties of the atrial endocardium make it a predilection site for lesions. In acute bacteremias—especially in young fish, those due to aeromonads and vibrios—toxic endocardial and atrial myocardial necroses are not uncommon

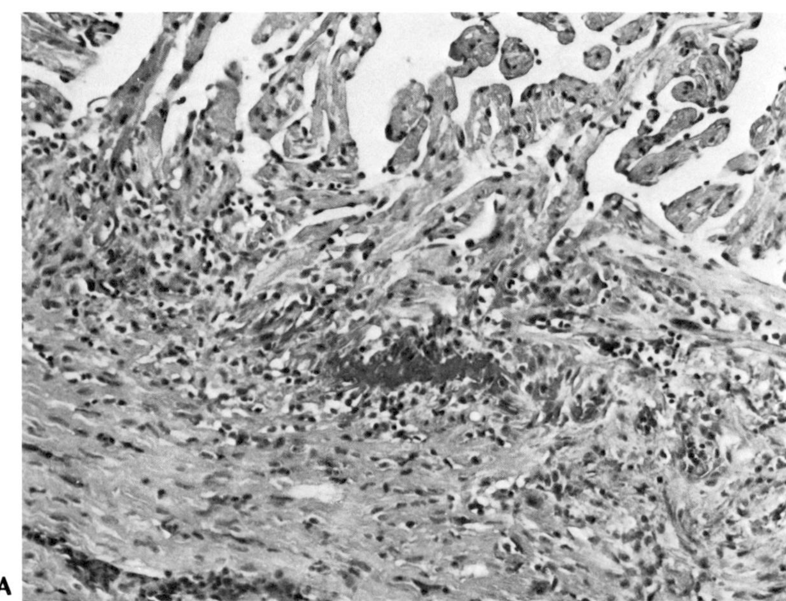

A

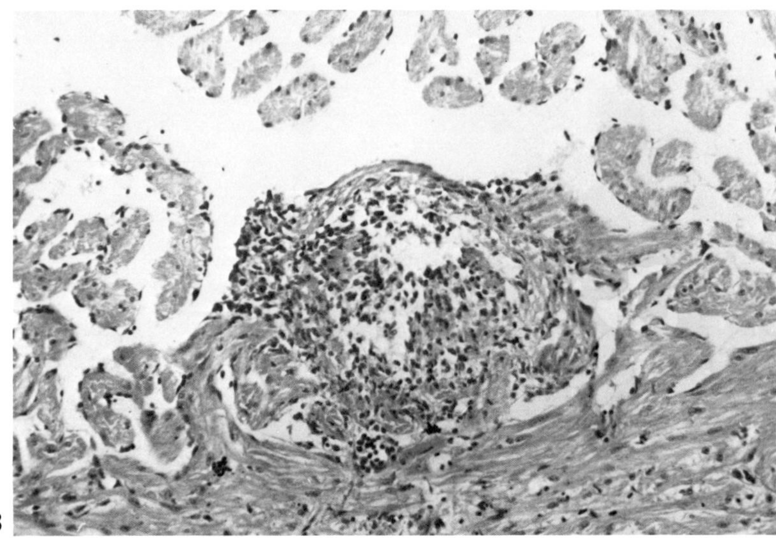

B

Fig. 6.5. A. Brown trout with severe postspawning hypoproteinemia and anemia showing fibrinoid necrosis of ventricular myocardium at compact-spongy interface. **B.** Rainbow trout with bacterial kidney disease showing typical pyogranulomatous focus at periphery of spongy layer.

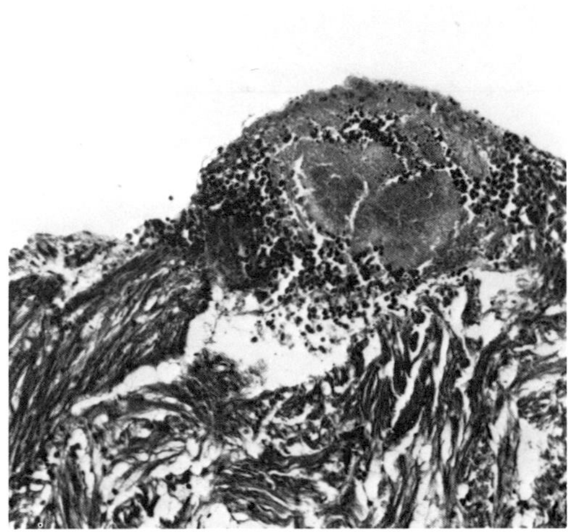

Fig. 6.6. Creek chub (*Semotilus atromaculatus*) with furunculosis. A large bacterial colony surrounded by a mixed inflammatory cell infiltrate is situated just beneath the epicardium. Indeed, organisms are already within the epicardium, and the bulging lesion gives the appearance of imminent rupture.

findings. The atrial endocardial cells are swollen, pyknotic, and often slough, with consequent degeneration and necrosis of the underlying myocardium. In these cases, cardiac failure may partly explain the associated edema, clinically manifested by swelling of the abdomen, exophthalmia, and softening of the skeletal musculature. Roberts (1978) suggests that the process may be complicated by concomitant failure of renal and branchial circulation, leading to loss of ion and fluid control, although any drop in renal perfusion pressure should enhance renin-angiotensin production in an attempt to compensate. This affinity of particulates for the heart is borne out experimentally in rainbow trout, in which species a rapid rise in water temperature has been shown to increase the numbers of bacteria that localize there.

The epicardium also is a common site for lesions associated with parasites (Fig. 6.7) or microbial agents, possibly because of the lymph draining into the pericardial sac. (Hemopoietic tissue and scattered thyroid follicles are not infrequent findings in this location, and it is important, therefore, not to overinterpret hypercellularity as inevitable epicarditis.) Grossly obvious adhesions of epicardium to pericardium, or cardiac tamponade due to excess fluid in the pericardial sac are, however, uncommon, except sometimes as a sequel to blood sampling by cardiac puncture (Fig. 6.8), especially in those species with coronary vessels, which presumably may be punctured by the needle. A constrictive pericarditis is occasionally seen in severe fibrogranulomatous responses to metacercariae of digeneans, with high mortality a consequence (Fig. 6.9), although even here adhesions to the pericardial sac are not prominent. A serofibrinous exudate is seen in the pericardial sac in Atlantic salmon suffering from Hitra disease, a condition reported by some to have a basis in vitamin E deficiency (others feel that a low-temperature *Vibrio–V. salmonicida*—is more important) and similar in this respect to mulberry heart disease of pigs. Other major similarities to mulberry heart disease include associated vascular lesions.

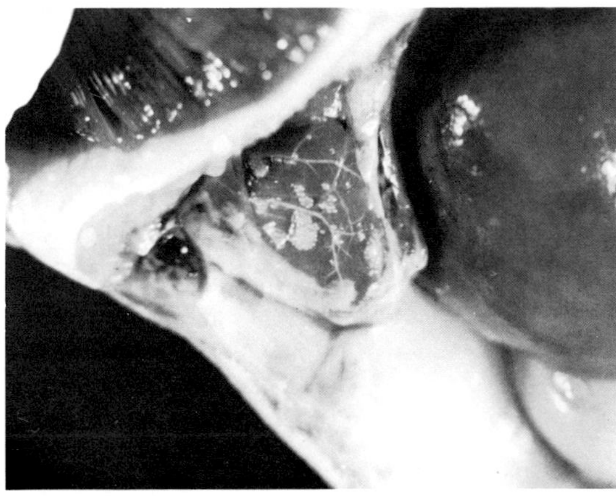

Fig. 6.7. Brown trout with pericardium partially removed to reveal numerous metacercariae of the digenetic trematode *Cotylurus* situated on epicardium. Reaction here is minimal and epicardial surface is still shiny.

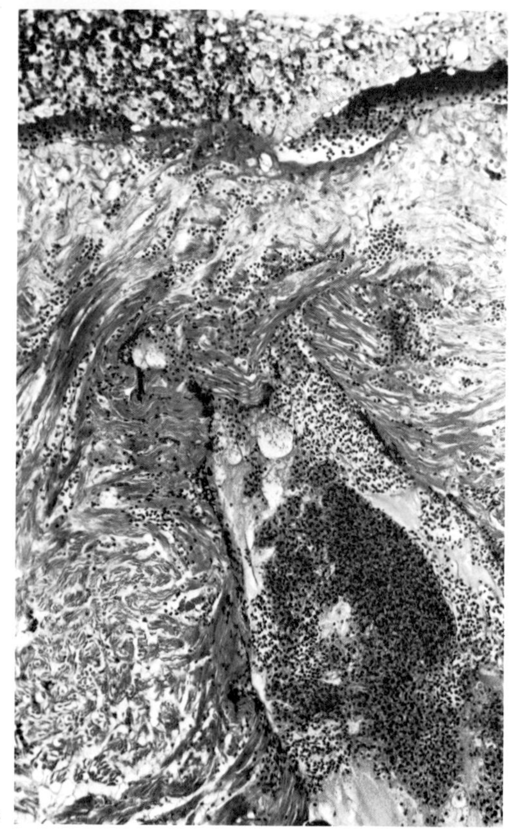

A

Fig. 6.8. Heart from tilapia that had been repeatedly blood sampled by cardiac puncture with hypodermic needle. Different ages of lesions are present in the same fish corresponding to separate sampling times. Despite apparently severe lesions, the fish appeared clinically normal and interested in mating, which merely emphasizes the impressive sex drive of these fish! (*Material courtesy of M. Singh*) **A.** Most recent puncture wound showing blood accumulation within cavity so produced, and extravasation onto epicardial surface. **B.** Organizing thrombus that has been sequestered within ventricle associated with semilunar valves. **C.** Repairing puncture wound with increased quantities of fibrous tissue.

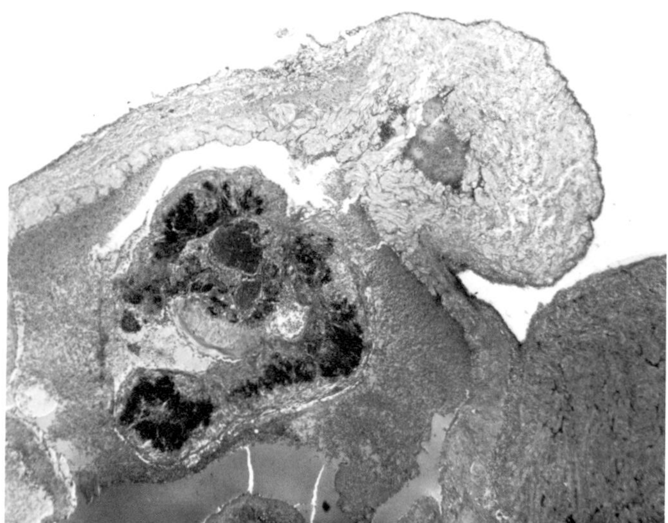

B

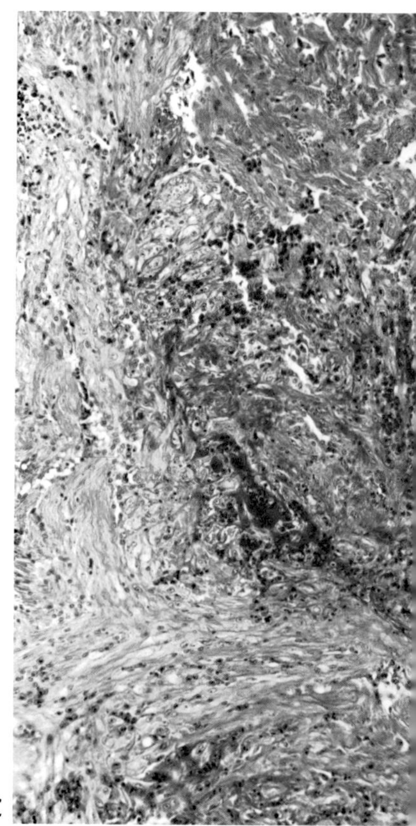

C

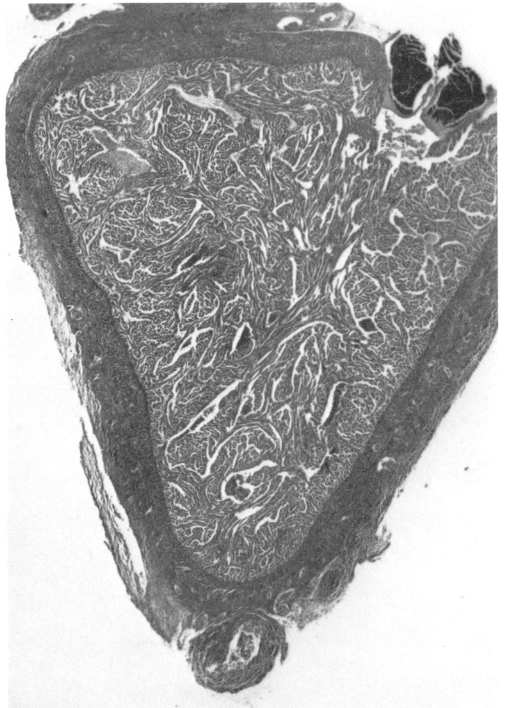

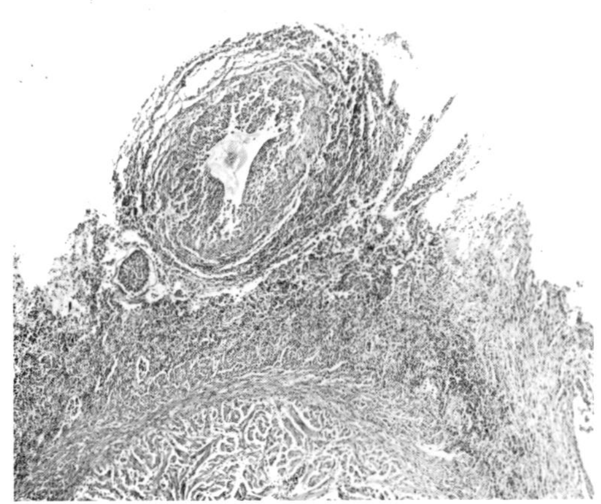

Fig. 6.9. A and **B.** Sea-caged rainbow trout with constrictive epicarditis due to pronounced granulomatous response to metacercariae of the digenetic trematode *Stephanostomum baccatum.* The consequences of such a response for an animal that relies heavily on an elevated stroke volume for increasing its cardiac output will probably depend on the demand placed upon it in any particular environment.

Valvular lesions are also uncommon, compared to other vertebrates, but they are occasionally encountered in bacterial diseases causing vegetative valvular endocarditis mainly of the semilunar valves between the bulbus arteriosus and ventricle (Fig. 6.10). Any damage to endocardium can promote thrombosis. This is frequently seen in the bacterial or nutritionally based heart lesions.

Consequences. Cardiac lesions are frequently encountered in both farmed and wild fish, but their significance to the host is hard to assess. In only a few cases are there obvious systemic responses, such as edema or anoxic changes, that can be specifically ascribed to impaired cardiac function. There is little information on the mechanisms of degeneration and repair in the fish heart. It is unknown, for instance, whether the atrial endocardial phagocytes that slough off are replaced by recruitment from the circulation, from adjacent still-viable cells, or from the "reserve" subendocardial cells. Severe diffuse myocardial degeneration, seen in some nutritional myopathies, repairs with a reduction in spongy architecture, leading to a much more compact arrangement of myocardial trabeculae. The consequences of this for the animal are unknown, but it is important to remember that as a group, fish place a heavier reliance on stroke volume than on heart rate for increasing cardiac output; while heart function is probably compromised therefore, differences from mammals should be anticipated. Certainly, anything inhibiting proper expansion could be expected to impair output, and indeed a restrictive epicarditis due to digenetic metacercarial infection of rainbow trout has been shown in vitro to cause a 20% to 40% reduction in output compared to healthy controls, while heart rate remains essentially the same (Tort et al. 1987).

Whenever examining and interpreting changes in relative proportions of myocardial tissue, it is also important to remember

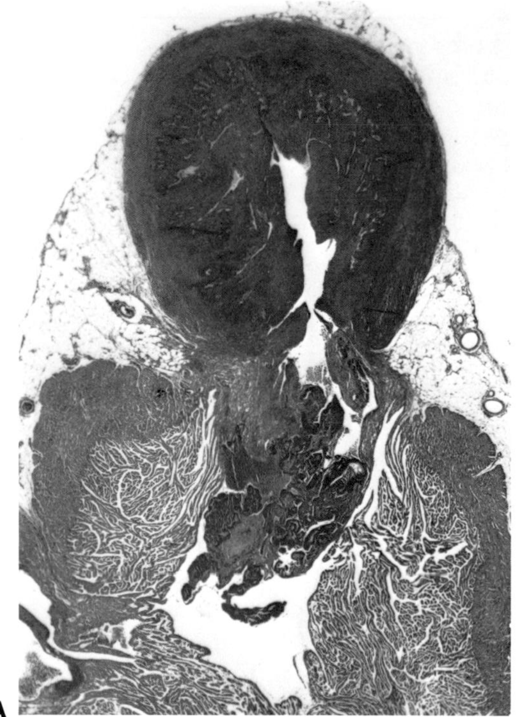
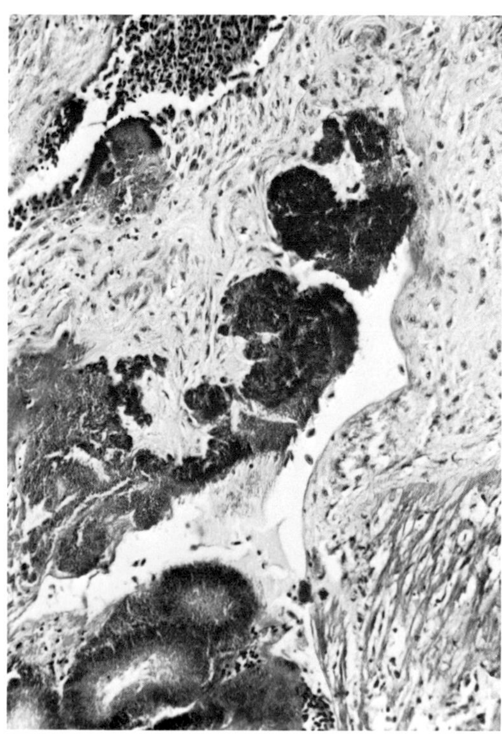

Fig. 6.10. A. Rainbow trout with septic vegetative valvular endocarditis of semilunar valves between bulbus arteriosus and ventricle. **B.** High power of lesion showing bacteria dissecting up through valves into bulbus. *Lactobacillus* sp. were isolated.

that morphological alterations naturally occur with aging. For example, salmon parr have a much higher proportion of spongy to compact myocardium than do smolts, which experience a rapid increase in the proportion of compact muscle to facilitate the increased demands placed on the heart for swimming at this stage in the life cycle of the fish. The particular nutritional demands placed upon the fish to accommodate such changes are poorly understood.

Specific Conditions. The heart is a target organ in several bacterial diseases, notably those due to *Aeromonas salmonicida* and *Renibacterium salmoninarum*. The presence of large colonies of small bacilli within and on the trabeculae is almost pathognomonic for furunculosis in salmonids in fresh water (Fig. 6.11), although other lesions are often also present, especially a fibrinous epicarditis. This latter lesion is frequently also seen in bacterial kidney disease (BKD) and in more chronic cases can be severe enough to suggest cardiac embarrassment from constriction (Fig. 6.12). Indeed, the granulation tissue can in some cases be so extensive, with pronounced neovascularization, that the epicardium can be thicker than the compact myocardium. More commonly, however, the typical necrotizing and pyogranulomatous foci seen in this disease are found scattered throughout the myocardium, often at the compact-spongy interface. Cardiac lesions are also seen in the other disseminated granuloma-inducing bacterial diseases, notably mycobacteriosis, and although granulomas can be present anywhere in the heart (Fig. 6.13), the

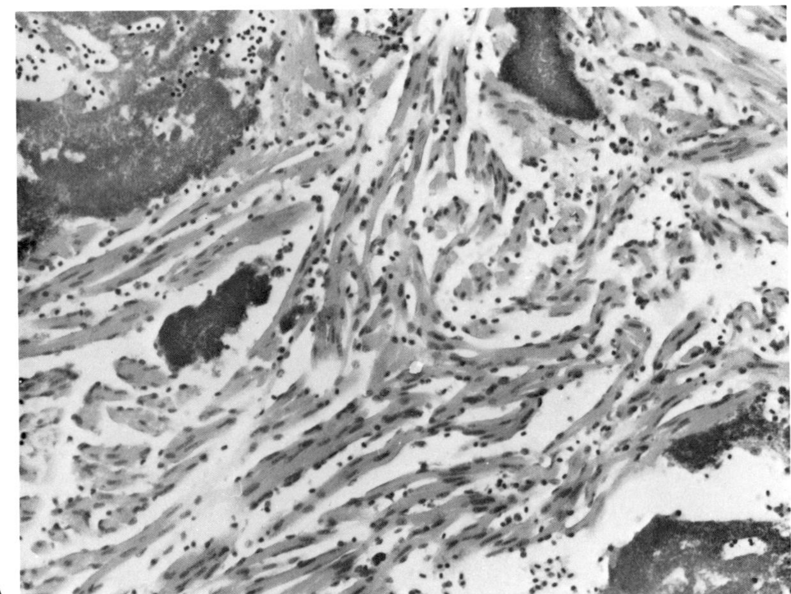

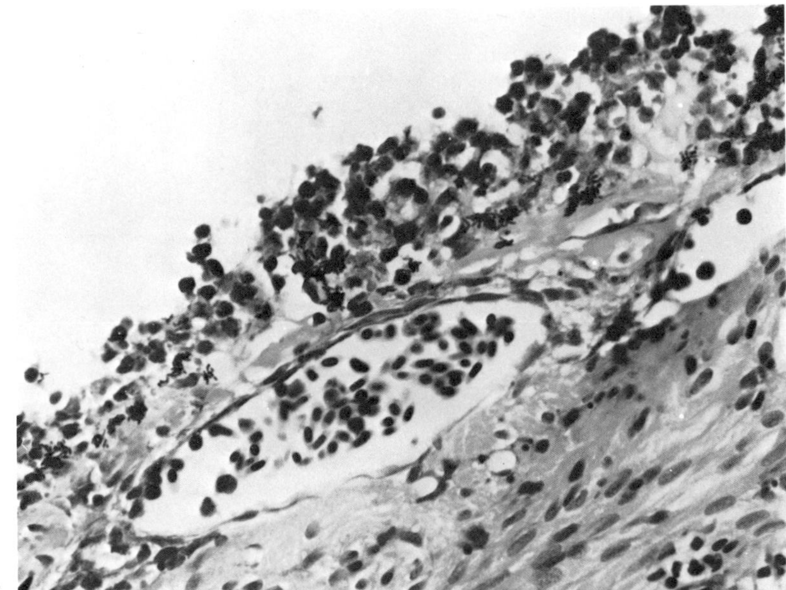

Fig. 6.11. Brown trout with furunculosis.
A. The presence of large colonies of bacteria within myocardial trabeculae and with this typical appearance is almost pathognomonic in fresh water for *Aeromonas salmonicida* in a variety of species. Note the relative absence of surrounding myocardial necrosis. This is not postmortem growth of bacteria.
B. An early fibrinous epicarditis is present. Numerous small bacilli are trapped within the response.

epicardium is a favored location. In these bacterial diseases, septic embolism and subsequent localization to the capillary beds of the gills and elsewhere are doubtless important consequences of the heart's colonization.

Lesions associated with viral diseases are seen in the generalized rhabdovirus infections such as viral hemorrhagic septicemia (VHS). Although many changes, such as hemorrhage and edema, probably are based on anoxia as a result of vessel damage, VHS can also cause multifocal myocarditis. Lymphocystis virus can attack the epi- and pericardial fibroblasts, although lesions here are less common than on the skin. The heart

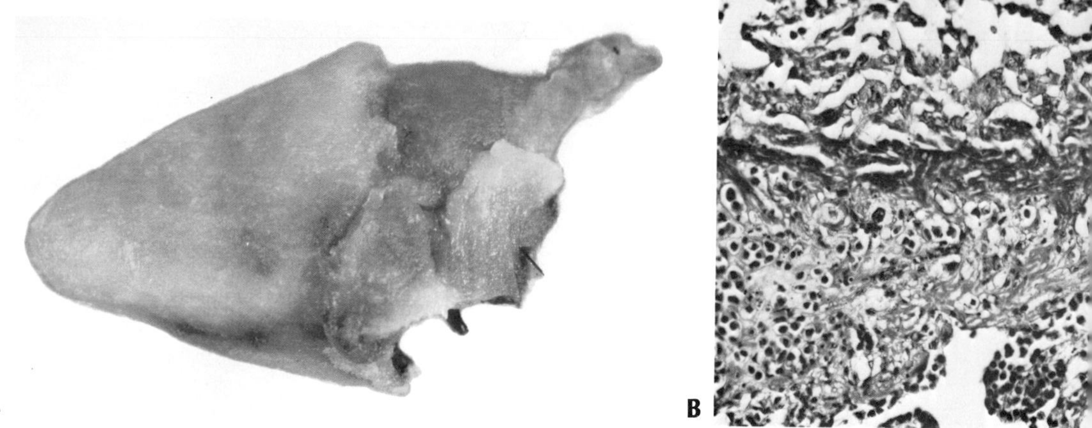

Fig. 6.12. A. Severe fibrinous epicarditis in rainbow trout due to *Renibacterium salmoninarum.* The lesion extends over entire heart (atrium is largely missing in this specimen) and is probably constrictive. **B.** Bacterial kidney disease in young coho salmon. There is marked necrosis of the entire compact layer and an extensive pyogranulomatous inflammatory response on the epicardium.

is a target organ for the fungus *Ichthyophonus hoferi,* which can cause a granulomatous epicarditis and myocarditis in several species, including pleuronectids, herring, and rainbow trout.

Many parasites localize on or within the heart, and although they are usually considered incidental findings, their significance is mostly unknown. Examples include various species of myxosporeans present within the myocardium, epicardium, or bulbus; some of these are considered pathogenic. Others include metacercariae of digeneans, notably *Cotylurus* spp. (Fig. 6.7) and *Stephanostomum* spp. (Fig. 6.9), with the latter causing death due to constrictive epicarditis. Potentially serious are the copepods, such as *Lernaeocera branchialis* or *Cardiodectes* spp., which can penetrate heart tissue with their attachment organs and cause hypertrophy of bulbus or ventricle. For the most part, however, these parasites do not penetrate to the lumen and rarely are associated with thrombosis.

Nutritional cardiomyopathy is seen in a variety of fish, especially under conditions of intensive culture, and notably in salmonids and turbot that are fed diets that may or may not be deficient in vitamin E and selenium. Lesions include an increase in eosinophilia and loss of striation, with fragmentation later (Fig. 6.14). Mineralization is not a feature. Repair is dominated by

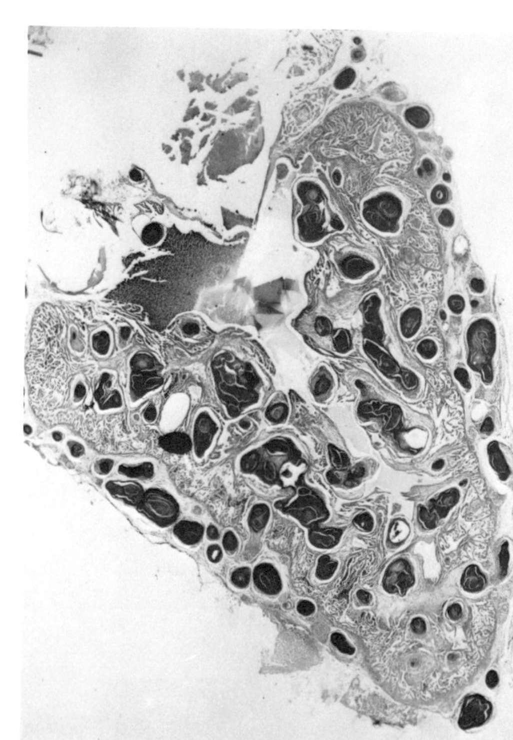

Fig. 6.13. Mycobacteriosis in cichlid. Large numbers of granulomas are scattered throughout ventricle and on its surface. It is difficult to imagine that such extensive involvement would not have clinical significance, even if only due to septic embolism.

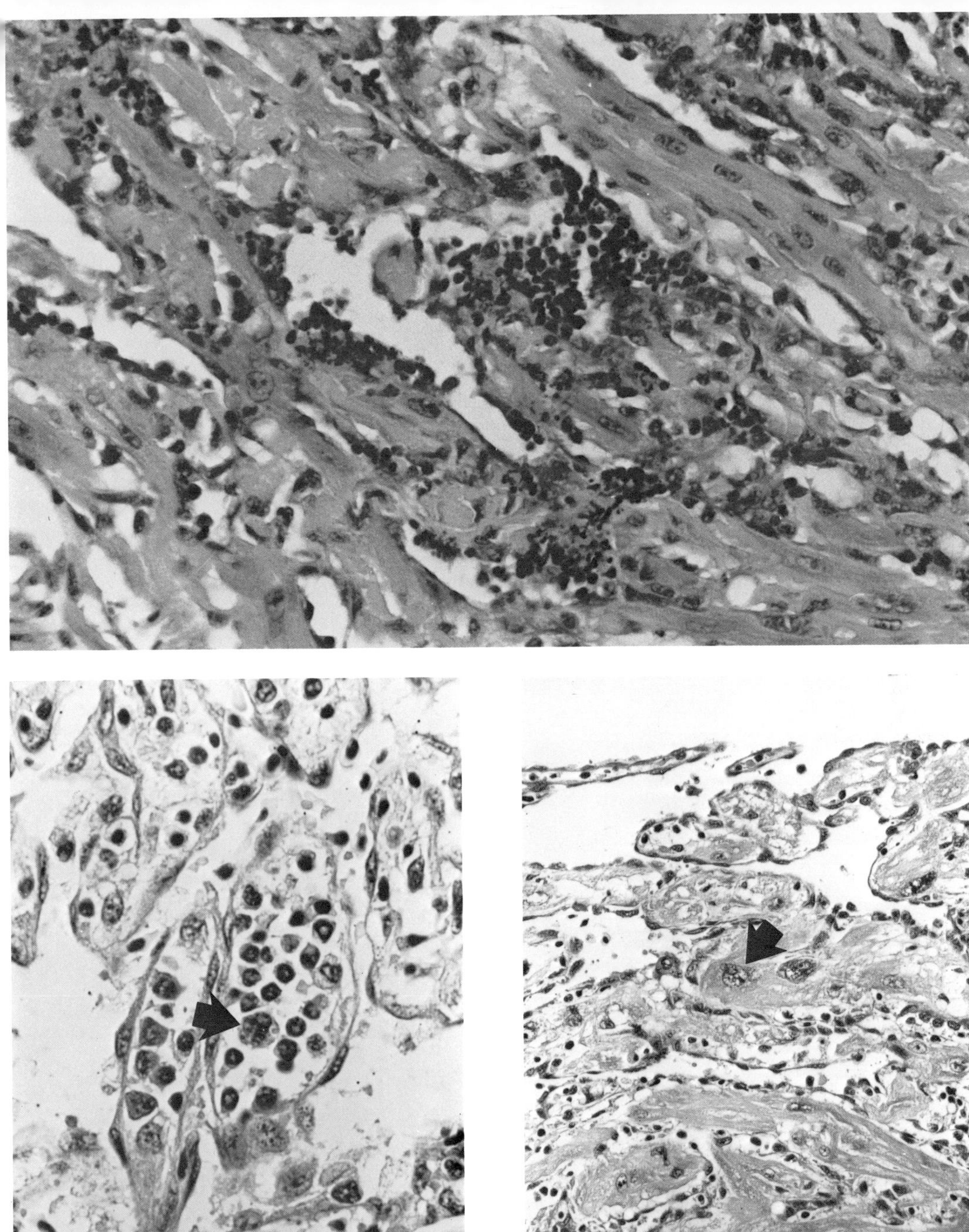

Fig. 6.14. A. Ventricle from sea-caged Atlantic salmon with severe degenerative cardiomyopathy (so-called pancreas disease). There is increased eosinophilia and loss of striation of cardiac myocytes plus multifocal mural thrombosis. **B.** Severe degenerative cardiomyopathy in broodstock Atlantic salmon. Macrophages (*arrow*) have "reamed out" necrotic myofiber. **C.** Fish recovering from cardiomyopathy showing hypertrophy (compensatory?) of surviving myocyte nuclei (*arrow*). (*Material for* **B** *and* **C** *courtesy of T. Poppe*)

hypertrophy of surviving myocytes, with hyperplasia of endocardium and a reduction in spongy matrix so that the ventricle appears more compact. Fibrosis follows but may be relatively minor.

The Vessels

Vascular lesions are quite common in fish. One of the most frequently observed is peripheral hyperemia and hemorrhage encountered in toxemias and bacteremias. This is most easily seen at the base of the fins, around the eyes, and occasionally (classically but infrequently in enteric redmouth disease) around the mouth. In many bacteremias, especially those associated with *Aeromonas salmonicida, Vibrio* spp., and *Renibacterium salmoninarum,* careful examination will frequently reveal septic thrombosis and occasional infarcts in a variety of tissues (Fig. 6.15). Some fish with *Renibacterium* infection may have extensive vascular involvement, lesions comprising necrosis of media and endothelium, with fibrin exudation or thrombosis as a consequence.

Most of the rhabdoviruses of fish also have an affinity for vascular endothelium, and consequently, the clinicopathological picture in such diseases is dominated, especially in the early stages, by punctate hemorrhaging. Mycotic invasion of vessels is reported in branchiomycosis of cyprinids in which the fungus *Branchiomyces sanguinis* grows into the lumen of the branchial vessels. Other fungi, including *Saprolegnia,* are occasionally seen in, or closely associated

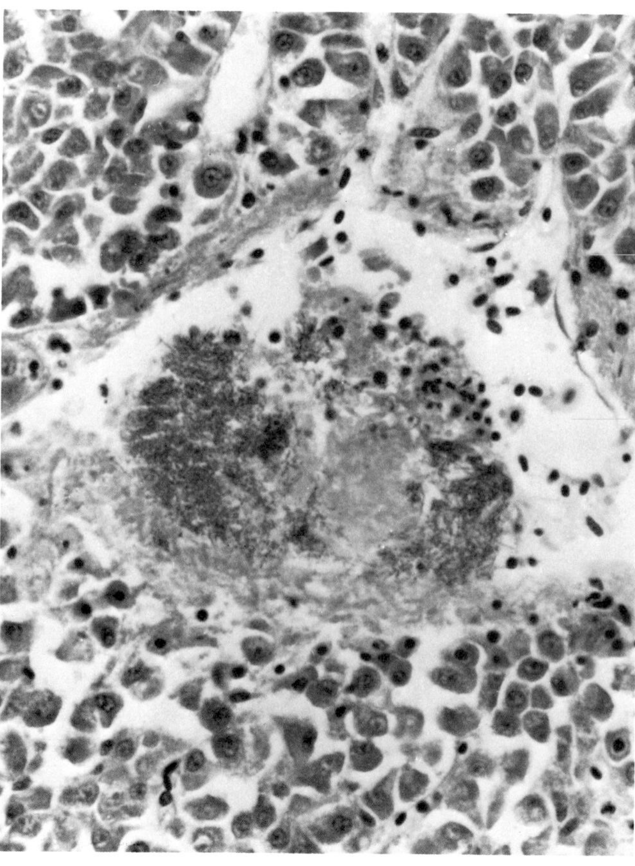

Fig. 6.15. Septic thrombus in central vein of liver from brown trout with furunculosis.

with, blood vessels. This is notable in the gastrointestinal tract of small fish, but any vessel may be involved, including the aorta.

Aneurysms of branchial vessels (sometimes referred to as telangiectasis) are a common finding in a variety of circumstances, especially acute toxic exposure (see Fig. 2.8), but they are also easily produced by rough handling and euthanasia procedures. The aneurysms thrombose, fibrose, and are eventually repaired. Dissecting aneurysms are occasionally encountered, especially in the bulbus arteriosus, possibly as a consequence of the marked stress and strain normally imposed on this highly distensible structure. Causes, or at least associated organisms, include bacteria (Fig. 6.10B) and digenetic metacercariae such as *Ascocotyle tenuicollis,* as reported in the mosquito fish (*Gambusia affinis*).

Fibrinoid necrosis of the afferent and efferent glomerular arterioles, similar in some respects to the Shwartzman reaction in mammals, is reported in white perch (*Morone americane*) (Fig. 6.16). Although associated with subendothelial electron-dense deposits, suggesting the possibility of an immune-complex-mediated lesion, the cause and pathogenesis of the condition remain unknown.

Degeneration and medial necrosis of peripheral vessels are also encountered in the vitamin E/selenium deficiency syndromes of intensively cultured fish. Such changes,

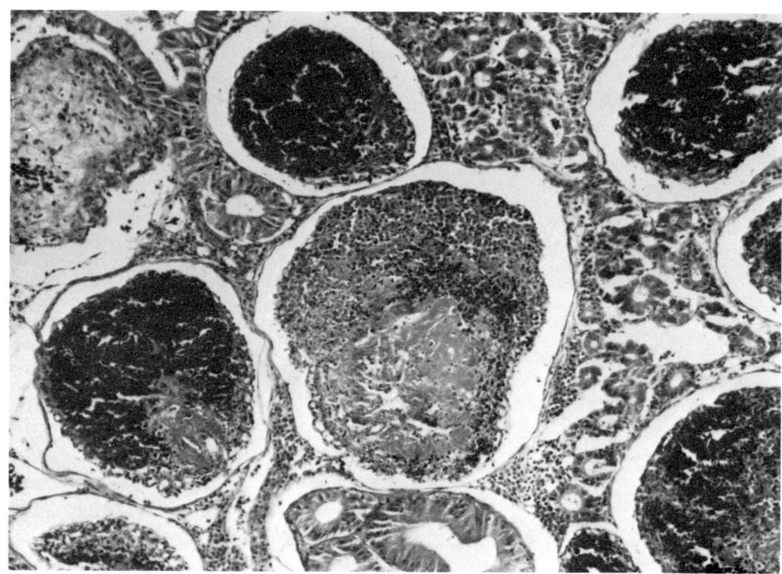

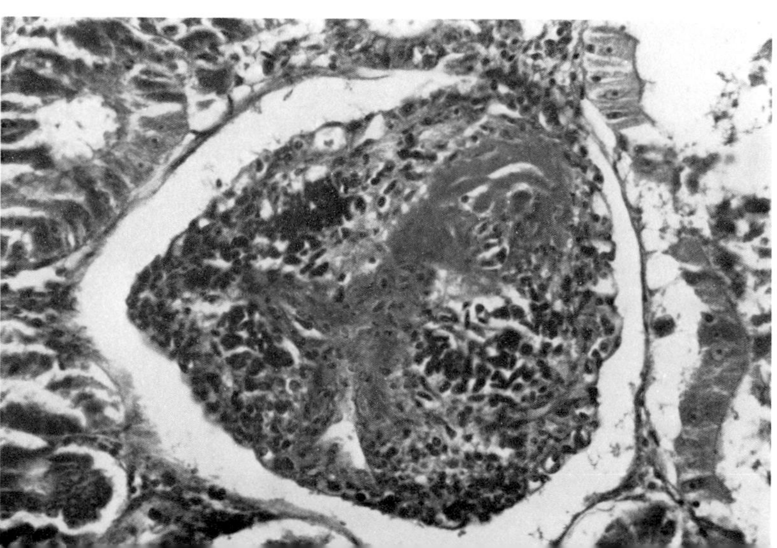

Fig. 6.16. A and **B.** Severe acute necrotizing glomerulonephritis in white perch. There is marked fibrinoid necrosis of afferent and efferent glomerular arterioles and aneurysm of some tuft capillaries with hemorrhage as a consequence. Pathogenesis of the lesion, which may affect most glomeruli in a fish and most fish in a population within a given location, is unknown. Subendothelial deposits are also present (see Fig. 4.7), but whether these are immune-complexes or fibrin is also unknown.

along with thrombosis, are reported in "Hitra disease" of farmed Atlantic salmon (Salte et al. 1987). The cause of this "hemorrhagic syndrome" is currently unknown although favored hypotheses include vitamin E/selenium deficiency, a low temperature *Vibrio* infection (*V. salmonicida*), or possibly some combination of the two. Certainly *Vibrio* seem to be frequently involved.

Aortic thrombosis and stenosis are seen in salmonids, possibly associated with extrusion of disc material via the vertebral arteries. Lesions are commonest at the main flexure point, just below the dorsal fin (Fig. 6.17). As expected, posterior incoordination and paralysis are the clinical presentation, together with erosion of skin and tail fins. The cause of the condition, which may affect large numbers of fish, is unknown.

Compared to mammals, teleost capillaries are highly permeable. The reason for this is unknown, although the relatively lower blood pressures found in these animals would not seem to demand such tight intercellular junctions to prevent leakage as are required in mammals. Whatever the reason, one of the consequences is that blood proteins, including the large IgM-type molecules, which by and large are the major ones found in teleosts, have relatively little difficulty in leaving the circulation. Thus when skin damage is extensive, from trauma or severe parasitism, even if there is little actual hemorrhage, hemodilution or hemoconcentration are distinct possibilities, depending on whether the fish is in fresh or salt water. Certainly it has been shown in the later stages of "ulcerative dermal necrosis" (UDN) in Atlantic salmon (the fish are in fresh water when this disease develops) that as the animal becomes progressively overhydrated, serum protein levels can fall by 60%. In fresh water, the dermal lymphatics would speed such a response, rapidly removing the inflowing water. Similar changes have been observed in brown trout suffering from extensive saprolegniasis.

Arteriosclerosis of coronary vessels in spawning steelhead trout, Atlantic and Pacific salmon is a transitory lesion that is considered to be largely under the control of the massive hormonal changes occurring in these anadromous species. Lesions include intimal plaques and cartilagenous or fibrous metaplasia. Spontaneous intimal proliferations are also described in pike.

In proliferative kidney disease (PKD) of salmonids, macrophages and lymphocytes cluster around the presumed myxosporean parasites. This response can occur within vessels, especially in the kidney but also elsewhere, and the cellular aggregates, closely apposed to or possibly coming through the endothelium, can be so extensive as to suggest obliteration of the lumen (Fig. 6.18). The microsporidian *Loma* has caused endarteritis of epizootic proportions in juvenile chinook salmon. Pathological changes include acute inflammation, necrosis, and eventual obliteration of the vessel lumen leading to infarction of tail muscles. Gill vessels may also be affected (Fig. 6.19).

Adult sanguinicolid flukes, for example *Sanguinicola inermis* residing in the bran-

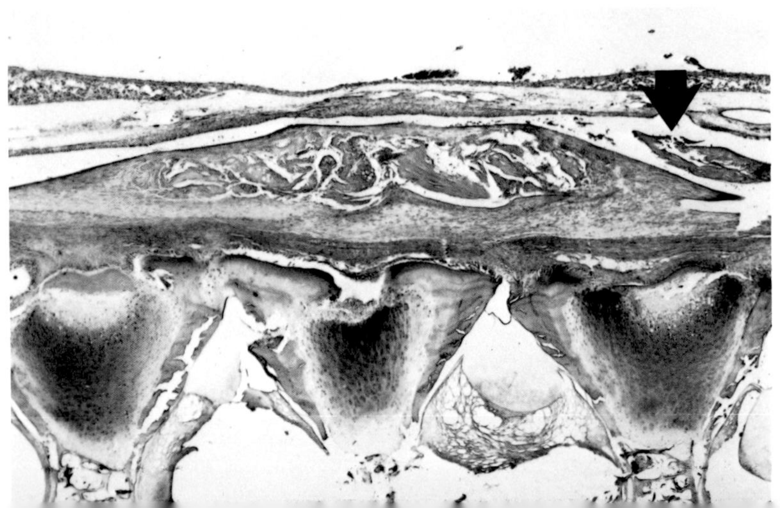

Fig. 6.17. Aortic stenosis and thrombosis in rainbow trout. Greatly thickened wall of aorta almost obliterates the lumen of the vessel. Note the tail of a large thrombus "downstream" from the lesion (*arrow*). (*Material courtesy of B. Hicks*)

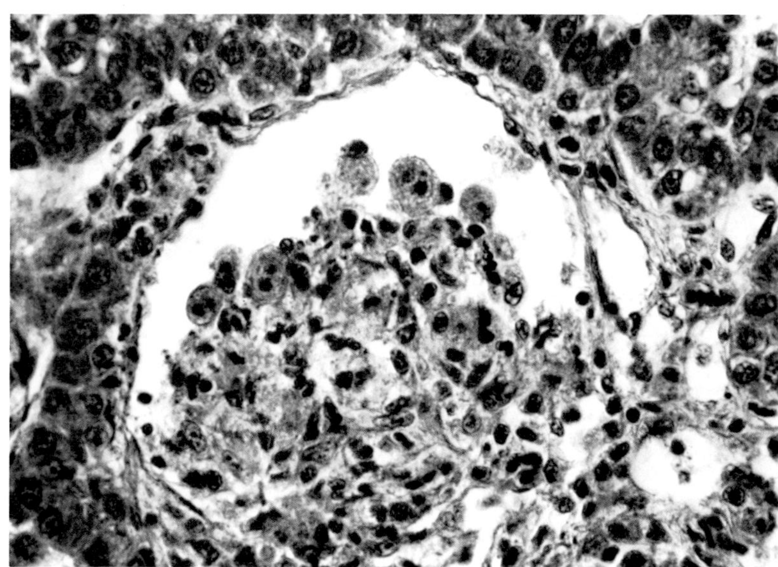

Fig. 6.18. Proliferative kidney disease in trout showing parasites and accompanying inflammatory cells (largely macrophages and lymphocytes) forming stenotic cluster within hepatic vessel.

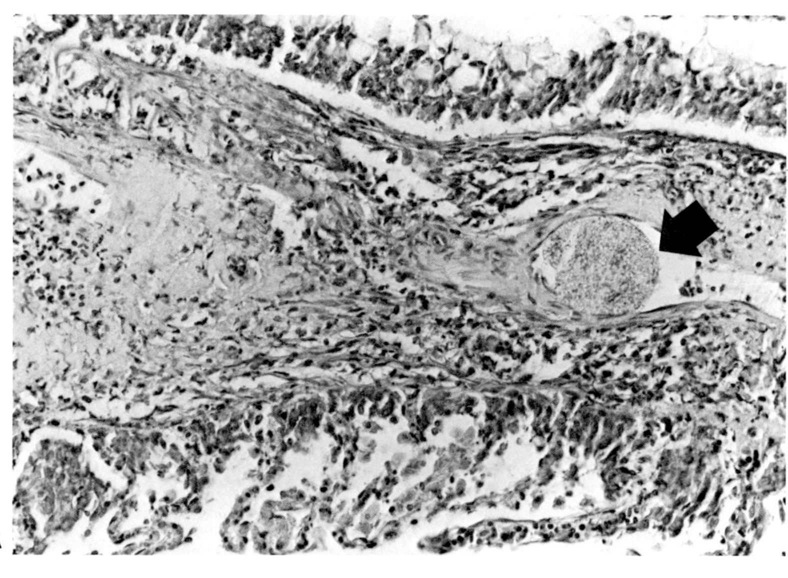

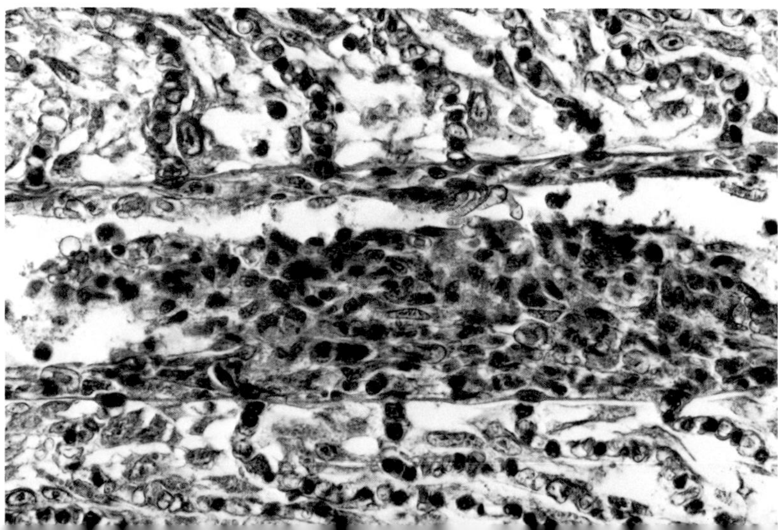

Fig. 6.19. A and **B.** Gills from chinook salmon showing severe vasculitis and thrombosis within filamental vessels. These lesions are associated with microsporidian parasites, probably *Loma* (*arrow*).

chial and other vessels (Fig. 6.20), release eggs that hatch in the gill capillaries. These then migrate out of the vessels, causing hemorrhage and possibly death. Low water temperatures are reported to slow hatching, possibly resulting in endothelial proliferation and encapsulation. Eggs may also be found in the kidney where they may cause renal dysfunction.

The "sand-grain grub" is the metacercaria of the heterophyid trematode *Apophallus brevis*. It is found in yellow perch, causing bony cysts within the lumen of the peripheral vasculature, especially of the axial musculature, but also of the extrinsic muscles of the eye. Of interest are the concentric rings within the bone suggesting annual growth increments as seen in scales and other bony structures in fish.

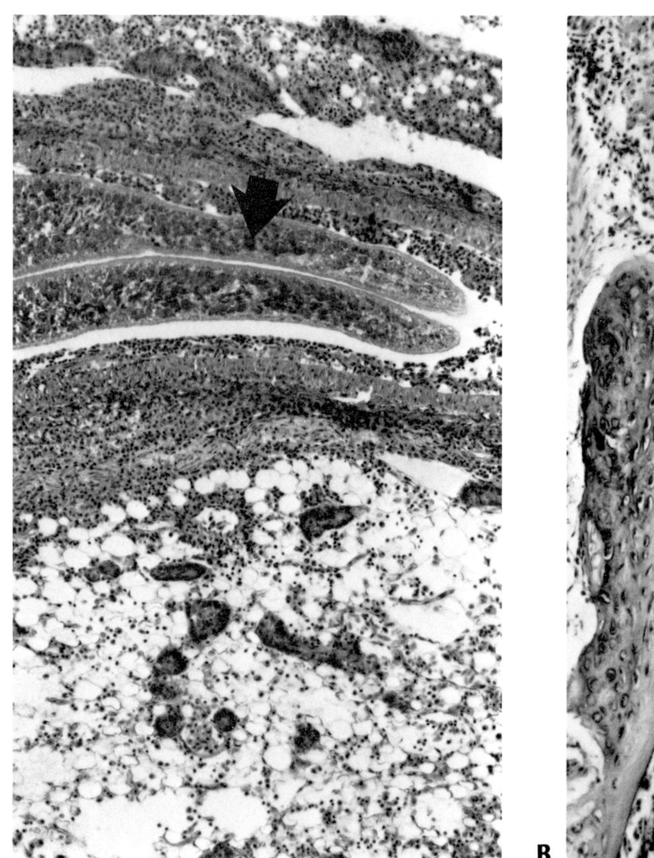

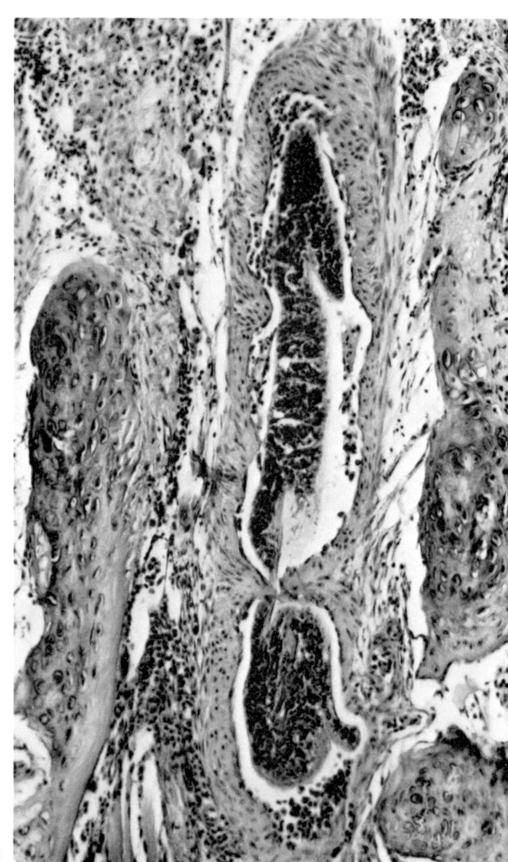

Fig. 6.20. A and **B.** White suckers with flukes (probably *Sanguinicola*) within major mesenteric (**A**) and branchial (**B**) blood vessels. In **A,** the parasite (*arrow*) is associated with moderate inflammation of surrounding mesenteric fat and pancreas.

References

Carlsten, A., I. Poupa, and R. Volkmann. 1983. Cardiac lesions in poikilotherms by catecholamines. *Comp. Biochem. Physiol.* 76:567–81.

Castric, J., and P. de Kinkelin. 1984. Experimental study of the susceptibility of two marine fish species, sea bass (*Dicentrarchus labrax*) and turbot (*Scophthalmus maximus*), to viral haemorrhagic septicaemia. *Aquaculture* 41:203–12.

Driedzic, W. R., J. M. Stewart, and G. McNairn. 1985. Control of lactate oxidation in fish hearts by lactate oxidase activity. *Can. J. Zool.* 63:484–87.

Farrell, A. P. 1984. A review of cardiac performance in the teleost heart: intrinsic and humoral regulation. *Can. J. Zool.* 62:523–36.

______. 1987. Coronary blood flow in a perfused rainbow trout heart. *J. Exp. Biol.* 129:107–23.

Farrell, A. P., and C. L. Milligan. 1986. Myocardial intracellular pH in a perfused rainbow trout heart during extracellular acidosis in the presence and absence of adrenaline. *J. Exp. Biol.* 125:347–59.

Farrell, A. P., K. R. MacLeod, and B. Chancey. 1986. Intrinsic mechanical properties of the perfused rainbow trout heart and the effects of catecholamines and extracellular calcium under control and acidotic conditions. *J. Exp. Biol.* 125:319–45.

Feller, G., R. Bassleer, G. Goessens, and G. Hamoir. 1983. Relative size and myocardial structure of the heart of an Antarctic fish devoid of haemoglobin and myoglobin, *Channichthys rhinoceratus. J. Zool. Lond.* 199:51–57.

Ferguson, H. W. 1975. Phagocytosis by the endocardial lining cells of the atrium of plaice (*Pleuronectes platessa*). *J. Comp. Pathol.* 85:561–69.

Ferguson, H. W., D. A. Rice, and J. K. Linas. 1986. Clinical pathology of myodegeneration ("pancreas disease") in Atlantic salmon (*Salmo salar*). *Vet. Rec.* 119:297–99.

Hargens, A. R., R. W. Millard, and K. Johansen. 1974. High capillary permeability in fishes. *Comp. Biochem. Physiol.* 48A:675–80.

Hauck, A. K. 1984. A mortality and associated tissue reactions of chinook salmon, *Oncorhynchus tshawytscha* (Walbaum), caused by the microsporidan *Loma* sp. *J. Fish Dis.* 7:217–29.

Karttunen, P., and R. Tirri. 1986. Isolation and characterization of single myocardial cells from the Perch, *Perca fluviatilis. Comp. Biochem. Physiol.* 84:181–88.

Lanctin H. P., L. E. McMorran, and W. R. Driedzic. 1980. Rates of glucose and lactate oxidation by the perfused isolated trout (*Salvelinus fontinalis*) heart. *Can. J. Zool.* 58:1708–11.

Leknes, I. L. 1980. Ultrastructure of atrial endocardium and myocardium in three species of Gadidae (Teleostei). *Cell Tissue Res.* 210:1–10.

______. 1981. Ultrahistochemical studies on the moderately electron dense bodies in teleostean endocardial cells. *Histochem.* 72:211–14.

______. 1986. Fine structure and endocytic properties of subendothelial macrophages in the bulbus arteriosus of two bony fish species. *Acta Histochem.* 79:155–60.

Lemanski, L. F., E. P. Fitts, and B. S. Marx. 1975. Fine structure of the heart in the Japanese Medaka, *Oryzias latipes. J. Ultrastruct. Res.* 53:37–65.

Mayberry, L. F., J. R. Bristol, D. Sulimanovic, N. Fuan, and Z. Petrinec. 1986. *Rhabdospora thelohani*: epidemiology of and migration into *Cyprinus carpio* bulbus arteriosus. *Fish Pathol.* 21:145–50.

Midttun, B. 1980. Ultrastructure of atrial and ventricular myocardium in the pike *Esox lucius* L. and mackerel *Scomber scombrus* L. (Pisces). *Cell Tissue Res.* 211:41–50.

Muller, R. 1983. Coronary arteriosclerosis and thyroid hyperplasia in spawning coho salmon (*Onchorrhynchus kisutch*) from Lake Ontario. *Acta Zoologica Pathologica Antverpiensia* 77:3–12.

Priede, I. G. 1974. The effect of swimming activity and section of the vagus nerves on heart rate in rainbow trout. *J. Exp. Biol.* 60:305–19.

______. 1976. Functional morphology of the bulbus arteriosus of rainbow trout (*Salmo gairdneri* Richardson). *J. Fish Biol.* 9:209–16.

Rasheed, V., C. Limsuwan, and J. Plumb. 1985. Histopathology of bullminnows, *Fundulus grandis* Baird & Girard, infected with a non-haemolytic group B *Streptococcus* sp. *J. Fish Dis.* 8:65–74.

Roberts, R. J. 1978. The pathophysiology and systemic pathology of teleosts. In *Fish Pathology,* ed. R. J. Roberts, 55–91. London: Boilliere Tindall.

Rumyantsev, P. P. 1979. Some comparative aspects of myocardial regeneration. In *Muscle Regeneration,* ed. A. Mauro et al. 335–55. New York: Raven Press.

Salte, R., P. Nafstad, and T. Asgard. 1987. Disseminated intravascular coagulation in "Hitra Disease" (Hemorrhagic Syndrome) in farmed Atlantic salmon. *Vet. Pathol.* 24:378–85.

Sanchez-Quintana, D., and J. M. Hurle. 1987. Ventricular myocardial architecture in marine fishes. *Anat. Rec.* 217:263–73.

Santer, R. M. 1985. Morphology and innervation of the fish heart. Adv. in *Anat. Embryology & Cell Biol.* vol. 89. Berlin: Springer-Verlag.

Sidell, B. D., W. R. Driedzic, D. B. Stowe, and I. A. Johnston. 1987. Biochemical correlations of power development and metabolic fuel preferenda in fish hearts. *Physiol. Zool.* 60:221–32.

Smialowska, E., and W. Kilarski. 1983. The fine structure of endothelial cells of white-blooded fish *Chaenocephalus aceratus. Z. mikrosk.-anat. Forsch.* 97:967–78.

Stewart, J. M., and W. R. Driedzic. 1986. Kinetics of lactate dehydrogenase from heart and white muscle of ocean pout: a single isozyme system. *Can. J. Zool.* 64:2665–68.

Tort, L., J. J. Watson, and I. G. Priede. 1987. Changes in *in vitro* heart performance in rainbow trout, *Salmo gairdneri* Richardson, infected with *Apatemon gracilis* (Digenea). *J. Fish Biol.* 30:341–47.

Ueno, S., H. Yoshikawa, Y. Ishida, and H. Mitsuda. 1986. Electrocardiograms recorded from the body surface of the carp, *Cyprinus carpio. Comp. Physiol.* 85:129–33.

Vogel, W. O. P. 1985. The caudal heart of fish: not a lymph heart. *Acta Anat.* 121:41–45.

Wood, C. M. 1977. Cholinergic mechanisms and the response to ATP in the systemic vasculature of the rainbow trout. *J. Comp. Physiol.* 122:325–45.

7 Gastrointestinal Tract, Pancreas, and Swimbladder

Gastrointestinal Tract

Normal. The great variety of ecological niches occupied by teleosts is reflected in the diversity of patterns of organization of the gastrointentinal (GI) tract. Compared to mammals, the teleost GI tract is histologically simple, probably because it is so easy to provide an aqueous vehicle for the digestive products and also because, in some species at least, the rate of digestion can be slow. Hence, less complex digestive glands and a less well-developed muscular apparatus are needed.

The oral cavity and pharynx are lined by a thin stratified squamous epithelium containing abundant mucus-secreting cells. Fungiform and filiform papillae may be found, and most species have teeth, which vary greatly in shape. Some species have isodont dentition (all teeth alike in shape), while others are heterodonts. The tooth consists of an enamel coating, a dentine layer, and a pulp core.

The esophagus is short and thick walled, the muscularis comprising interweaving skeletal muscle fibers that may extend as far as the stomach. The stratified cuboidal or columnar epithelium may be ciliated, and it contains numerous goblet cells and occasionally taste buds. In addition, multicellular serous or cardiac glands may be found posteriorly. The mucosa is thrown into longitudinal folds that end at the stomach, giving way to rugae. The serosa contains prominent nerve fibers of the vagus.

The stomach itself is highly variable and indeed is absent entirely in most larvae and in about 15% of all adult species (a good example of this is the cyprinids). Where present, the gastric mucosa typically contains both fundic and pyloric glands. The submucosa of some species contains large numbers of coarse eosinophilic granular cells, in addition to lymphatics, blood vessels, and nerve fibers.

As in mammals, carnivores tend to have shorter intestines than herbivores, and in these latter species there is extensive coiling and possibly large cecal pouches designed for fermentative microbial digestion (Rimmer and Wiebe 1987). Many species have pyloric caeca, blind-ending diverticula with multifolded intestinal-type epithelium, located at the anterior end of the intestine. Some species have longitudinal folding of the mucosa, but only a few have anything resembling the intestinal villi of mammals. In general, the mucosa is not easily divisible into regions on the basis of histology; i.e., the small and large intestines of mammals are not present, although a rectum is present, and in some species this is separated by a connective tissue annulus ("intestinal valve"). There may, however, be some functional differentiation, with the proximal, middle, and distal segments specializing in lipid, protein, and ionic regulation, respectively. The mucosal lining is of simple columnar type, the cells possessing microvilli (forming the brush border) and in some species, occasional cilia. Lymphocytes and rodlet cells are not uncommon in the mucosa, nor are apoptotic cells. The latter represent effete cells phagocytosed by adjacent viable ones, the remnants appearing as brightly eosinophilic spherical inclusions; they are prominent during periods of starvation. The lamina propria and submucosa of several species contain large numbers of eosinophilic granular cells and variable quantities of lymphoid tissue. In general, many of those species without a stomach tend to have greater quantities of lymphoid tissue than those with a stomach (an attempted compensation for increased antigenic exposure?).

Disease. Severe diseases that specifically target the gastrointestinal tract are less common in fish than in mammals. There are, therefore, relatively few examples of severe degenerative or necrotizing enteritides and no information on anything similar to progenitor versus functional cell compartment damage (progenitor zones may be found in some teleosts at the base of longitudinal mucosal folds). At a diagnostic level, interpretation is complicated by frequently poor fixation (few people bother to open the intestine to expose the mucosal surface) and by the atrophic changes that are found normally in temperate species during winter, when feeding is reduced.

MOUTH. Oral congestion and hemorrhages are encountered in many bacteremias and viremias. The classic example is the ecchymotic hemorrhaging frequently but not invariably seen in salmonids with the enteric redmouth bacterium, *Yersinia ruckeri*. Oral hemorrhage also may be found in poor water quality conditions, notably nitrite toxicity, although in this case, the hemorrhages tend to be more brown than red, due to methemoglobin formation. Diffuse oral congestion and hemorrhage are seen in feral European cyprinids infected with metacercariae of *Bucephalus polymorphus,* which are found in large numbers in the subcutis of the mouth, and elsewhere in the fish where, however, they evoke less inflammation.

Necrotic stomatitis is frequently seen in *Flexibacter*-like infections, particularly at warmer temperatures (Fig. 7.1). In aquarium species, the condition is known popularly as cotton-wool mouth; fungi may also become secondarily involved. The infection may be extensive and ulcerative, involving the opercula, teeth, maxilla, and mandible, as well as the spongy bones of the head.

ESOPHAGUS. Esophageal lesions are rarely reported (it is rarely examined!), but they may be expected in any generalized myopathy involving striated muscle (such as in "pancreas disease" of Atlantic salmon) or in systemic infections. If severe, they may impede swallowing or constrict the pneumatic duct. Mycotic esophagitis is often seen in salmonids and probably other fry. Although the pathogenesis is unknown, it may relate to impaction of the esophagus by an incorrect particle food size with secondary fungal involvement (Fig. 7.1). The pneumatic duct and swimbladder are readily colonized from this location.

STOMACH. In those species that have a stomach, and many do not, gastritis is not uncommon, although its significance is usually hard to determine. Involvement is expected in viremias and bacteremias (Fig. 7.2) although infarctive lesions associated with gram-negative sepsis, as seen in many domestic mammals, are not reported. The well-vascularized gastric submucosa is, however, a frequent site for lesions in the disseminated granulomatous infections such as mycobacteriosis. Smaller granulomas, with

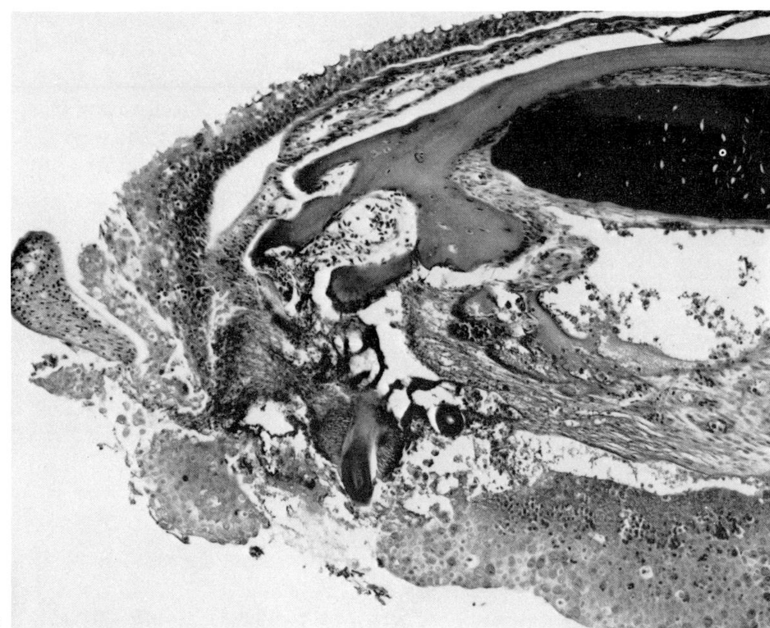

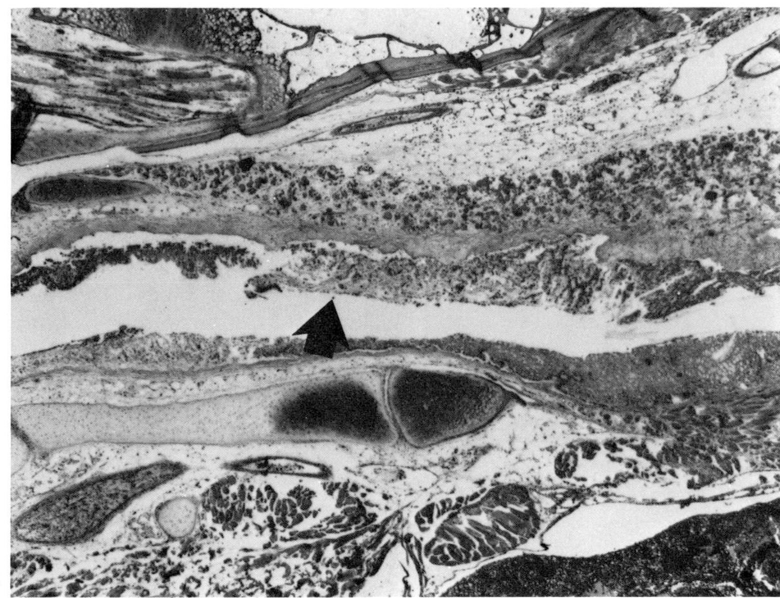

Fig. 7.1. A. Necrotic stomatitis in young cultured walleye caused by *Flexibacter columnaris.* Similar lesions are common in aquarium species, the bacteria often growing to produce a grossly visible mat, the so-called cotton-wool mouth. **B.** Mycotic esophagitis in Atlantic salmon fry. Mucosa is virtually replaced by fungal hyphae (*arrow*).

mineral at the core instead of caseonecrotic debris, are commonly encountered in intensively reared salmonids and are considered part of the syndrome nephrocalcinosis (Fig. 7.3). They are rarely very large and are probably of little significance per se, although they may be an indicator of stress. Certainly, just as the stomach can act as a signpost to stress in mammals, the fish stomach can also react to social stress with atrophy of mucosal glands and replacement by connective tissue. The full significance of such responses, especially in intensive culture conditions, is as yet unclear.

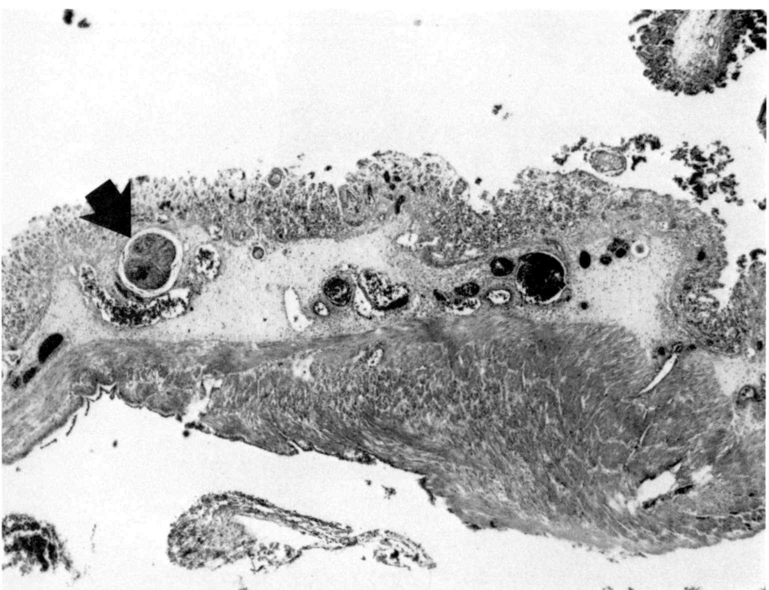

Fig. 7.2. Stomach wall from fish with bacteremia. Note the submucosal edema plus congestion and thrombosis (*arrow*) of the vessels.

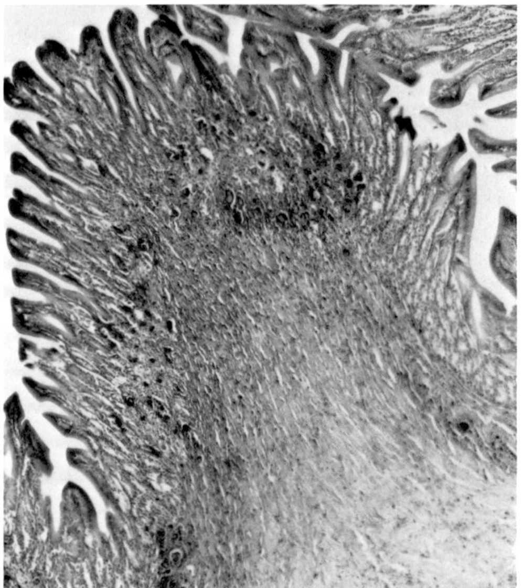

Fig. 7.3. Stomach wall from cage-reared Atlantic salmon with numerous foci of mineral in lamina propria. Such deposits may elicit a granulomatous response and are frequently seen with nephrocalcinosis.

Necrosis and inflammation of the gastric mucosa are seen in intensively reared salmonid sac-fry, associated with *Saprolegnia* infections. The stomach may be dilated and edematous, and mortality is usually high; the pathogenesis of the condition, however, is unclear. Another fungal disease of salmonids, that due to *Exophiala salmonis,* may cause an eosinophilic gastritis and enteritis, although other organs, notably kidney and liver, are more severely affected. The fungus *Phoma herbarum* and the yeast *Candida sake* have been associated with severe gastric dilation and necrosis in chinook salmon and amago (*Oncorhynchus rhodurus*), respectively. As in mammals, however, dietary influences predisposing to primary mucosal damage should be considered in cases such as these, with the fungi present as secondary invaders.

Severe gastric dilation is also seen in cichlids, particularly in private aquaria, and is known as cichlid bloat. Lesions include submucosal granulomas and severe diffuse necrotizing loss of gastric epithelium (Fig. 7.4). Whether these different lesions represent a progression of the same entity, or are due to separate causes, is unknown. Clostridia and flagellated protozoa, possibly

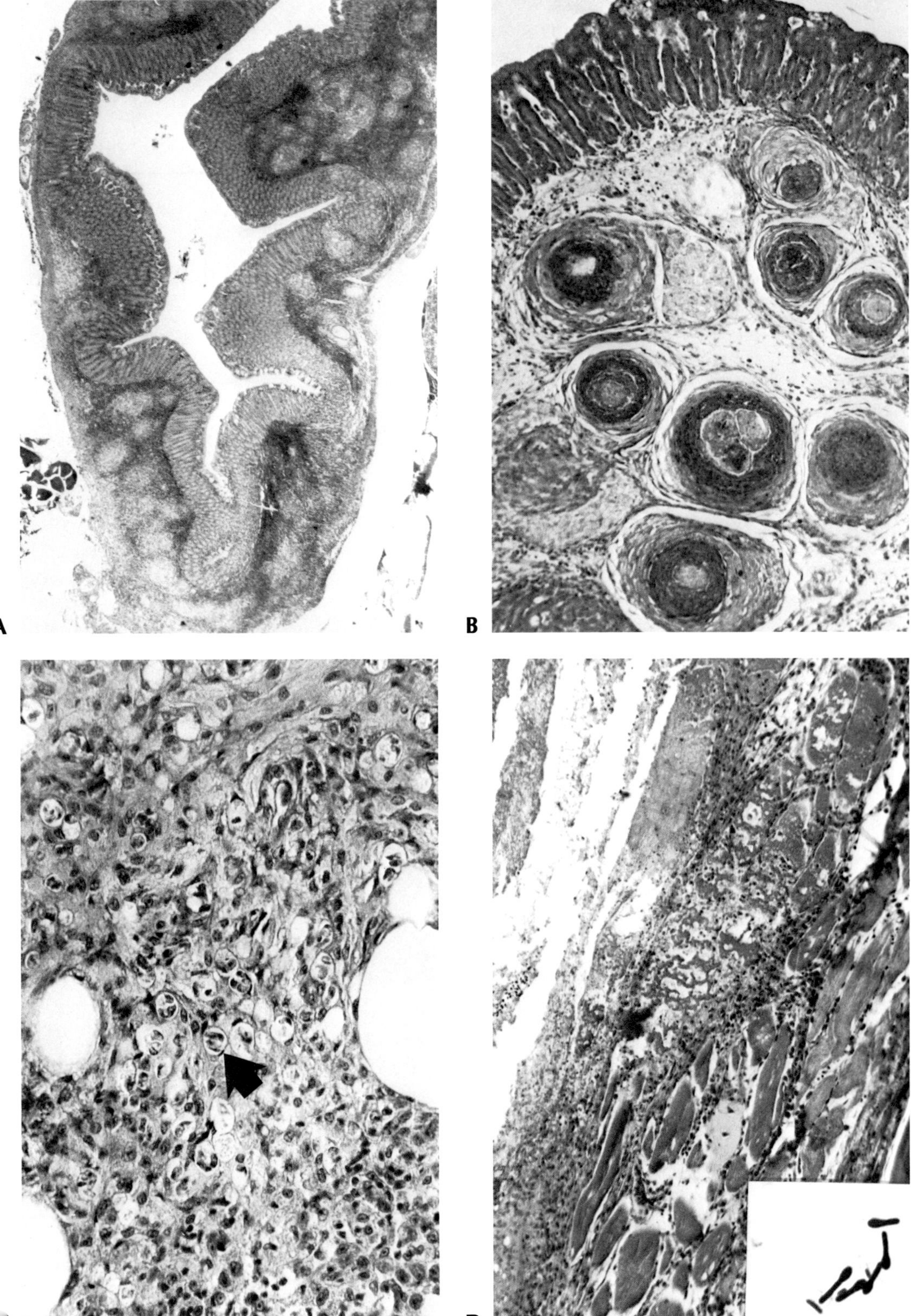

Fig. 7.4. Cichlid bloat in *Tropheus duboissii.* **A.** Stomach with multifocal granulomas distending lamina propria. **B.** Another fish with submucosal granulomas with caseonecrotic cores. **C.** Some gastric granulomas associated with flagellated protozoa (possibly *Cryptobia*) (*arrow*). **D.** Marked distension of stomach, eventually leading to perforation with necrosis of the body musculature. Inset shows spore-forming *Clostridium hastiforme* present in peritoneal cavity.

Cryptobia spp., are both associated with the condition that may eventually lead to gastric perforation, peritonitis, and even full thickness necrosis of the body-wall musculature.

Cryptobia spp. are reported to invade the gastric epithelium of European cyprinids and flounders causing submucosal granulomatous inflammation or edema, respectively. A further example of severe gastritis of tropical fish, which may also lead to perforation, is that seen in hexamitiasis in Siamese fighting fish. The flagellates are associated with a full-thickness gastritis (Fig. 7.5) and may also penetrate blood vessels, leading to dissemination. A peritonitis frequently accompanies severely affected animals, largely due to the bacterial contamination resulting from the perforated ulcer.

Other parasites associated with the stomach wall include some species of myxosporeans, the developing stages of which may be found in the brush border before penetrating the epithelium, or the more common space-occupying cysts as seen, for example, with *Myxobolus dermatobia* in wild eels. The microsporidian *Glugea anomala* may be found anywhere within the mesenchymal cells of the three-spined stickleback. Host responses include xenoma formation within the gastric wall, resulting in glandular atrophy. The hemogregarine *Haemogregarina sachai* also causes pressure atrophy of gastric glands in intensively reared turbot (*Scophthalmus maximus*) due to space-occupying submucosal granulomas. *Anisakis simplex* infection of cod is an example of helminth-associated gastritis; craterous granulomatous lesions may be present, possibly an immune response inhibiting larval penetration.

INTESTINE AND PERITONEAL CAVITY. The intestine is involved in several generalized virus diseases, notably infectious pancreatic necrosis (IPN), infectious hematopoietic necrosis (IHN), and channel catfish virus disease (CCVD). First described as a catarrhal enteritis, IPN disease in salmonids is reported to cause increased numbers of rounded eosinophilic degenerate cells in the epithelium (so-called McKnight cells; Roberts 1978). Such a finding is not uncommon in a variety of conditions, including starvation, and it may well be that the cells are no more than apoptotic remnants of necrotic cells. Submucosal hemorrhage is seen in CCVD and viral hemorrhagic septicemia of trout. Necrosis of the submucosal eosinophilic granular cells is considered pathognomonic for IHN, but this lesion is not invariable. An adenoviruslike disease of anorexic cultured sturgeons targets the epithelium of the intestine and spiral valve although the liver is also affected; intranuclear inclusions are present (Hedrick et al. 1985).

Diseases such as nocardiosis or mycobacteriosis frequently result in the formation of multifocal granulomas in the lamina propria or sub-mucosa. These are rarely severe enough however to suggest major functional impairment, although some protein leakage and malabsorbtion probably do occur. Similarly, acute bacteremias often cause congestion or petechial hemorrhaging of the intestinal wall; in diseases such as enteric septicemia of catfish (*Edwardsiella ictaluri*), furunculosis, or vibriosis, this can be severe, especially of the rectum, and prolapse may even occur. In our experience, enteric disease associated with lumenal overgrowth and adherence of bacteria to the brush border is not uncommon as isolated cases in marine tropicals (especially those fed high protein diets), although the identity of the organism(s) in these cases is rarely established.

Fig. 7.5. Disseminated hexamitiasis in Siamese fighting fish. **A.** Stomach showing severe full-thickness gastritis and perforation with large granuloma on serosal surface. **B.** High-power view of serosal response showing necrotic core. **C.** Kidney from same fish showing flagellates within renal parenchyma (*arrows*). (See also Fig. 4.20.) **D.** Giemsa-stained smear of peritoneal exudate showing large numbers of *Hexamita*-like flagellates (*arrows*) as well as numerous bacteria, erythrocytes, and other cells.

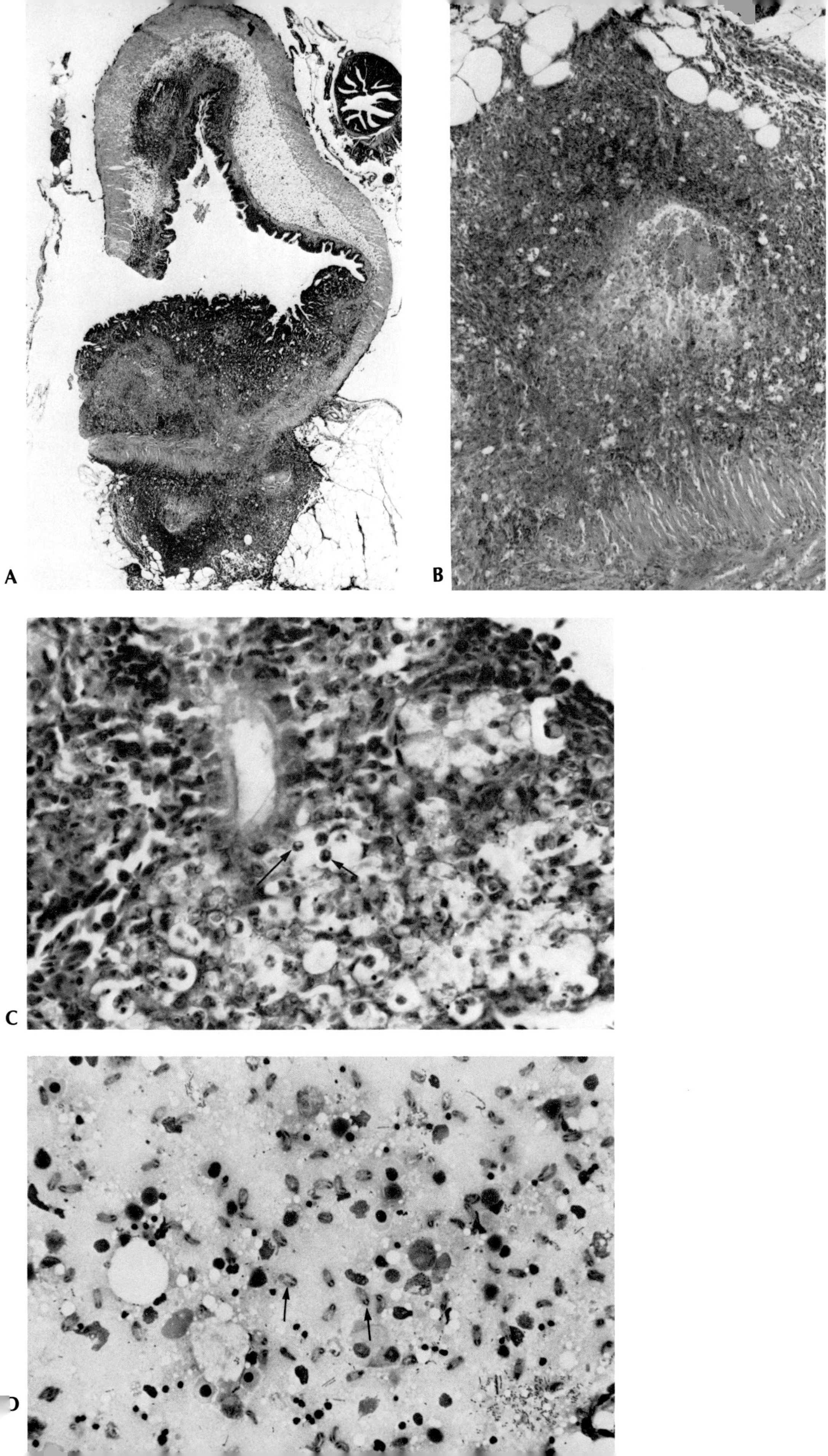

A

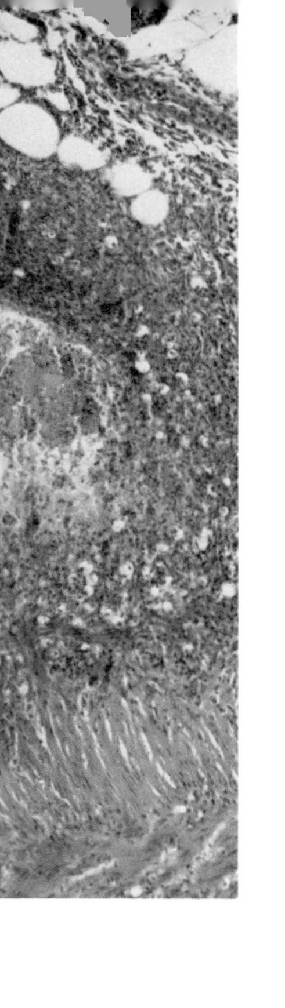

B

C

D

Rearing larval fish is notoriously difficult, and mortality can be especially high when the yolk-sac reserves become depleted (several days after hatching, depending on water temperatures) unless the fish have started feeding. Even then, the type of food may also determine survival. For example, larval grayling fed on zooplankton have been shown to accumulate excessive quantities of fat in enterocytes that eventually lyse (Eckmann 1987). It is suggested this is due to a disturbance in the fat transport process, possibly because of a deficit in lipoprotein, which is needed for the formation of chylomicrons.

There is surprisingly little information on toxic damage to gut epithelium, although heavy metals such as cadmium and lead are reported to disrupt intestinal brush border with alterations in ionic fluxes as a consequence (Crespo et al. 1986). Similarly, acute gastrointestinal inflammation is seen in cases of drug overdose. The impact and significance of such responses in treating fish and in cases of environmental or feed-associated toxicity remain to be properly investigated.

Protozoan diseases of the intestine are quite common, and some are associated with moderately severe pathological changes. Various coccidia affect many species with developmental stages in both epithelium and lamina propria. *Eimeria subepithelialis* is the cause of nodular coccidiosis in European carp; the mucosa is damaged when the parasites are released. *Cryptosporidia* have been reported in fish (Fig. 7.6), but their significance is unknown. They must be carefully differentiated from other coccidia or even myxosporea, some of which also have developmental stages in the brush border and which can therefore superficially resemble *Cryptosporidia.*

The flagellate *Hexamita* is a protozoa whose pathogenicity has been questioned; in rainbow trout, *H. salmonis* is found in the upper intestine and pyloric ceca. Nevertheless, they are associated with disease and often high mortality, and treatment to remove them does result in dramatic recovery. Grossly, the fish may have exophthalmia and an abdomen distended by a clear peritoneal fluid (Fig. 7.7). Despite close apposition to the brush border (Fig. 7.8), there is virtually no discernible pathological change in the underlying epithelium except cytoplasmic blebbing, a reaction of the vertebrate enterocyte common in protozoan disease. There are, however, increased numbers of apoptotic

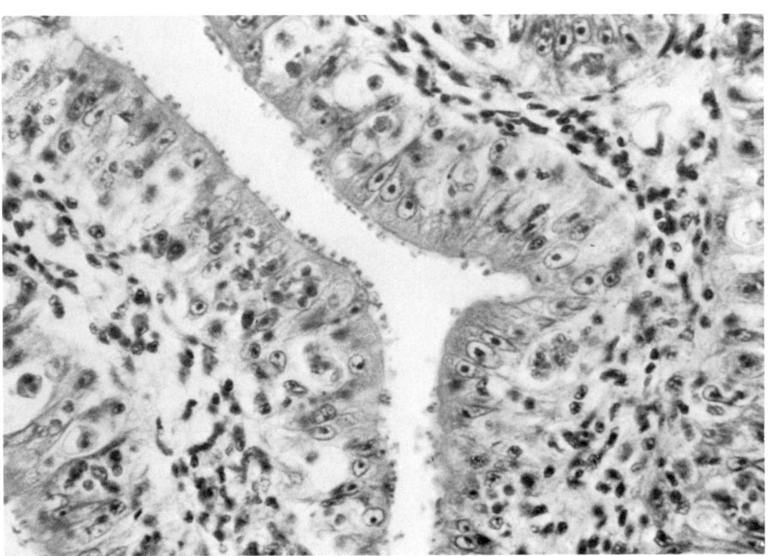

Fig. 7.6. Marine tropical fish showing numerous blebs on brush border of intestine as well as coccidia within epithelium. Such parasites may be *Cryptosporidium,* but these must be differentiated from other coccidia such as *Epieimeria* or some myxosporea, which also have developmental stages within the brush border.

Fig. 7.7. Hexamitiasis in rainbow trout fry showing exophthalmia and abdominal distension.

bodies in the epithelium, representing dead enterocytes phagocytosed by adjacent viable cells. Focal hepatic necrosis may also be present, although the cause is unknown. Another hexamitid parasite, *Spironucleus,* is frequently encountered in tropical freshwater angel fish. When large numbers are present, it is associated with moderately severe enteritis, particularly, it would appear, if the nematode *Capillaria* is also present. The ciliate *Balantidium* is associated with disease in cyprinids, although it is often regarded as a commensal. A variety of microsporidians and myxosporeans may be encountered, forming cysts within the intestinal wall. Of note are *Glugea* and *Thelohanellus kitauei,* the cause of intestinal giant cystic disease of carp.

There is a long list of metazoan parasites found in the gastrointestinal tract of fish. Most are well adapted and seem to cause little harm, or evoke much in the way of an inflammatory response. Exceptions include the attachment point of acanthocephalans, which may be necrotic and can even perforate, leading to peritonitis. The pseudophyllidean tapeworm *Bothriocephalus gowkongensis* (*acheilognathi*) causes in cyprinids in various parts of the world necrosis, hemorrhage, and inflammation at attachment points, as well as space-occupying distension of the intestine and possible perforation. Mortality can be high. The larval cestode *Diphyllobothrium dentriticum* can cause severe granulomatous enteritis in trout (Fig. 7.9) and may also lead to peritonitis with visceral adhesions and death.

Prolapsed rectum is occasionally encountered with large space-occupying visceral lesions or bacteremias. Intussusception is rarely seen in fish, and when present, it may be a terminal event.

PERITONEUM. In addition to a perforated stomach or intestine, causes of peritonitis in salmonids include postspawning infections with alpha-hemolytic *Streptococcus* and *Lactobacillus piscicola,* possibly largely the result of trauma at the hands of inexperienced strippers! Peritoneal and visceral involvement is common in bacteremias and viremias, as well as mycotic disease in most species. Examples of the latter include *Saprolegnia* and *Fusarium solani* (Fig. 7.10). The myxosporean *Ceratomyxa shasta* causes granulomatous peritonitis in young

Fig. 7.8. A. Pyloric caeca from fish shown in Figure 7.7 showing large numbers of flagellates on mucosal surface. **B.** High-power view of mucosal surface showing several pear-shaped flagellates and several large blebs (*arrow*). The cracking of the brush border is artifact. **C** and **D.** Low and high power of parasites within lumen of pyloric caeca. Note that despite close apposition to the mucosal surface, the brush border appears absolutely normal, as do the terminal web, mitochondria, and other details of the enterocytes. **E.** Virtually the only pathological change accompanying this infection is blebbing of the enterocytes into the lumen, representing ballooning dilatation of one or several microvilli. Compare with **B. F.** Section of pyloric caeca showing remnants of two necrotic cells (*arrows*) that have been phagocytosed by adjacent viable enterocytes. Between them can be seen a theliolymphocyte.

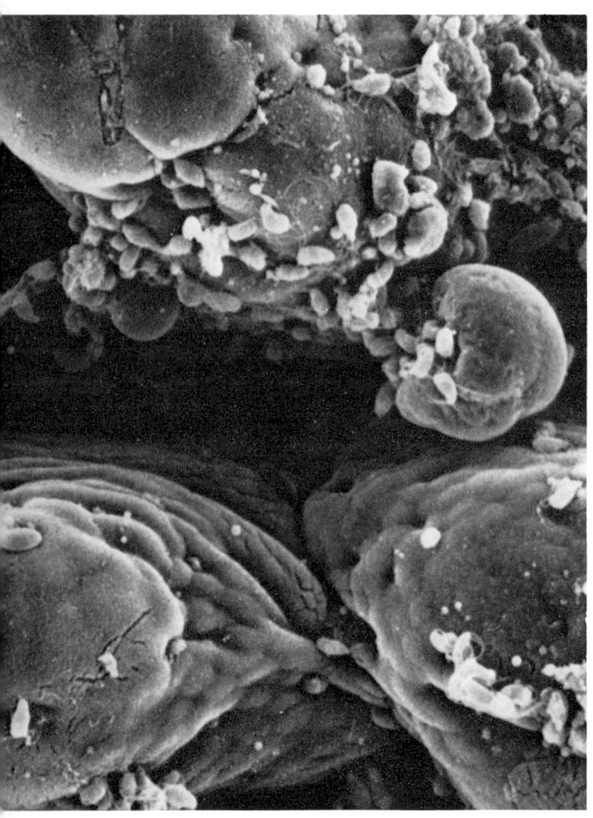

A

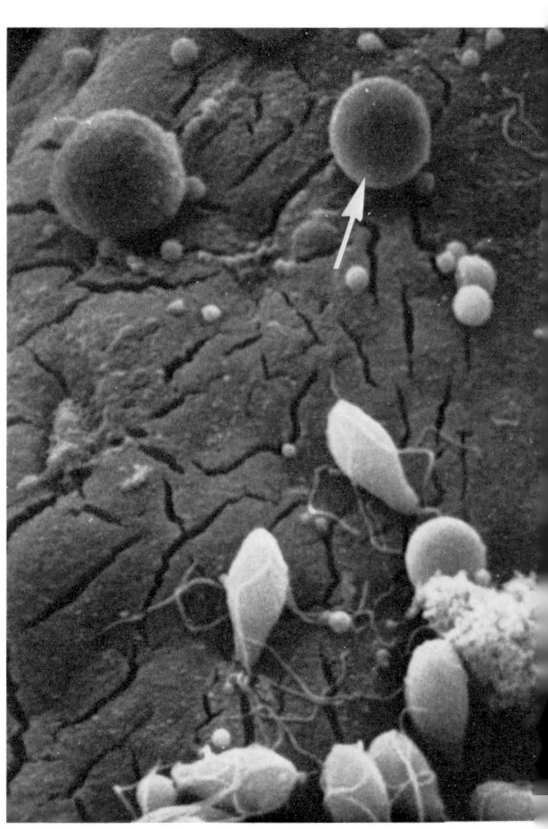

B

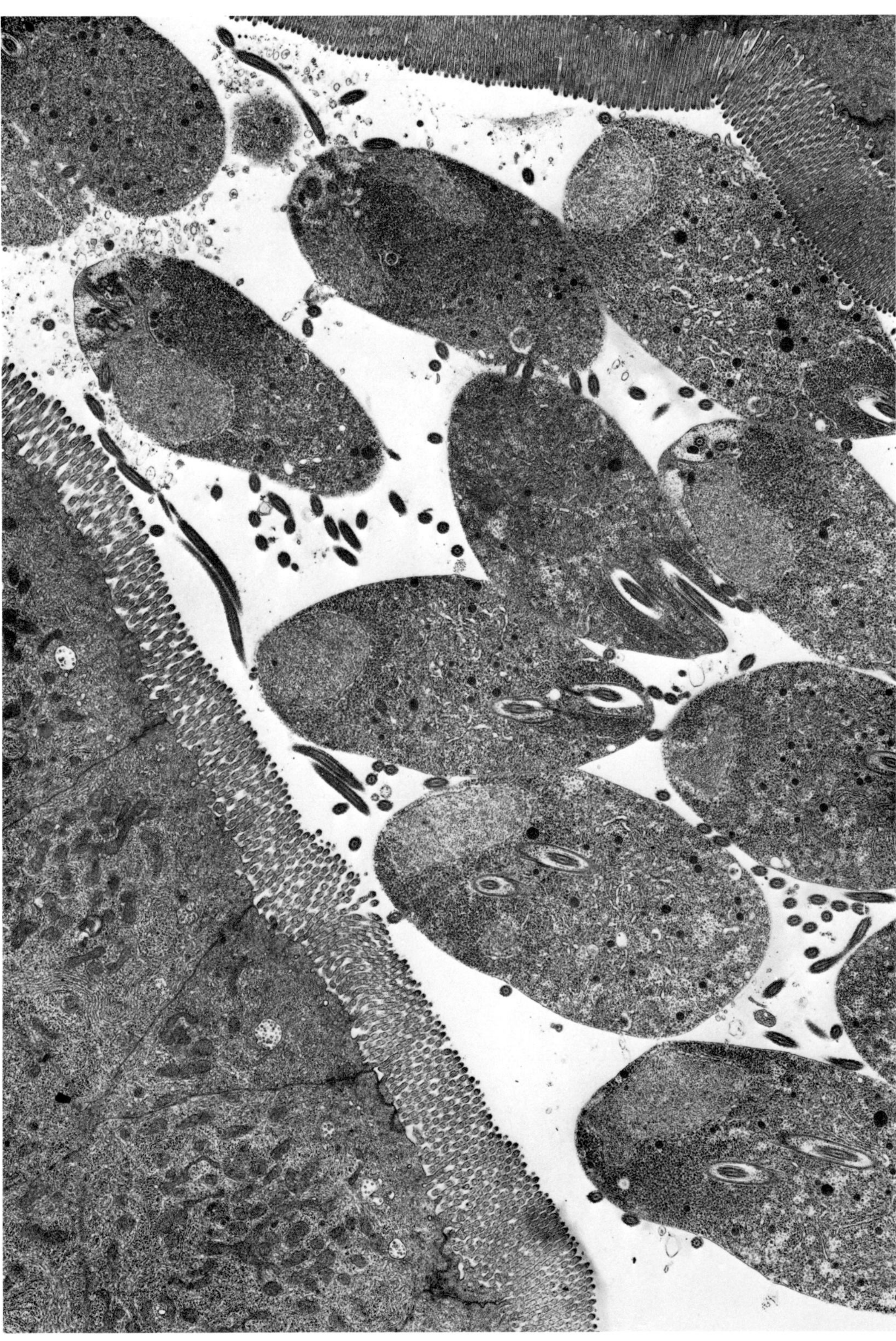

C

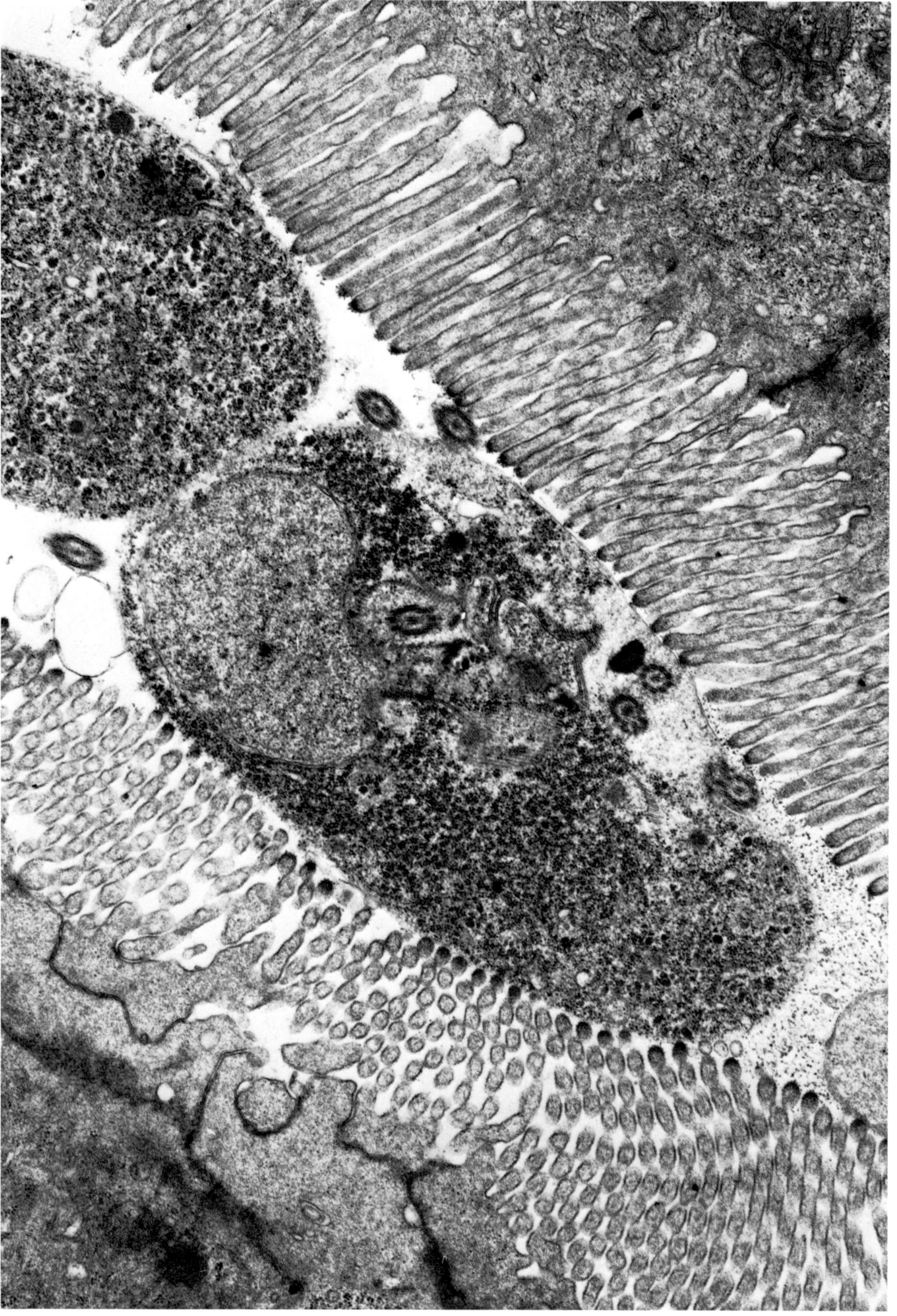

D

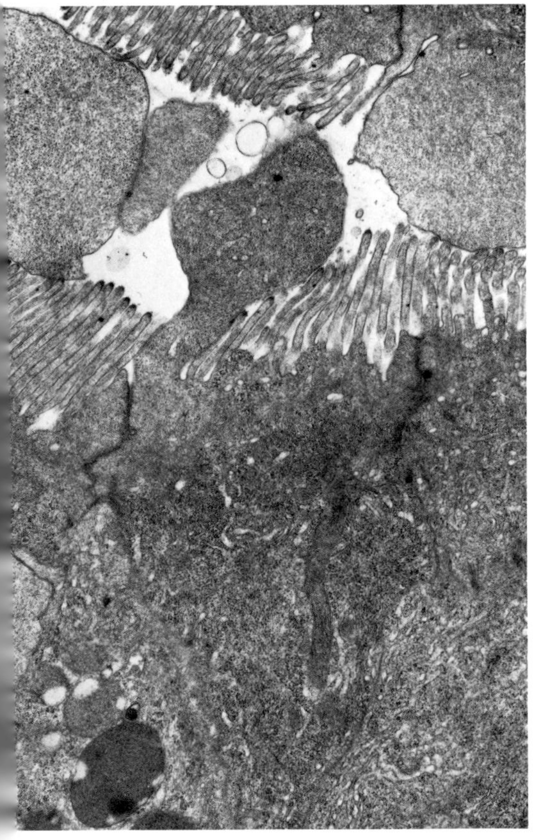

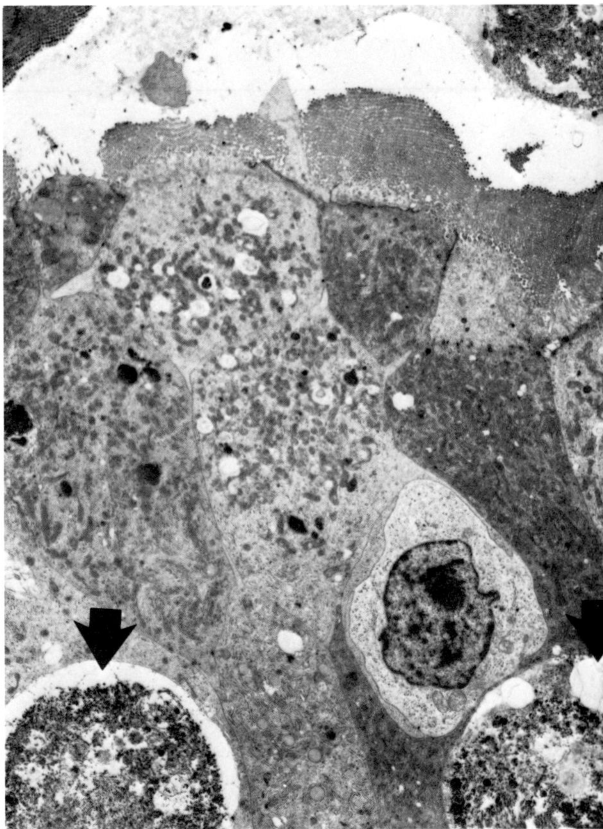

F

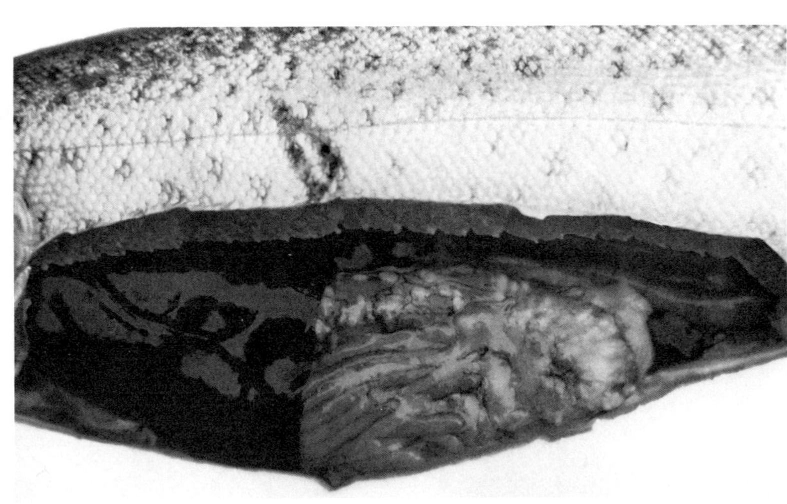

Fig. 7.9. Brown trout with multifocal granulomatous enteritis due to larval *Diphyllobothrium dentriticum.*

salmonids (mainly chinook and coho salmon), but there is involvement of most other visceral organs and muscle of the body wall as well. Nodular lesions may be found in the gut wall or there may be diffuse thickening; occasionally some of these lesions may perforate causing septic peritonitis. The geographically restricted nature of this disease, mainly to the Columbia River basin, has suggested to some the possibility of an intermediate host.

Metacercariae of digeneans may encyst in the abdominal cavity; the resulting black spot with those species of parasite that attract melanin should not be confused with the often pronounced melanization of the peritoneum found normally in many fish. Peritonitis with visceral adhesions is pro-

A

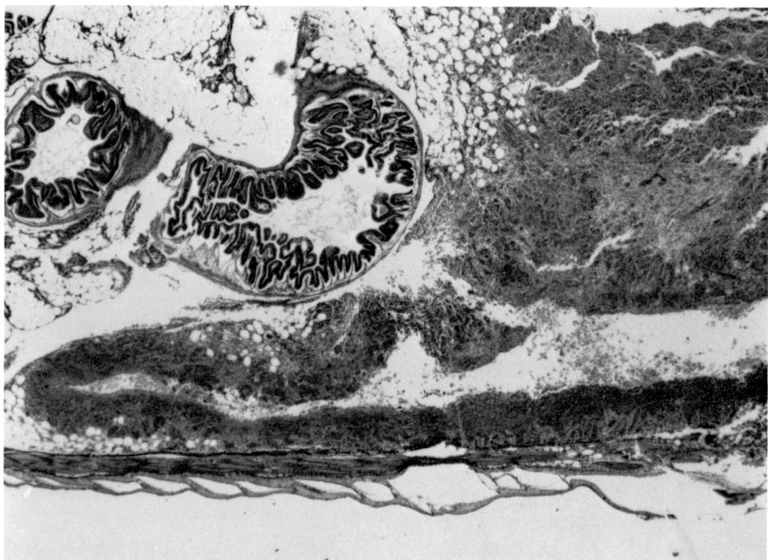

B

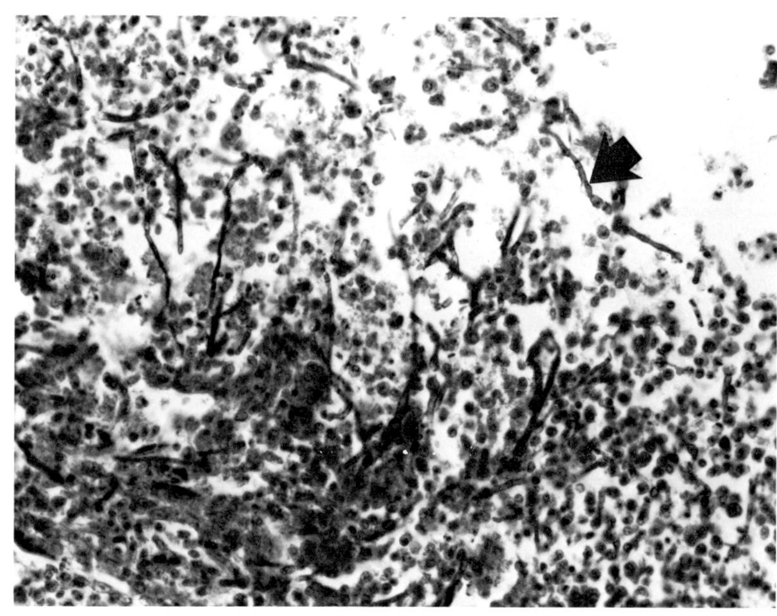

Fig. 7.10. Mycotic peritonitis in desert pupfish caused by *Fusarium solani.* **A.** Granulomatous response extending ventrally along floor of abdomen. **B.** PAS stain reveals fungi (*arrow*) to be branching and septate.

nounced with some larval tapeworms such as *Ligula intestinalis* and *Proteocephalus ambloplitis,* the bass tapeworm. The nematode *Philonema agubernaculum* encysts in the peritoneal cavity of salmonids causing extensive adhesions. The space-occupying parasites, especially *Ligula,* cause pressure atrophy of liver, gonads, and body-wall musculature. A distended abdomen can cause a reduction in swimming speed as well as scale protrusion, thereby predisposing to trauma and secondary bacterial or fungal invasion.

A granulomatous peritonitis is sometimes encountered in yolk-sac fry, in which the reaction centers around droplets of the yolk itself. Such a "yolk peritonitis" may be associated with a bacterial or fungal infection, but often the cause is unknown.

Pancreas

Location of pancreatic tissue varies within and between species. Common sites include the mesenteric fat interspersed among the pyloric caeca (as in salmonids) or surrounding the portal vessels entering the liver to form an hepatopancreas. In some species, acinar distribution is more diffuse, and they may also be found within the splenic capsule and elsewhere. Typically, interlobular ducts merge to form the large pancreatic ducts that run alongside the bile duct before joining the intestine. Although there are few data to support the idea, severe pancreatic disease probably results in exocrine insufficiency and maldigestion. The acinar tissue is involved in most of the economically important virus diseases of salmonids, notably IPN, IHN, and VHS, but also in *Herpesvirus salmonis.* Necrosis of acinar cells is seen in all these diseases except *H. salmonis,* in which acinar cell syncytia are formed: this is considered to be pathognomonic. Other lesions in this geographically restricted condition (United States) include renal interstitial and tubular necrosis, with focal necrosis in myocardium, liver, and posterior gut, which may lead to cast formation.

IPN virus replicates largely within pancreatic acini and causes pyknosis and karyorrhexis with a moderate inflammatory infiltrate. Interpretation of such lesions however is frequently hampered by poor fixation, especially at a routine level. When it is appreciated that IPN virus is easily recovered from apparently healthy fish, including nonsalmonids, a reliable diagnosis of the *disease* IPN can sometimes be surprisingly difficult to make unless lesions are correlated with the presence of viral antigen by means of fluorescent or other antibody tests. Other lesions reported in IPN include a catarrhal enteritis, with possible sloughing of gut mucosa, hyaline degeneration of skeletal muscle, and necrosis of renal hemopoietic elements. Indeed, recent reports suggest that the latter lesion is more common than pancreatic necrosis, which is in fact the exception rather than the rule. The presence of a mucoid gel in the stomach and anterior intestine is considered by some to be highly suggestive of IPN although we have associated this also with high ammonia levels.

Pancreas disease is a syndrome of unknown etiology affecting farmed Atlantic salmon in sea-caged sites particularly in Scotland, Ireland, and Norway. Lesions include pronounced atrophy and fibrosis of the pancreas (Fig. 7.11) with some hemorrhage. Corresponding with these changes in some cases is severe degenerative myopathy affecting mainly heart and red skeletal muscle (see Figs. 12.2 and 12.3). Recovery of the pancreas may correlate with cardiac repair. The syndrome is associated with plasma and tissue vitamin E levels dramatically lower than control healthy fish, but whether this represents the cause or the result of the disease is unknown. It is also unknown whether the pancreatic lesions and the myopathy represent different conditions or merely different ends of the same spectrum.

There are also other syndromes with a possible basis in vitamin E deficiency and in which the pancreas is involved, notably pan-

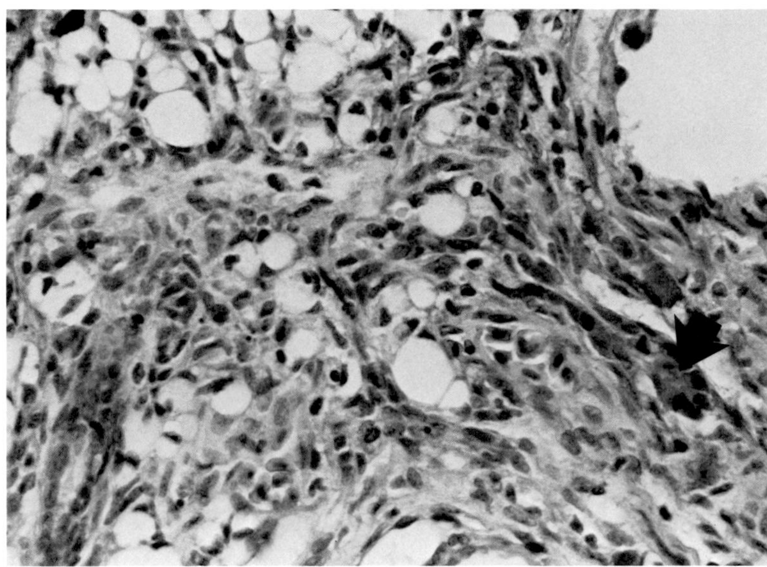

Fig. 7.11. Pancreas disease in Atlantic salmon. There is almost complete fibrous replacement and atrophy of normal acini, accompanied here by a mild cellular infiltrate. *Arrow* shows acinar remnants. (See also Fig. 6.14.)

steatitis in rainbow trout in which there is inflammation of the fat surrounding the pancreas and elsewhere, and a granulomatous steatitis in Sunapee trout affecting pancreas and viscera with giant cell formation and adhesions (Fig. 7.12). Pancreatic atrophy is also described in channel catfish with vitamin C deficiency, although the major lesion in this disease is osteoporosis.

The pancreas is a not infrequent site for colonization in bacteremias such as furunculosis and bacterial kidney disease (BKD) (Fig. 7.13), and in granulomatous diseases such as mycobacteriosis. Parasitic involvement includes visceral coccidia such as *Calyptospora* (*Eimeria*) *funduli,* which can almost obliterate the organ in killifish (Fig. 7.14). Included too are the myxosporeans such as *Myxobolus osburni,* which causes a pancreatitis in pumpkinseeds, microspori-

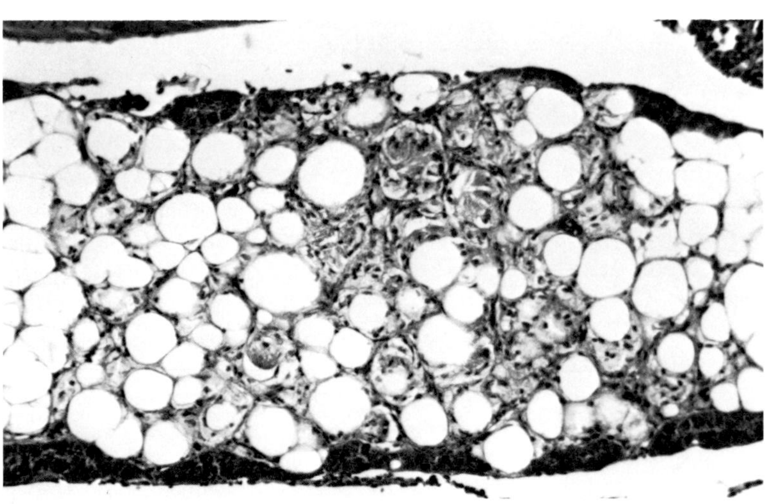

Fig. 7.12. Granulomatous pancreatitis of unknown cause in rainbow trout. Although the response involves mainly the fat between the pyloric caeca, there is some compression involvement of acini.

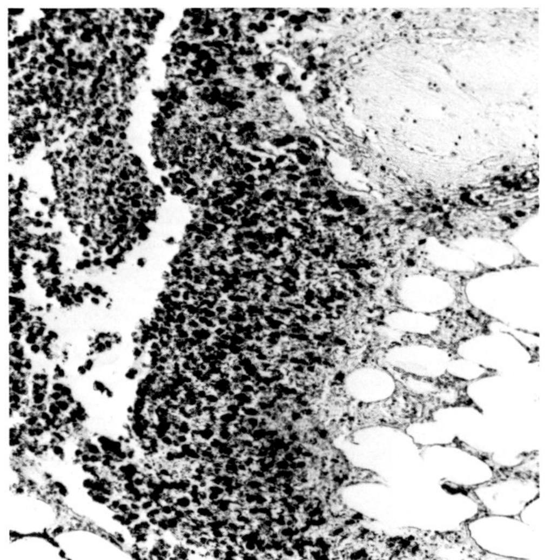

Fig. 7.13. Severe granulomatous peritonitis and pancreatitis in minnow due to experimental infection with *Renibacterium salmoninarum*. This section has a Gram stain to show massive number of black organisms.

dians such as *Glugea,* the ubiquitous digenetic metacercariae, and sanguinicolids, which may sometimes be found in mesenteric vessels, the accompanying inflammation extending into the surrounding pancreas (see Fig. 6.20). Degeneration and necrosis of pancreas are sometimes seen in toxic exposure, an example being crude oil spills, although in this case, branchial hyperplasia is the more significant lesion.

Swimbladder

Present in most teleosts, the swimbladder is connected to the esophagus by a pneumatic duct, which is patent and functional (physostomes) in most soft-rayed fish (such as salmonids) but closed in spiny-rayed fish (physoclists). The bladder may be subdivided into two or even three chambers, and may have a well-developed rete mirabile that enhances the blood supply to the organ. In some species, the swimbladder wall may even have respiratory value.

Diseases of the swimbladder are not commonly reported although it is involved in many systemic diseases, especially bacteremias and viremias with petechial and ecchymotic hemorrhaging frequently seen. This is particularly obvious with the endotheliotropic rhabdoviruses. The specific, commercially important syndrome, swimbladder inflammation (SBI) of common carp prevalent in Eastern European

Fig. 7.14. A. Large number of coccidia in common shiner replacing pancreas. **B.** Granulomatous pancreatitis in sunfish due to digenetic metacercariae.

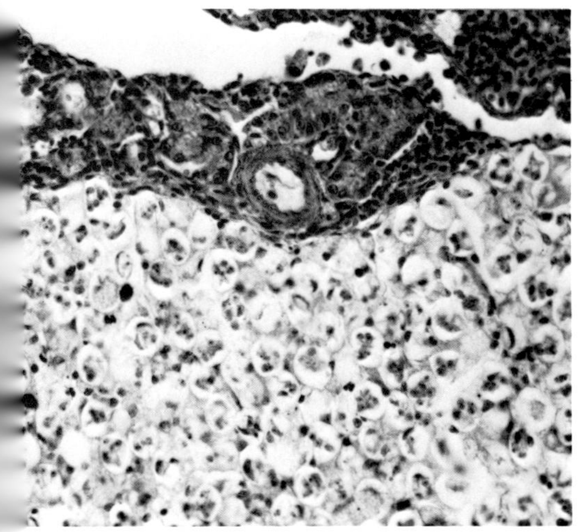

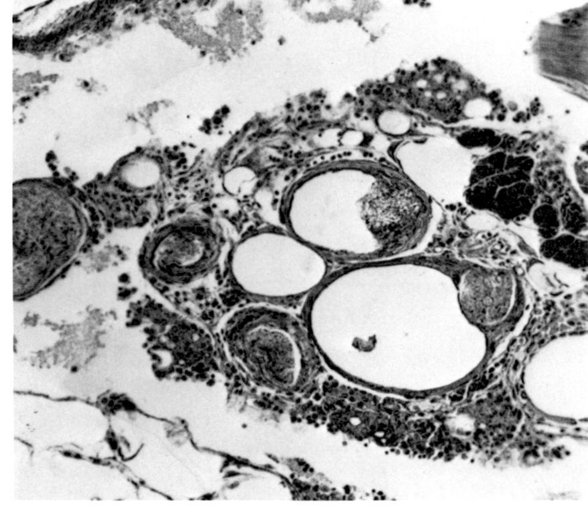

B

countries and once considered to have a bacterial and then a rhabdoviral etiology, has now been shown to be associated with developing presporogonic stages of the myxosporean parasite *Sphaerospora angulata,* the spores of which are found within the renal tubules. Grossly, the swimbladder wall is cloudy, opaque, and hemorrhagic in acutely affected fish, while in more advanced stages, there is fibrin deposition. Histologically, many developing protozoa are present within the well-vascularized thickened wall.

Coccidian parasites of the genus *Goussia* are found in several species, notably cichlids and gadoids. They are present within the epithelial lining, causing hypertrophy and sloughing of individual cells (Landsberg and Paperna 1985). Metazoan parasites are occasionally observed in the swimbladder. The best known of these is the nematode *Cystidicola,* and *C. stigmatura,* which infests lake trout and Arctic char, may be found in 100% of fish in a single location, some fish harboring worm burdens exceeding 1000. Pathogenicity is reportedly low. *Anguillicola* is another nematode found in the swimbladder of eels. Again pathogenicity is low, although large numbers cause thickening of the wall and may increase susceptibility to other diseases. Also found is the adult digene *Acetodextra.*

Fungal infections are occasionally seen, especially in salmonids that are physostomes (duct patent). The ducts of small fish may trap fine particulate feed and then subsequently become colonized by opportunistic fungi or bacteria. Examples include the fungi *Phoma herbarum,* which may progress to systemic involvement, and *Saprolegnia.* Another systemic mycotic disease of Atlantic salmon parr is that caused by *Phialophora* sp. in which the swimbladder is the target organ. Necrosis and hemorrhage are seen although cellular infiltration is minimal. Fusion of the bladder wall with the dorsally situated renal capsule may also occur.

Severe hemorrhage or even rupture of the swimbladder may be seen in deep or midwater species that are hauled to the surface too quickly after capture or in any species subjected to underwater explosions. Such an occurrence results from an inability to adjust quickly enough to the changes in pressure associated with these procedures.

A common but rarely described condition of intensively reared rainbow trout is overinflation of the swimbladder so that it

Fig. 7.15. Grossly overinflated swimbladder in rainbow trout ventrally displaced at its caudal end. Such a lesion must be differentiated from cystic dilatation of a ureter due to hypoplasia or acquired blockage (rare in our experience).

grossly distorts the lateral musculature (Fig. 7.15). Affected fish swim on the surface, often on their side, and may eventually die, apparently from exhaustion. Morbidity may be high. The precise cause of this condition is unknown although it appears to be associated with overstocking and high carbon dioxide levels. Hence, thinning the fish out leads to recovery within 24 hours; if left too long, however, the fish will not recover. A similar syndrome, also of unknown pathogenesis, has been observed in pantothenic-acid-deficient rainbow trout.

Failure of the swimbladder to inflate is reported in a variety of species of commercial importance. This is usually encountered in young larval fish as an irreversible condition that is often but not invariably incompatible with life. Spinal deformities are a frequent accompaniment. The cause of the condition is unknown.

References

Adair, B. M., and H. W. Ferguson. 1981. Isolation of infectious pancreatic necrosis (IPN) virus from non-salmonid fish. *J. Fish Dis.* 4:69–76.

Bennett, R. O., J. N. Kraeuter, L. C. Woods, III, M. M. Lipsky, and E. B. May. 1987. Histological evaluation of swimbladder inflation in striped bass larvae *Morone saxatilis. Dis. Aquat. Org.* 3:91–95.

Bergeron, T., and B. Woodward. 1983. Ultrastructure of the granule cells in the small intestine of the rainbow trout (*Salmo gairdneri*) before and after stratum granulosum formation. *Can. J. Zool.* 61:133–38.

Bucke, D. 1971. The anatomy and histology of the alimentary tract of the carnivorous fish the pike *Esox lucius* L. *J. Fish. Biol.* 3:421–31.

Burnstock, G. 1959. The morphology of the gut of the brown trout (*Salmo trutta*). *Q. J. Microsc. Sci.* 100:183–98.

Caceci, T. 1984. Scanning electron microscopy of goldfish, *Carassius auratus,* intestinal mucosa. *J. Fish Biol.* 25:1–12.

Coley, T. C., A. J. Chacko, and G. W. Klontz. 1983. Development of a lavage technique for sampling *Ceratomyxa shasta* in adult salmonids. *J. Fish Dis.* 6:317–19.

Cone, D. K. 1983. A *Lactobacillus* sp. from diseased female rainbow trout, *Salmo gairdneri* Richardson, in Newfoundland, Canada. *J. Fish Dis.* 5:479–85.

Crandall, T. A., and P. R. Bowser. 1981. A microsporidian infection in a natural population of mosquitofish, *Gambusia affinis* (Baird and Girard). *J. Fish Dis.* 4:317–24.

Crespo, S., G. Nonnotte, D. A. Colin, C. Leray, L. Nonnotte, and A. Aubree. 1986. Morphological alterations induced in trout intestine by dietary cadmium and lead. *J. Fish Biol.* 28:69–80.

Dykova, I., and J. Lom. 1978. Tissue reaction of the three-spined stickleback *Gasterosteus aculeatus* L. to infection with *Glugea anomala* (Moniez, 1887). *J. Fish Dis.* 1:83–90.

Eckmann, R. 1987. Pathological changes in the midgut epithelium of grayling, *Thymallus thymallus* L., larvae reared on different kinds of food, and their relation to mortality and growth. *J. Fish Dis.* 10:91–99.

Ferguson, H. W. 1979. Scanning and transmission electron microscopical observations on *Hexamita salmonis* (Moore, 1922) related to mortalities in rainbow trout fry *Salmo gairdneri* Richardson. *J. Fish Dis.* 2:57–67.

Ferguson, H. W., and R. D. Moccia. 1980. Disseminated hexamitiasis in Siamese fighting fish. *J. Am. Vet. Med. Assoc.* 177:854–57.

Ferguson, H. W., S. Rosendal, and S. Groom. 1985. Gastritis in Lake Tanganyika cichlids (*Tropheus duboisii*). *Vet. Rec.* 116:687–89.

Fouchereau-Peron, M., M. Laburthe, J. Besson, G. Rosselin, and Y. Le Gal. 1980. Characterization of the vasoactive intestinal polypeptide (VIP) in the gut of fishes. *Comp. Biochem. Physiol.* 65:489–92.

Gardiner, C. H., and R. M. Bunte. 1984. Granulomatous peritonitis in a fish caused by a flagellated protozoan. *J. Wildl. Dis.* 20:238–40.

Hayunga, E. G. 1979. Observations on the intestinal pathology caused by three caryophyllid tapeworms of the white sucker *Catostomus commersoni* Laecpede. *J. Fish Dis.* 2:239–48.

Hedrick, R. P., J. Speas, M. L. Kent, and T. McDowell. 1985. Adenovirus-like particles associated with disease of cultured white sturgeon *Acipenser transmontanus. Can. J. Fish. Aquat. Sci.* 42:1321–25.

Hoffman, G. L. 1965. *Eimeria aurati* n. sp. (Protozoa:Eimeriidae) from goldfish (*Carassius auratus*) in North America. *J. Protozool.* 12:273–75.

______. 1980. Asian tapeworm, *Bothriocephalus acheilognathi* Yamaguti, 1934, in North America. *Fisch und Umwelt* 8:69–75.

Hoover, D. M., F. J. Hoerr, W. W. Carlton, E. J. Hinsman, and H. W. Ferguson. 1981. Enteric cryptosporidiosis in a naso tang, *Naso lituratus* Bloch and Schneider. *J. Fish Dis.* 4:425–28.

Humbert, W., R. Kirsch, and M. F. Meister. 1984. Scanning electron microscopic study of the oesophageal mucous layer in the eel, *Anguilla anguilla* L. *J. Fish Biol.* 25:117–22.

Jilek, R., J. L. Crites. 1982. Intestinal histopathology of the common bluegill, *Lepomis macrochirus* Rafinesque, infected with *Spinitectus carolini* Holl, 1928 (Spirurida:Nematoda). *J. Fish Dis.* 5:75–77.

Landsberg, J. H., and I. Paperna. 1985. *Goussia cichlidarum* n. sp. (Barrouxiidae, Apicomplexa), a coccidian parasite in the swimbladder of cichlid fish. *Z. Parasitenkd.* 71:199–212.

McKnight, I. J., and R. J. Roberts. 1976. The pathology of infectious pancreatic necrosis. I. The sequential histopathology of the naturally occurring condition. *Br. Vet. J.* 132:76–85.

Mitchell, L.G., J. Ginal, and W. C. Bailey. 1983. Melanotic visceral fibrosis associated with larval infections of *Posthodiplostomum minimum* and *Proteocephalus* sp. in bluegill, *Lepomis macrochirus* Rafinesque, in central Iowa, U.S.A. *J. Fish Dis.* 6:135–44.

Molnar, K. 1982. Nodular coccidiosis in the gut of the tench, *Tinca tinca* L. *J. Fish Dis.* 5:461–70.

Molnar, K., and M. Reinhardt. 1978. Intestinal lesions in grasscarp *Ctenopharyngodon idella* (Valenciennes) infected with *Balantidium cteno pharyngodonis* Chen. *J. Fish Dis.* 1:151–56.

Munro, A. L. S., A. E. Ellis, A. H. McVicar, H. A. McLay, and E. A. Needham. 1984. An exocrine pancreas disease of farmed Atlantic salmon in Scotland. *Helgolander Meeresunters* 37:571–86.

Noga, E. J. 1986. Diet-associated systemic granuloma in African cichlids. *J. Am. Vet. Med. Assoc.* 189:1145–48.

Nohynkova, E. 1984. A new pathogenic *Cryptobia* from freshwater fishes: a light and electron microscopic study. *Protistol.* 20:181–95.

Ostberg, Y., R. Fange, A. Mattisson, and N. W. Thomas. 1976. Light and electron microscopical characterization of heterophilic granulocytes in the intestinal wall and islet parenchyma of the hagfish, *Myxine glutinosa* (Cyclostomata). *Acta Zool.* (Stockh.) 57:89–102.

Pankhurst, N. W., and P. W. Sorensen. 1984. Degeneration of the alimentary tract in sexually maturing European *Anguilla anguilla* (L.) and American eels *Anguilla rostrata* (LeSueur). *Can. J. Zool.* 62:1143–49.

Paperna, I. 1983. A *Chloromyxym*-like myxosporean infection in the stomach of cultured *Sparus aurata* (L.). *J. Fish Dis.* 6:85–89.

Rimmer, D. W., and W. J. Wiebe. 1987. Fermentative microbial digestion in herbivorous fishes. *J. Fish Biol.* 31:229–36.

Roberts, R. J. 1978. The pathophysiology and systemic pathology of teleosts. In *Fish Pathology,* ed. R. J. Roberts, 55–91. London: Bailliere Tindall.

Roberts, R. J., and R. H. Richards. 1978. Pansteatitis in farmed rainbow trout *Salmo gairdneri* Richardson. *Vet. Rec.* 103:492–93.

Rombout, J. H. W. M., H. W. J. Stroband, and J. J. Taverne-Thiele. 1984. Proliferation and differentiation of intestinal epithelial cells during development of *Barbus conchonius* (Teleostei, Cyrpinidae). *Cell Tissue Res.* 236:207–16.

Rombout, J. H. W. M., C. H. J. Lamers, M. H. Helfrich, A. Dekker, and J. J. Taverne-Thiele. 1985. Uptake and transport of intact macromolecules in the intestinal epithelium of carp (*Cyprinus carpio* L.) and the possible immunological implications. *Cell Tissue Res.* 239:519–30.

Scott, A. L., and J. M. Grizzle. 1979. Pathology of cyprinid fishes caused by *Bothriocephalus gowkongensis* Yea, 1955 (Cestoda:Pseudophyllidea). *J. Fish Dis.* 2:69–73.

Shostak, A. W., and T. A. Dick. 1986. Intestinal pathology in northern pike, *Esox lucius* L., infected with *Triaenophorus crassus* Forel, 1868 (Cestoda:Pseudophyllidea). *J. Fish Dis.* 9:35–45.

Sis, R. F., P. J. Ives, D. M. Jones, D. H. Lewis, and W. E. Haensly. 1979. The microscopic anatomy of the oesophagus, stomach and intestine of the channel catfish, *Ictalurus punctatus*. *J. Fish Biol.* 14:179–86.

Smith, M. W. 1983. Membrane transport in fish intestine. *Comp. Biochem. Physiol.* 75:325–35.

Swanson, R. N., and J. H. Gillespie. 1979. Pathogenesis of infectious pancreatic necrosis in Atlantic salmon (*Salmo salar*). *J. Fish. Res. Board Can.* 36:587–91.

Trust, T. J., and R. A. H. Sparrow. 1974. The bacterial flora in the alimentary tract of freshwater salmonid fishes. *Can. J. Microbiol.* 20:1219–28.

Van Noorden, S., and G. J. Patent. 1980. Vasoactive intestinal polypeptide-like immunoreactivity in nerves of the pancreatic islet of the teleost fish, *Gillichthys mirabilis*. *Cell Tissue Res.* 212:139–46.

Wierzbicka, J., and T. Einszporn-Orecka. 1986. Flagellates *Spironucleus mobilis* sp.n. in eel, *Anguilla anguilla* (L.). *Acta Protozool.* 25:75–80.

Yamamoto, T., and J. E. Sanders. 1979. Light and electron microscopic observations of sporogenesis in the myxosporida, *Ceratomyxa shasta* (Noble, 1950). *J. Fish Dis.* 2:411–28.

Youson, J. H., and R. M. Langille. 1981. Proliferation and renewal of the epithelium in the intestine of young adult anadromous sea lampreys, *Petromyzon marinus* L. *Can. J. Zool.* 59:2341–49.

8 Liver

IN MANY SPECIES the liver is a discrete organ in the anterior portion of the abdominal cavity, whereas in others it is split into lobes that interdigitate with the intestine and may run the length of the abdominal cavity. The histology is essentially similar to mammals, although lobulation is less distinct and less organized, and typical portal triads are not obvious.

Fish liver also serves *metabolic functions* similar to those in mammals, and despite earlier assertions to the contrary, teleosts do indeed possess drug metabolizing enzymes including phase 1 (activation), phase 2 (inactivation of reactive groups), and phase 3 (conjugation) pathways (Balk 1985). Considering the increasing number of environmental xenobiotics (a chemical foreign to the biological system) and the rapid efficient uptake of these compounds by gut, gills, and skin of the fish, the presence of these pathways is of vital importance to the adaptability and success of the group. It is interesting to note that enzymic activity in other tissues may be as high if not higher than those seen in the liver. Such tissues include anterior and posterior kidney, gut, and gills. Cytochrome p448-type enzymes are a major mixed-function oxidase (MFO) enzyme system in fish. Although they do possess the p450-type enzymes as found in mammals, rainbow trout (and probably other teleosts) seem to be refractory to the usual mammalian p450 inducers (such as phenobarbitone).

Hepatic lesions are not as well recognized in fish as in their mammalian counterparts, especially in infectious disease. This may partly be explained by the absence of a well-developed Kupffer cell system lining the sinusoids (Fig. 8.1), and hence a relatively poor capacity to trap circulating antigens. Some species have pancreatic tissue surrounding the major portal vessels, while others have melanomacrophage centers (MMCs) scattered throughout the parenchyma (Fig. 8.2); most have a biliary system (Fig. 8.3), which in some species is described as extending into hepatocytes, not just between hepatocytes, as seen in mammals. It is interesting to note that in the lamprey, a primitive cyclostome fish that parasitizes other fish, metamorphosis of the larvae to adults correlates with disappearance of the bile duct system and gallbladder.

Hepatocyte appearance varies greatly between species and also with season, nutritional status, age, sex, and pollutant exposure, especially with regard to fat and endoplasmic reticulum content. For example, hepatocytes from starved milkfish are smaller than normal, and they have smaller nuclei with condensed chromatin. There is loss of stored glycogen but an increase in

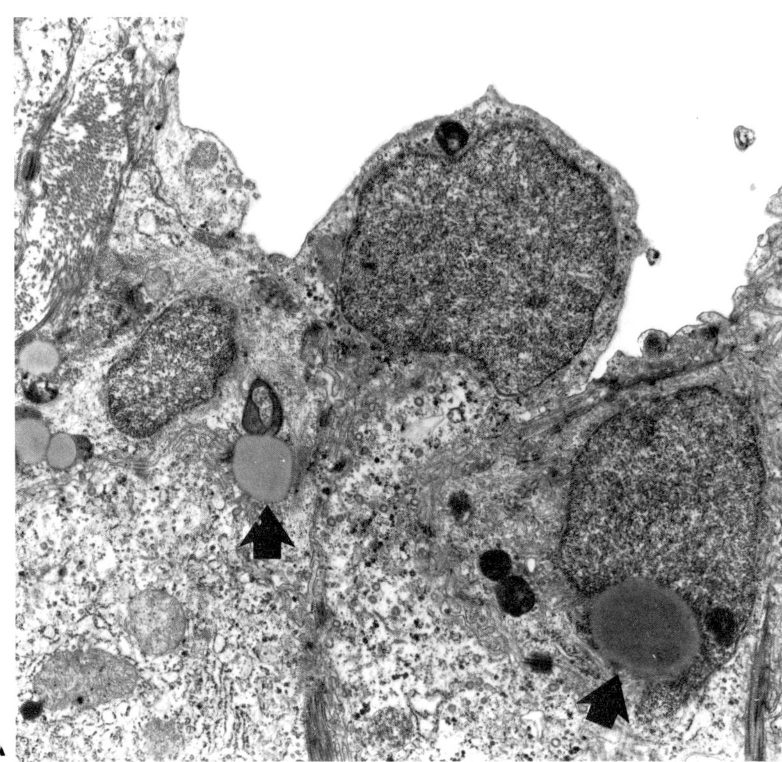

Fig. 8.1. A. Sinusoid from plaice showing endothelial cell nucleus projecting out into lumen. Beneath this cell are two fat-storing cells showing fat globules in their cytoplasm (*arrows*). No Kupffer cells are present.
B. Plaice injected with carbon and killed 6 hours later showing particle accumulation in fat-storing-cell (FSC) cytoplasm (*arrow*). The fenestrated nature of the endothelial lining is not apparent in this section. Trapping ability of FSC is probably limited to particles small enough to traverse endothelial fenestrations.

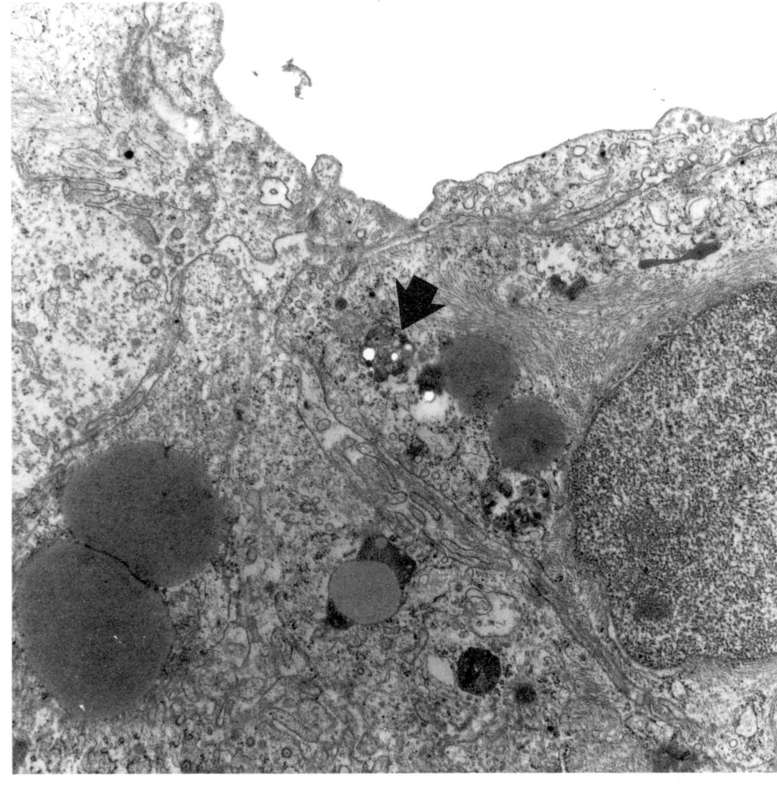

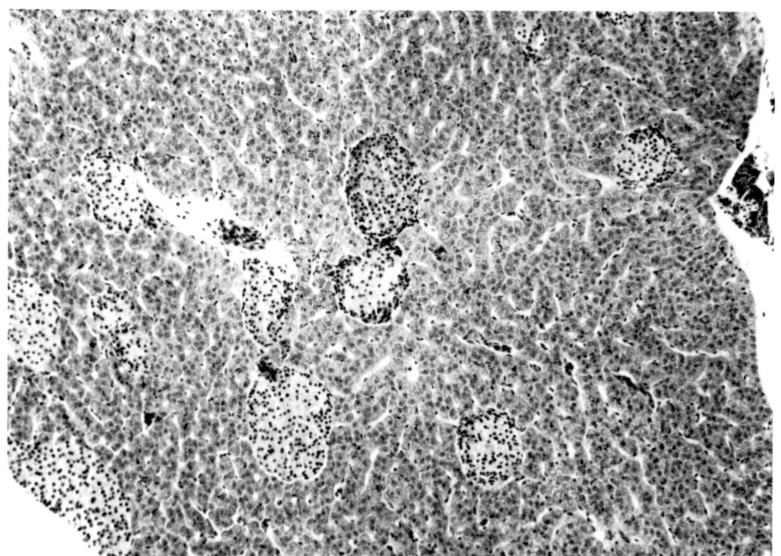

Fig. 8.2. Marine flatfish with numerous melanomacrophage centers throughout parenchyma.

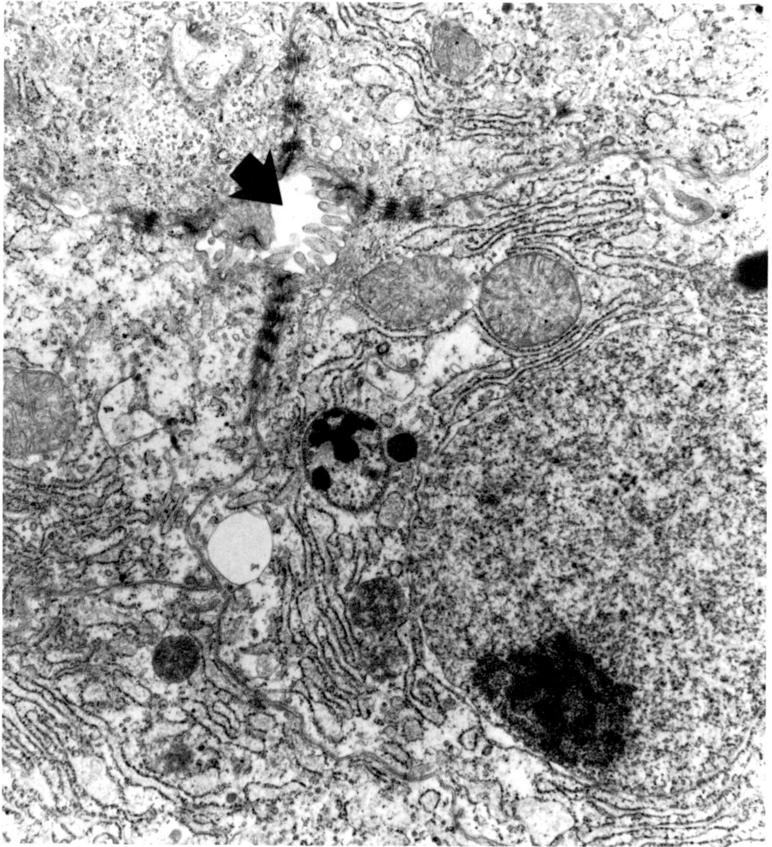

Fig. 8.3. Four hepatocytes contributing to intercellular bile canaliculus (*arrow*). Note tight junctions and microvilli.

mitochondrial size and iron-containing dense bodies. Thus great care should be exercised whenever interpreting such changes in fish liver, especially fatty change, as the liver in several species normally stores large quantities of lipid. Despite all these considerations, fatty degeneration is occasionally encountered and may be either diffuse or patchy.

Fish such as trout utilize dietary carbohydrates poorly, and an excess can accumulate in hepatocytes as glycogen. As well as leading to hepatomegaly, this results in impaired liver function as demonstrated by reduced clearance of sulphobromophthalein (BSP) from the plasma (Hilton and Dixon 1982).

Hepatic siderosis is sometimes seen in aquarium fish, as in other vertebrates, and may be associated with syncytial cell formation. Whether this condition represents an iron storage problem or a metabolic defect is unknown. Progressive accumulation of copper within hepatocytes is described in a variety of vertebrates including dogs (Bedlington and West Highland White terriers) and man (Wilson's disease). High hepatic copper levels have recently been reported in feral white perch from the Chesapeake Bay area, the accumulations being associated with peribiliary fibrosis and inflammation, enlarged MMCs, and disruption of architecture in older fish (Bunton et al. 1987). Whether the copper was responsible for the cholangiohepatitis or whether high levels (often greater than 1,000 μg/g wet weight) are normal in this species remains to be determined. Certainly other species occupying the same body of water had much lower levels, similar to those encountered in fish in the Great Lakes (often of the order of 2.0 to 10.0 μg/g wet weight).

Jaundice is not a commonly observed clinical feature in fish, although in toxic liver failure and in hemolytic diseases, an increase in bile pigments in the serum is measured. In addition, there is a strange condition of northern pike recorded from both Sweden and Canada, in which there is a buccal or generalized green discoloration of the tissues. The significance or cause of this "green pike syndrome" is unknown.

Single-cell necrosis of hepatocytes is encountered quite frequently in septicemic and toxic disease. Early evidence of the change is heralded by rapid condensation of nuclear chromatin, followed by increased eosinophilia of the cytoplasm and shrinkage (individualization) from adjacent viable hepatocytes. Remnants of the cell may be phagocytosed by surrounding cells (apoptosis). These necrotic changes may be accompanied by an inflammatory cell infiltrate.

Severe patchy or zonal necrosis of hepatocytes is one of the major lesions seen in a recently encountered syndrome (salmon anemia syndrome) in farmed Norwegian Atlantic salmon (Fig. 8.4). Associated lesions include anemia and fragmentation degeneration of striated muscle. The cause of this condition is unknown as is its relationship to pancreas disease and other syndromes with some component of rhabdomyopathy.

Zonal changes are less common in fish than in mammals, although they are seen around central veins in metabolic disturbances. They are also seen in toxic exposure and in anoxia associated with severe gill disease or with gill netting when the opercula become progressively clamped shut, thereby preventing irrigation (Fig. 8.4). Passive venous congestion is encountered in severe cardiac disease (Fig. 8.4C) such as that seen in some of the salmonid cardiomyopathies, or in some cases of bacterial kidney disease (BKD). Fibrosis is occasionally a sequel.

Mitotic figures, binucleate cells, and cytomegaly are all relatively common observations in hepatocytes during repair, regeneration, or neoplastic processes. Cirrhosis is not commonly seen in fish although teleosts do have large numbers of fat-storing cells in the space of Disse between the sinusoid endothelium and hepatocytes (Fig. 8.1). In mammals these cells are responsible for collagen production in cirrhosis. Their involvement in granulomatous hepatitis is reported in

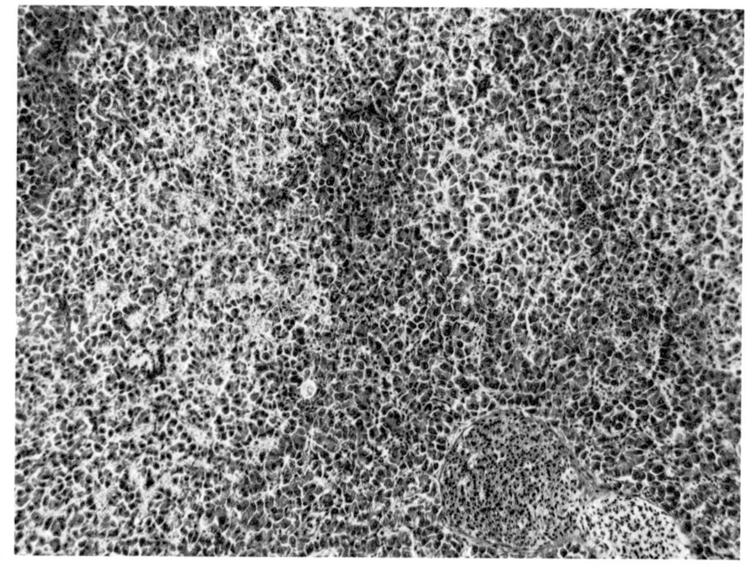

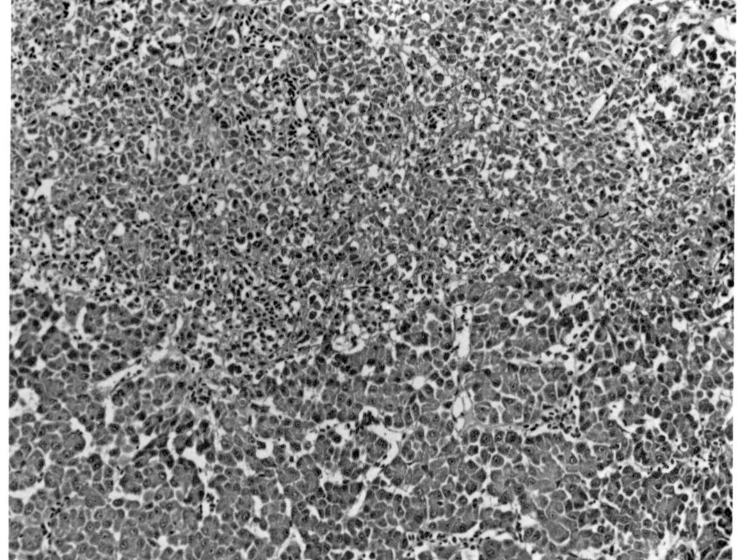

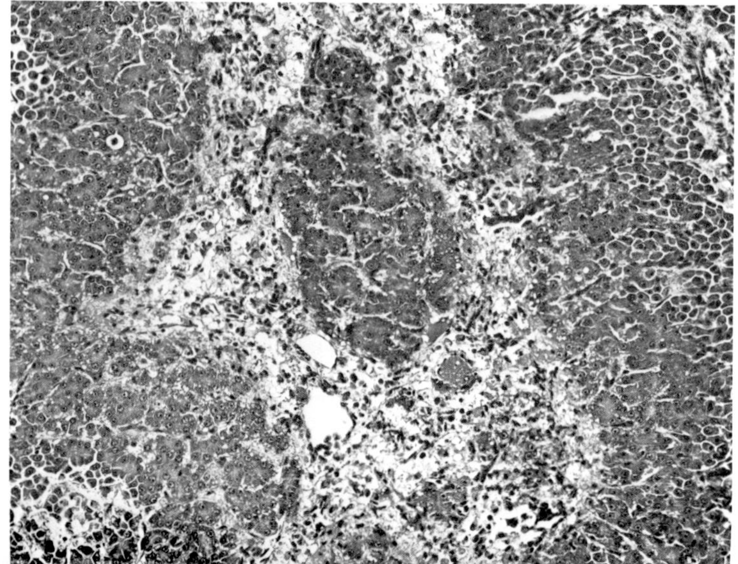

Fig. 8.4. A. Market-sized rainbow trout with severe bacterial gill disease showing zonal necrosis and degeneration of liver. **B.** Farmed Norwegian Atlantic salmon with patchy severe acute necrosis (top half of picture). The cause of this emerging syndrome is presently unknown. (*Material courtesy of T. Poppe*) **C.** Liver from farmed Atlantic salmon with severe degenerative cardiomyopathy. Chronic venous congestion has progressed to fibrosis, which is isolating still-viable islands of hepatocytes (cardiac fibrosis).

fish (Hawkins et al. 1981), and they are especially prominent in lipoid liver disease, often bulging into the sinusoid.

Cholangitis and cholangiohepatitis are both encountered. The associated inflammatory response is primarily lymphoid with some peribiliary fibrosis in the more advanced or chronic cases (Fig. 8.5). It is important to differentiate such an infiltrate from hemopoiesis, which is sometimes seen in this ancillary location in times of high demand or as a normal event in very young fish. Protozoa or metazoa are sometimes seen within bile ducts and gallbladder, and although they may provide a ready explanation for any associated inflammation, in many cases the cause of the response is unknown.

On the other hand, gouramis are prone to a systemic granulomatous disease associated with the presence of amoeboid protozoa; among other organs (kidney, spleen, submucosa of gut), the liver is a target, and in particular the bile ducts, which become surrounded by marked chronic fibroplastic and granulomatous inflammation. The presence of numerous parasites within the response makes a cause-effect relationship in this case a seemingly reasonable deduction.

Fish in heavily polluted areas may have prominent cholangitis with increased numbers of ducts as well as metaplastic changes in the duct epithelium or even frank necrosis. It has been suggested that the often high incidence of associated bile duct and hepatocyte carcinomas in these fish is not merely coincidence but a direct result of such pollution. A dramatic example of peribiliary fibrosis in farmed fish is seen in "hepatorenal syndrome" of marine flatfish.

Lesions in the gallbladder often parallel those found in the bile ducts, but the organ

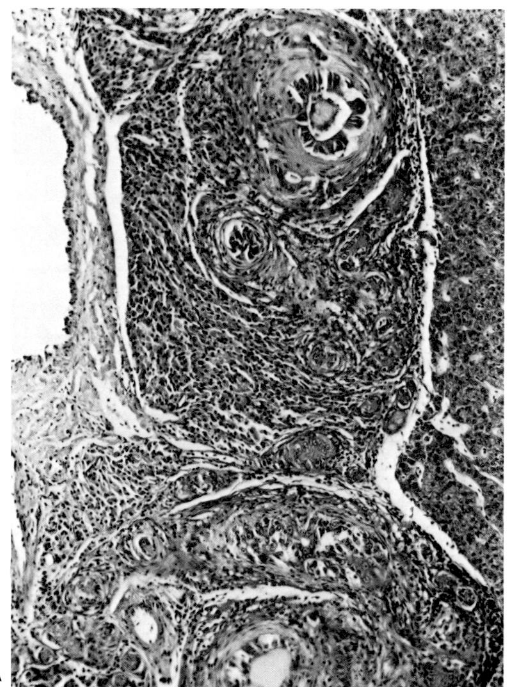

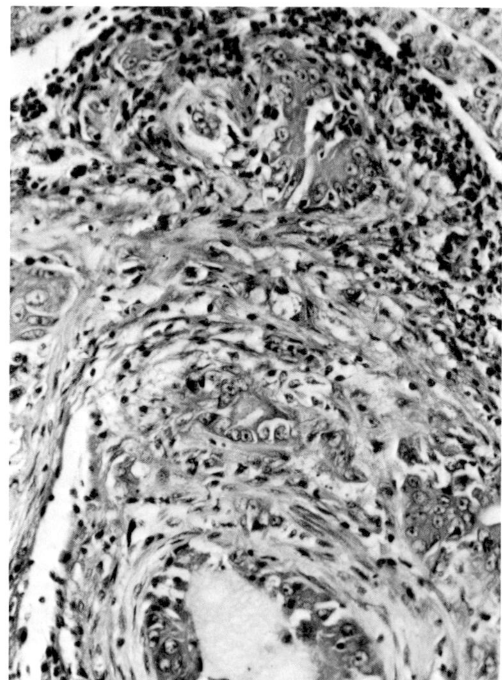

Fig. 8.5. **A** and **B.** White sucker showing cholangiohepatitis. There is a marked hyperplasia of bile ducts and an infiltration of lymphocytes and macrophages plus early fibroplasia.

should not be overlooked, especially on gross examination as it can give an indication of how the fish was feeding. For example, the gallbladder in starved Atlantic salmon increases to 20% of liver weight by 36 hours, and the bile changes from straw colored to green to blue after 1, 4, and 6 days, respectively. Enlargement of the gallbladder and thickening of its wall are seen in various myxosporean infections, which may also involve bile ducts and sometimes extend out into the hepatic parenchyma. For example, *Myxidium oviformis* in Atlantic salmon causes severe cholangiohepatitis and inflammation of the gallbladder with emaciation of the host. The myxosporean *Ceratomyxa* may cause damage severe enough to result in total loss of the gallbladder. Virtually the only gross lesion seen with experimental arsenic toxicity in rainbow trout is fibrosis of the gallbladder wall. (In green sunfish, there is an increased number of electron-dense bodies within both nucleus and cytoplasm.)

Gallstones are occasionally encountered in a variety of species. One report describes them in tilapia (*Oreochromis mossambicus*) subjected to adverse environmental conditions. The significance or precise etiology of the condition is rarely determined.

Multifocal hepatitis is seen in bacteremias, but not as frequently or as severely as in mammals. Septic thrombi within sinusoids and central veins are also seen (see Fig. 6.15), and these may cause ischemic lesions. Infarction as a result of such lesions, or necrotic tracts due to parasitic migration, is sometimes encountered. These foci of necrotic cells can create a microenvironment suitable for the growth of anaerobes, and large bacilli can therefore occasionally be encountered within these lesions (Fig. 8.6). Focal hepatic necrosis is also seen in several virus diseases such as infectious hematopoietic necrosis (IHN) of salmonids and channel catfish virus disease (CCVD) as well as the recently reported adenoviruslike disease of cultured sturgeon.

The disseminated granulomatous infections so common in aquarium fish often involve the liver, although usually to a much lesser degree than the kidney or spleen. Similarly, the salmonid liver is involved in diseases such as BKD or proliferative kidney

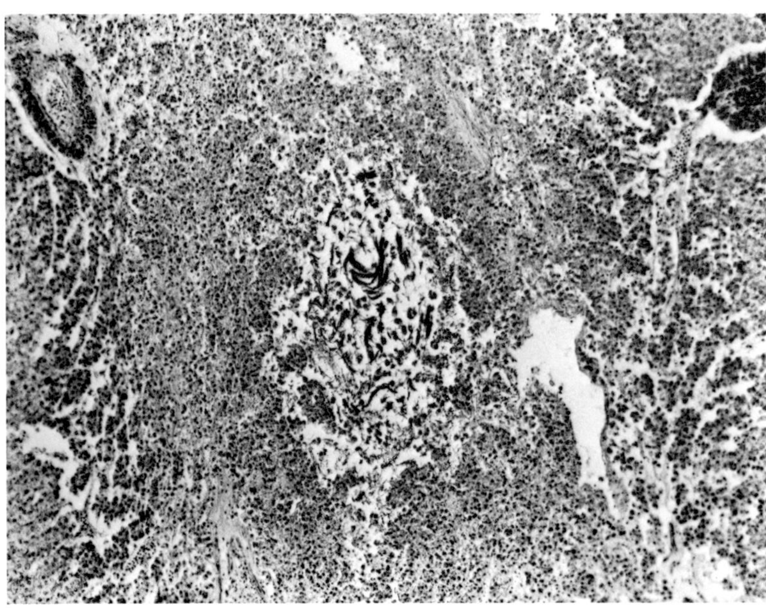

Fig. 8.6. Brown bullhead with large focus of necrosis caused by migrating metazoan parasites. In the center are many gram-positive bacilli.

disease (PKD), although once again to a lesser degree than other tissues (Fig. 8.7). Severe hepatic necrosis accompanies acute *Aspergillus* spp. infection in cultured tilapia, and multifocal hepatitis is also seen in Atlantic salmon associated with the fungus *Exophiala*. Similarly, any peritonitis of whatever cause will often involve the hepatic capsule, and the inflammatory response may be seen to extend into the underlying parenchyma.

Chronic granulomatous hepatitis is frequently encountered with various protozoan and metazoan parasites (Fig. 8.8). An example is hepatic coccidiosis of wild and cultured cyprinids and other species (Fig. 8.9). Although the parasites often evoke little more than a mild to moderate multifocal granulomatous response, in the killifish *Fundulus grandis, Calyptospora* (*Eimeria*) *funduli* can be present in numbers large enough to replace 85% of the liver and pancreas. A commercially important coccidian hepatitis is also reported in blue whiting.

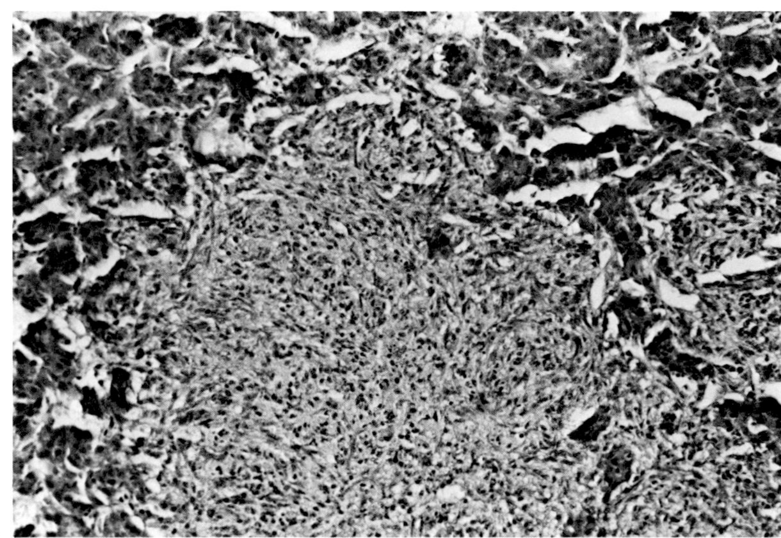

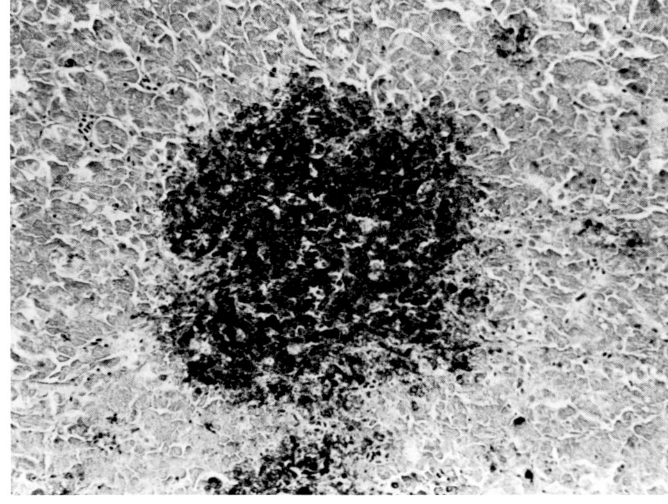

Fig. 8.7. A. Coho salmon with bacterial kidney disease showing focal granuloma compressing and invading normal parenchyma. **B.** Gram-stained section of similar but earlier lesion with large number of bacteria present.

Larval nematodes of the genus *Anisakis* can be present in numbers large enough to represent a space-occupying threat to normal functioning. Similarly, *Capillaria* spp. may cause hepatic destruction if the infection is massive, either by virtue of the parasites themselves or their eggs.

Hepatic pallor is seen in anemia and fat storage—which in extreme cases may also render the liver soft and friable, leading to hemorrhage of a spontaneous nature (a consequence of impaired production of clotting factors?)—or as a result of trauma (Fig. 8.10). Petechial or ecchymotic hemorrhages

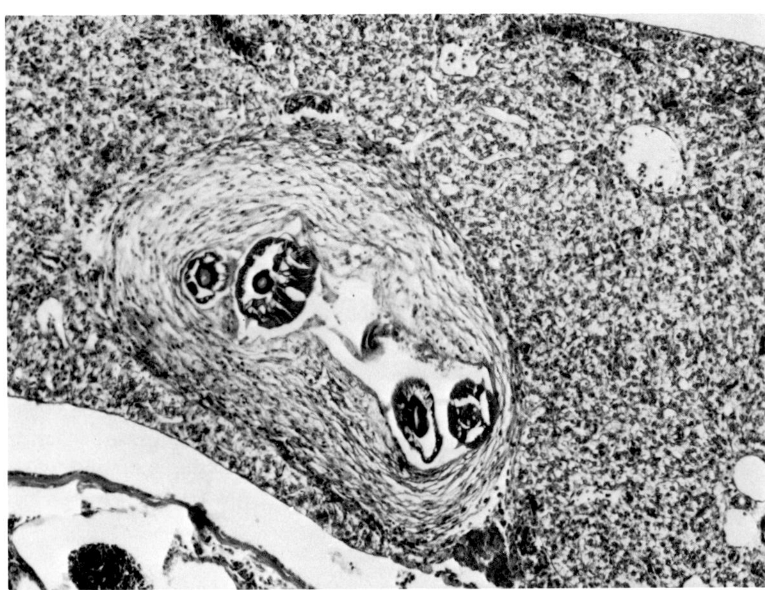

Fig. 8.8. Yellow perch with marked fibrosis surrounding nematodes. Lateral alae on parasites are quite distinct. Such findings are usually considered incidental.

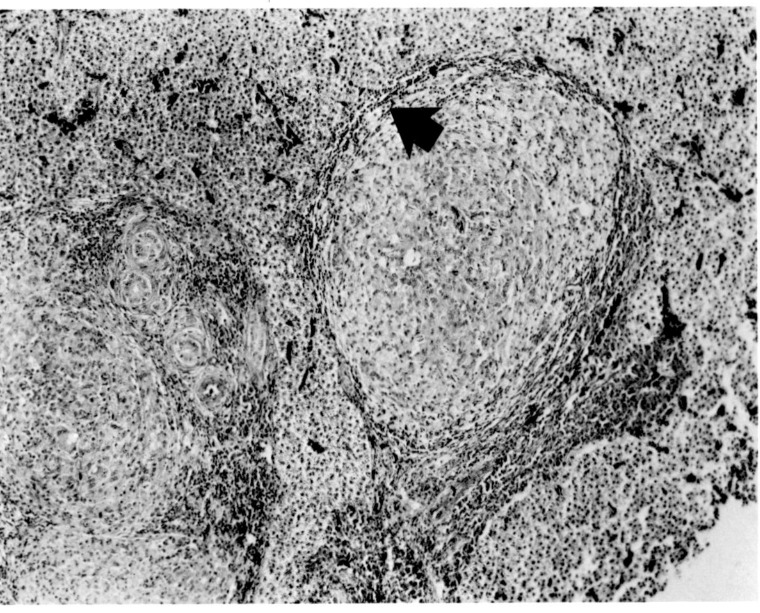

Fig. 8.9. Brown bullhead with multifocal hepatic granulomas due to coccidia. Large pale foci are comprised almost exclusively of macrophages. These in turn are separated from surrounding hepatic parenchyma by a layer of fibrous tissue and moderate numbers of lymphocytes (*arrow*).

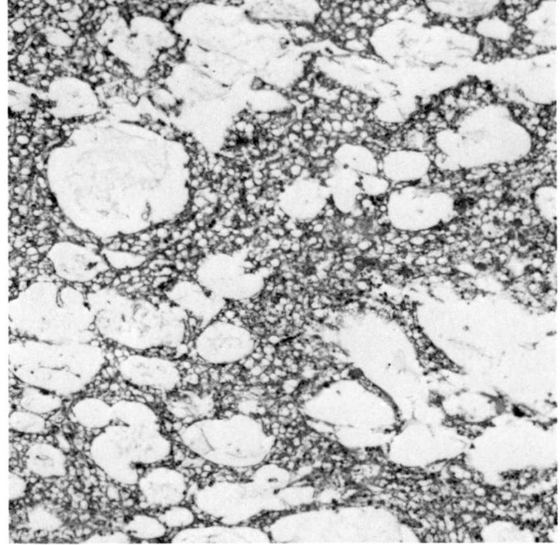

Fig. 8.10. Cichlid with severe fatty degeneration of liver. Intraperitoneal hemorrhage killed this animal.

are often seen in bacteremias and viremias as well as in acute exposure to toxins such as aflatoxin derived from the fungus *Aspergillus flavus,* which grows on oil seeds (chronic exposure, of course, leads to hepatoma; see Chapter 13).

A condition seen in salmonids, which renders the liver pale due to anemia as well as from the large patchy deposits of lipid metabolites, is *lipoid liver disease* (Fig. 8.11). This condition is seen most commonly in Atlantic salmon and rainbow trout (other species such as brown trout seem less susceptible) and is associated with feeding rancid fats. The high levels of polyunsaturated fats needed in fish diets render them prone to such oxidative change, especially if they are stored in moist and warm conditions. Antioxidants are usually added, but under severe conditions even their protection may be overcome.

Lipid products, mainly ceroid, accumulate within hepatocytes, fat-storing cells, and macrophages that may aggregate around the portal triads. The pigment autofluoresces under ultraviolet light and has typical membranous whorls ultrastructurally. Anoxic change due to the often severe anemia and the toxic action of the rancid fats themselves can complicate the histopathological interpretation. The anemia is a result of increased red cell fragility and therefore faster breakdown, plus a failure of maturation of polychromatocytes, probably a consequence of the vitamin E deficiency induced by the

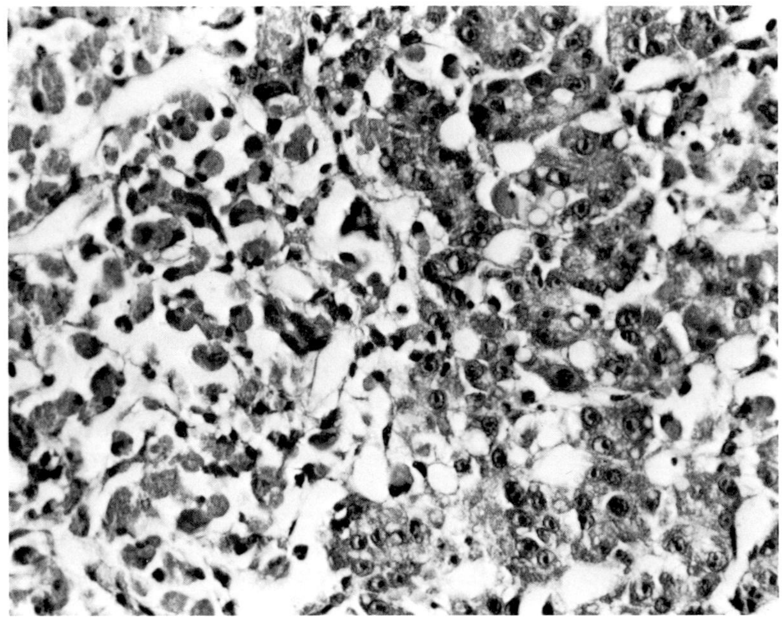

Fig. 8.11. Rainbow trout with lipoid liver disease. Abnormal hepatocytes (to the *left*) are rounded up and contain large quantities of ceroid (not obvious in an H and E–stained section).

rancid fats. Hemosiderin from the erythrocytes is deposited in large quantities in the spleen. Consumption of available endogenous antioxidants such as glutathione and vitamin E/selenium may predispose the hepatocytes to lipid peroxidation from other normally innocuous sources, thereby complicating the pathogenesis even more.

References

Balk, L. 1985. Characterization of xerobiotic metabolism in the ferol teleost northern pike (*Esox lucius*). Ph.D. thesis, Department of Biochemistry, University of Stockholm, Sweden.

Bruno, D. W., and A. E. Ellis. 1986. Multiple hepatic cysts in farmed Atlantic salmon, *Salmo salar* L. *J. Fish Dis.* 9:79–81.

Bunton,T. E., S. M. Baksi, S. G. George, and J. M. Frazier. 1987. Abnormal hepatic copper storage in a teleost fish (*Morone americana*). *Vet. Pathol.* 24:515–24.

Byczkowska-Smyk, W. 1968. Observation of the ultrastructure of the hepatic cells of the burbot (*Lota lota* L.). *Zool. Polon.* 18:287–94.

De Vos, R., C. De Wolf-Peeters, and V. Desmet. 1973. A morphologic and histochemical study of biliary atresia in lamprey liver. *Z. Zellforsch.* 136:85–96.

Elarifi, A. E. 1982. The histopathology of larval anisakid nematode infections in the liver of whiting, *Merlangius merlangus* (L.), with some observations on blood leucocytes of the fish. *J. Fish Dis.* 5:411–19.

Eurell, J. A., and W. E. Haensly. 1982. The histology and ultrastructure of the liver of Atlantic croaker *Micropogon undulatus* L. *J. Fish Biol.* 21:113–25.

Fujita, H., T. Tamaru, and J. Miyagawa. 1980. Fine structural characteristics of the hepatic sinusoidal walls of the goldfish (*Carassius auratus*). *Arch. Histol. Jpn.* 43:265–73.

Hacking, M. A., J. Budd, and K. Hodson. 1978. The ultrastructure of the liver of the rainbow trout: normal structure and modifications after chronic administration of a polychlorinated biphenyl Aroclor 1254. *Can. J. Zool.* 56:477–91.

Hawkins, W. E., M. A. Solangi, and R. M. Overstreet. 1981. Ultrastructural effects of the coccidium, *Eimeria funduli* Duszynski, Solangi and Overstreet, 1979 on the liver of killifishes. *J. Fish Dis.* 4:281–95.

Hendricks, J. D., L. J. Hunter, and J. H. Wales. 1976. Postmortem bile damage to rainbow trout (*Salmo gairdneri*) livers. *J. Fish. Res. Board Can.* 33:2613–16.

Heusequin, E. 1973. Sur les elements mis en evidence par la 3,3′-diaminobenzidine dans le foie de lebistes reticulatus. *Arch. Biol.* 84:243–79.

Hilton, J. W., and D. G. Dixon. 1982. Effect of increased liver glycogen and liver weight on liver function in rainbow trout, *Salmo gairdneri* Richardson: recovery from anaesthesia and plasma ^{35}S-sulphobromophthalein clearance. *J. Fish Dis.* 5:185–95.

Kranz, H., and N. Peters. 1985. Pathological conditions in the liver of ruffe, *Gymnocephalus cernua* (L.), from the Elbe estuary. *J. Fish Dis.* 8:13–24.

Maier, K. J. 1984. Gallstone induction in tilapia, *Oreochormis mossambicus* (Peters). *J. Fish Dis.* 7:521–24.

Mawdesley-Thomas, L. E., and D. H. Barry. 1970. Acid and alkaline phosphatase activity in the liver of brown and rainbow trout. *Nature* 227:738–39.

Mitchell, L. G., J. K. Listebarger, and W. C. Bailey. 1980. Epizootiology and histopathology of *Chloromyxum trijugum* (Myxospora:Myxosporida) in centrarchid fishes from Iowa. *J. Wildl. Dis.* 16:233–36.

Moccia, R. D., S. S. O. Hung, S. J. Slinger, and H. W. Ferguson. 1984. Effect of oxidized fish oil, vitamin E and ethoxyquin on the histopathology and haematology of rainbow trout, *Salmo gairdneri* Richardson. *J. Fish Dis.* 7:269–82.

Morrison, C. M., W. E. Hawkins. 1984. Coccidians in the liver and testis of the herring *Clupea harengus* L. *Can. J. Zool.* 62:480–93.

Nopanitaya, W., J. Aghajanian, J. W. Grisham, and J. L. Carson. 1979. An ultrastructural study on a new type of hepatic perisinusoidal cell in fish. *Cell Tissue Res.* 198:35–42.

Payne, J. F., L. F. Fancey, A. D. Rahimtula, and E. L. Porter. 1987. Review and perspective on the use of mixed-function oxygenase enzymes in biological monitoring. *Comp. Biochem. Physiol.* 86C:233–45.

Peek, W. D., E. W. Sidon, J. H. Youson, and M. M. Fisher. 1979. Fine structure of the liver in the larval lamprey, *Petromyzon marinus* L.; hepatocytes and sinusoids. *Am. J. Anat.* 156:231–50.

Pierce, K. V., B. B. McCain, and S. R. Wellings. 1980. Histopathology of abnormal livers and other organs of starry flounder *Platichthys stellatus* (Pallas) from the estuary of the Duwamish River, Seattle, Washington, U.S.A. *J. Fish Dis.* 3:81–91.

Poole, B. C., and T. A. Dick. 1984. Liver pathology of yellow perch, *Perca flavescens* (Mitchill), infected with larvae of the nematode *Raphidascaris acus* (Bloch, 1779). *J. Wildl. Dis.* 20:303–7.

Roald, S. O., D. Armstrong, and T. Landsverk. 1981. Histochemical, fluorescent and electron microscopical appearance of hepatocellular ceroidosis in the Atlantic salmon *Salmo salar* L. *J. Fish Dis.* 4:1–14.

Saez, L., R. Amthauer, E. Rodriguez, and M. Krauskopf. 1984. Effects of insulin on the fine structure of hepatocytes from winter-acclimatized carps: studies on protein synthesis. *J. Exp. Zool.* 230:187–97.

Saez, L., T. Zuvic, R. Amthauer, E. Rodriguez, and M. Krauskopf. 1984. Fish liver protein synthesis during cold acclimatization: seasonal changes of the ultrastructure of the carp hepatocyte. *J. Exp. Zool.* 230:175–86.

Smith, C. E. 1979. The prevention of liver lipoid degeneration (ceroidosis) and microcytic anaemia in rainbow trout *Salmo gairdneri* Richardson fed rancid diets: a preliminary report. *J. Fish Dis.* 2:429–37.

Sorensen, E. M. B. 1976. Ultrastructural changes in the hepatocytes of green sunfish, *Lepomis cyanellus* Rafinesque, exposed to solutions of sodium arsenate. *J. Fish Biol.* 8:229–40.

Storch, V., and J. V. Juario. 1983. The effect of starvation and subsequent feeding on the hepatocytes of *Chanos chanos* (Forsskal) fingerlings and fry. *J. Fish Biol.* 23:95–103.

Talbot, C., and P. J. Higgins. 1982. Observations on the gall bladder of juvenile Atlantic salmon *Salmo salar* L., in relation to feeding. *J. Fish Biol.* 21:663–69.

Tanuma, Y. 1980. Electron microscope observations on the intrahepatocytic bile canalicules and sequent bile ductules in the Crucian, *Carassius carassius*. *Arch. Histol. Jpn.* 43:1–21.

Weis, P. 1972. Hepatic ultrastructure in two species of normal, fasted and gravid teleost fishes. *Am. J. Anat.* 133:317–21.

Wunder, Von Wilhelm. 1976. Experimentelle untersuchungen uber Wundheilung und regeneration an der Leber der regenbogenforelle (*Salmo irideus* W. Gibb.). *Zool. Anz.* 196:357–83.

Yamamoto, T. 1965. Some observations on the fine structure of the intrahepatic biliary passages in goldfish (*Carassius auratus*). *Z. Zellforsch.* 65:319–30.

Youson, J. H., and E. W. Sidon. 1978. Lamprey biliary atresia: First model system for the human condition? *Experientia* 34:1084.

9 Nervous System

Brain

Normal. The teleost brain has the same basic components and arrangement as other vertebrate brains, but it has different proportions. The mammalian pathologist may therefore have some difficulty in recognizing the major parts.

The cerebral hemispheres (telencephalon or cerebrum) are much reduced by comparison with higher vertebrates and consist mainly of the olfactory lobes and olfactory bulbs. The bulbs are located anteriorly next to the olfactory sacs and are connected to the lobes by long tracts. Lateral ventricles within the telencephalon are not obvious.

The diencephalon (tween brain) is small but contains appendages such as the pineal body (in the epithalamus) and the pituitary gland (in the hypothalamus). The hypothalamus is the major part of the diencephalon and acts as a coordinator of olfactory and lateral-line information. Behind the infundibulum of the pituitary lies the saccus vasculosus, a choroid plexus producing cerebrospinal fluid for the third ventricle with which its lumen is continuous. (The saccus dorsalis on the roof of the epithalamus and continuous with the pineal gland is also a fluid-secreting plexus.)

Caudal to the olfactory lobes, the mesencephalon (optic lobes) is usually the largest portion of the brain and is the most obvious when the cranium is removed. It is divided into the optic tectum (divided into two lobes) and tegmentum, which form the roof and floor of the third ventricle, respectively.

The cerebellum (metencephalon) is not a very prominent portion of the teleost brain, although there is great interspecies variation. It lies behind the optic lobes, and although not folded into gyri as in mammals, there are easily recognizable molecular and granular layers, with Purkinje cells in between.

The medulla oblongata is continuous caudally with the spinal cord. The central canal is enlarged to form the fourth ventricle lined by ependymal cells and supplied with fluid from a choroid plexus in its roof.

In mammals, the immature ependyma, derived from neuroectoderm, is the origin of the glial cells (astrocytes and oligodendrocytes) as well as neurons and mature ependyma. In teleosts, which also have astrocytes and oligodendrocytes, it is reported that the immature ependyma is similarly pluripotential, but also that mature ependymal cells have retained the embryonic capacity for differentiation (Bernstein 1970). The inference is that damaged brain tissue can be *regenerated* and indeed this has been shown experimentally to be the case, although there are species differences. Evidence for ependymal reactivity sometimes can be encountered in routine diagnostic material where the cells may be seen streaming into the ventricular lumen in those relatively uncommon cases of severe encephalitis. Teleost ependymal cells possess large quantities of glycogen and numerous large mitochondria, suggesting specialized metabolic activity.

Mammalian microglia are mesodermal in origin and represent histiocytes. Whether fish possess microglia is unknown, and although cells with suspicious features (using mammalian criteria for identification) are seen in cases of encephalitis, their presence in teleosts awaits confirmation.

Disease. There is still relatively little information on the pathology of fish nervous tissue. Although basic responses are probably the same as in other vertebrates, a detailed study of general responses is nevertheless necessary. Indeed many pathologists even fail to examine the brain as a routine! In general, however, gliosis is readily recognized in situations where there is inflammation, with phagocytes of either microglial or astrocytic origin removing debris (Figs. 9.1B and 9.7A).

Associated with cranial ulceration, meningitis and encephalitis are reported in *Edwardsiella ictaluri* infection of channel catfish; the disease entity is termed enteric septicemia of catfish (ESC). The olfactory area is a particular target, and indeed it is suggested that once the disease is established here, the cranial ulceration is a direct result of dorsal extension of this lesion. Similar encephalitic lesions are also seen in *Streptococcus iniae* infection in cultured yellowtails. This latter granulomatous lesion extends into the ventricle and granular layer of the cerebellum and results in vertebral deformities. A similar granulomatous meningitis is occasionally seen in salmonids with severe bacterial kidney disease (BKD) or *Streptococcus* infections (Fig. 9.1). Epizootics of

Fig. 9.1. A and **B.** Rainbow trout with subacute suppurative meningitis due to *Renibacterium salmoninarum.* High-power micrograph in **B** shows intense gliosis (including several cells with microglial-like nuclei [*open arrows*]) and ongoing neuronophagia (*solid arrow*).

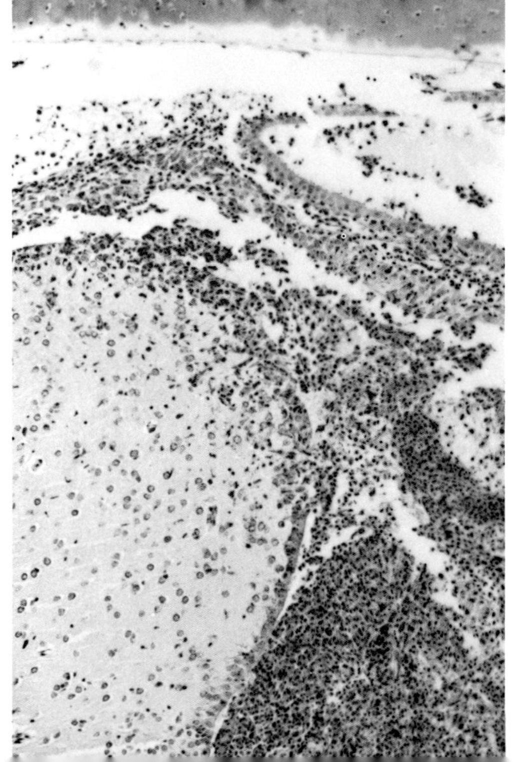

B

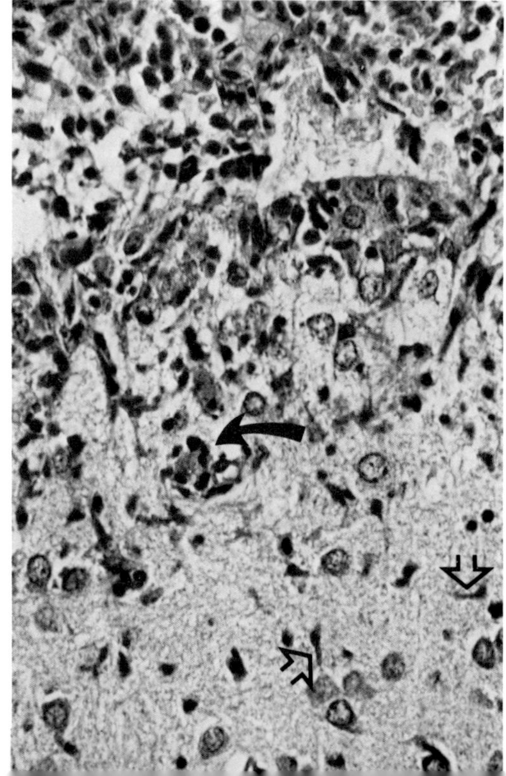

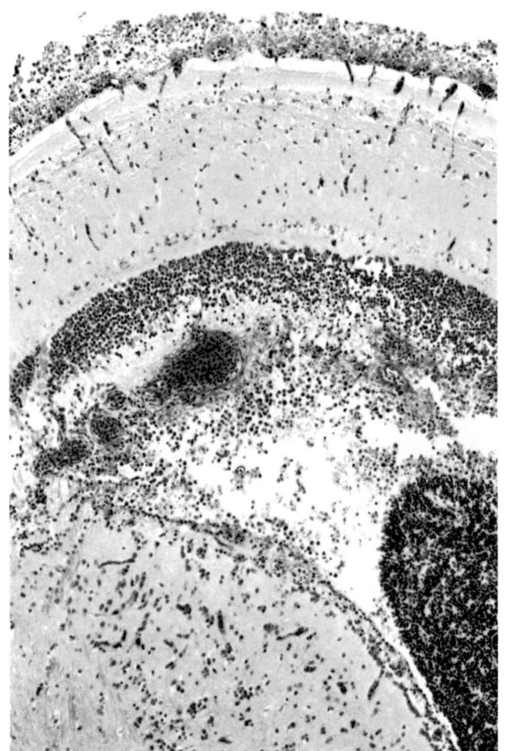

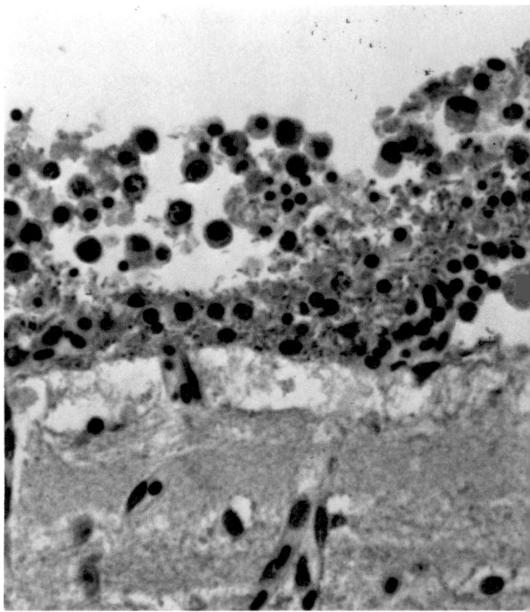

Fig. 9.2. A and **B.** Goldfish with severe acute suppurative meningitis and ventriculitis associated with large numbers of rod-shaped bacteria (**B**). Most blood vessels are prominent due to congestion and mild hypertrophy of endothelium.

meningitis in several species of estuarine fish in the Gulf of Mexico and Florida were associated with gram-positive filamentous anaerobes recovered from the brains of affected fish and identified as *Eubacterium tarantellus.* Other examples of subacute suppurative meningitis are occasionally encountered in routine diagnostic work (Fig. 9.2), although the precise cause is seldom established.

Viral encephalitides are rarely reported, although the herpesvirus of channel catfish can cause neuronal vacuolation and edema of both brain and spinal cord. Similarly pike fry rhabdovirus is said to cause hydrocephalus in young fish with massive fluid accumulation in the third ventricle, plus hemorrhage and congestion of the cranial vessels. The lesion manifests itself as a distinct "blip" on the head. It should be noted, however, that similar lesions have also been seen in rainbow trout and coho salmon, unrelated to any virus infection (Fig. 9.3). An iridovirus recovered from moribund redfin perch with hepatocellular and hematopoietic necrosis (the disease is known as epizootic haematopoietic necrosis [EHN] and represents the first fish virus disease reported from Australia) that was inoculated into Atlantic salmon produced vacuolar degeneration of optic lobe, optic nerve, and cerebellum. The lesions correlated well with clinical observations of blindness and depression (Langdon et al. 1986). Similar lesions (vacuolating encephalopathy and hematopoietic necrosis) have recently been reported from naturally

Fig. 9.3. Coho salmon fry each exhibiting a pronounced swelling on top of head due to cranial defect. Eventually there is pressure necrosis of the overlying skin, and the brain comes into direct contact with the water. The cause of this condition is unknown although we suspect teratogenic agents such as malachite green, which has been shown to specifically target cranial tissues and which is commonly used in hatcheries to control fungal growth on incubating eggs. (*Courtesy of R. D. Moccia*)

occurring outbreaks of disease in trout, and from which a morphologically similar iridovirus has also been recovered.

Exophiala salmonis has been associated with epizootic outbreaks of disease in farmed cutthroat trout (Carmichael 1966). The brain was the primary focus for the fungus, with later involvement of the eye and other tissues. As might be expected, the response was necrotizing and granulomatous and comprised of macrophages, giant cells, and lymphocytes. *Saprolegnia* may also on occasion penetrate to the central nervous system, as can *Ichthyophonus.*

Hoffman and Hoyme (1958) described the neuropathology of experimental infection of the stickleback (*Eucalia inconstans*) by cercariae of the trematode *Diplostomum baeri eucaliae.* Following skin penetration, the parasites migrated to the brain, localizing particularly in the choroid plexi which became hyperplastic. The associated granulomatous response was not reported to affect nervous function, although occasionally hemorrhage did cause death.

By contrast, metacercariae of another fluke, *Ornithodiplostomum ptychocheilus,* encysted within the brain of minnows have been shown to alter the behavior of the fish so that their schools were less compact, divided more frequently, and occupied positions closer to the water surface than did control schools. Such an effect would probably favor predation by the parasites' final host (Fig. 9.4). Similarly, paralysis has been

Fig. 9.4. A and **B.** Common shiner with cranium removed (**A**) to reveal slight bubbling appearance to meninges. This is due to massive numbers of fluke metacercariae replacing large portions of the brain as seen in **B.** Despite the extent of the involvement, there is relatively little inflammatory response.

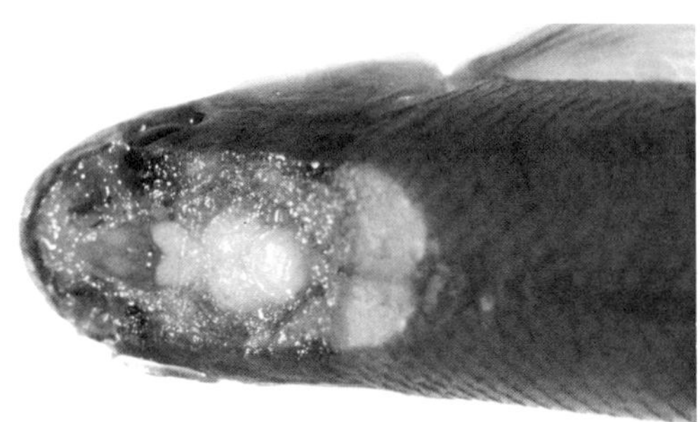

B

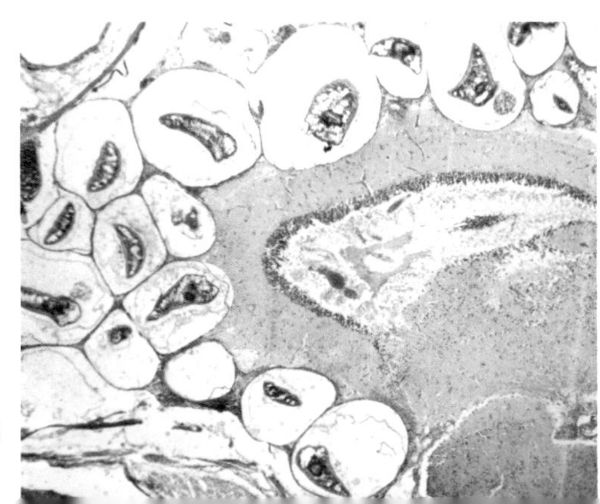

reported to be the result of severe cranial diplostomiasis in the Argentine fish (*Basilichthys microlepidotus*). Adult sanguinicolid flukes are also occasionally encountered in the meninges, for example *Sanguinicola idahoensis* in rainbow trout.

Other parasites found in the brain include myxosporeans and microsporidians. *Myxobolus hendricksoni* is found in the optic lobes and cerebellum as well as ventricles and meninges of the fathead minnow. Host response is minimal even though large areas of the brain may be replaced by the parasites. Spinal curvature may be a consequence of some myxosporean infections (see Chapter 12 for other causes of spinal damage). Such is the case with the encephalotropic species *Myxobolus buri* of yellowtail and probably also *Triangula percae* in redfin perch (Langdon 1988).

The microsporidian *Spraguea* (*Nosema*) *lophii* is found in the cerebrospinal ganglia of angler fish (*Lophius* spp.) where it evokes grossly visible xenoma formation. Clinical disease is not reported however. The ciliates *Ichthyophthirius multifiliis* and *Tetrahymena corlissi* (and their marine counterparts, *Cryptocaryon* and *Uronema*) may occasionally be found within meninges or deeper when infections are heavy, and especially when the epithelium of the nares or orbit is parasitized.

The pesticide carbaryl causes characteristic vacuolation of the optic tectum in salmonids. The insecticide endosulfan causes edema, meningitis, and encephalitis in tilapia. The response is characterized by an influx of eosinophilic granular cells and small foci of glial scarring in the long term. High ammonia levels induce convulsions or coma, probably due to interference with neurotransmitters, but there are no reported lesions similar to those seen in mammals, i.e., intramyelinic edema causing spongy change to white matter. Instead, a proliferation of the meninges has been observed in fathead minnows, although this lesion is not seen in salmonids (Smith 1984).

Thiamine deficiency has been reported to be the cause of polioencephalomalacic lesions in clinically affected herring with spongy degeneration of the periventricular ependyma, focal thalamic neuronal necrosis, and meningeal and ventricular hemorrhage (Roberts 1978). Despite similar clinical disease in rainbow trout, we have failed to confirm these findings in experimentally produced deficiency. Clinical evidence of severe nervous dysfunction in young cultured splake (lake trout–brook trout hybrid) was associated with marked hemorrhagic encephalomalacia (Fig. 9.5). Although of unknown cause, the lesions tended to have a laminar distribution at the white/gray matter interface in several parts of the brain. In mammals, possible causes would include thiamine "deficiency," lead poisoning, and "salt poisoning," whereas in poultry, vitamin E deficiency would be a possibility.

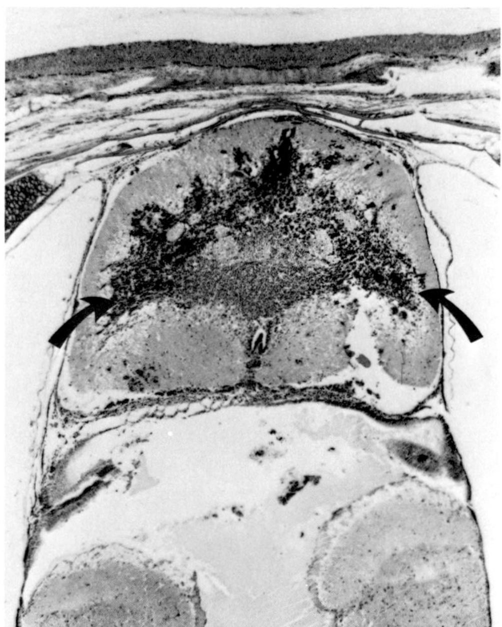

Fig. 9.5. Transverse section through head of young cultured splake showing marked hemorrhage with laminar distribution (*arrows*) within cerebellum.

Spinal Cord

Normal. Originating through the foramen magnum, the spinal cord runs down the vertebral column and terminates in a neurosecretory endocrine structure, the urophysis. On transverse section it can be seen that the cord is divided into both gray and white matter, the gray forming an inverted Y, so that there is a single dorsal horn and two ventrolateral ones. Just beneath the arms of the Y and ventral to the central canal lie the large myelinated Mauthner axons arising from the Mauthner neurons in the medulla. Excitation results in a powerful tail flip, the so-called Mauthner-initiated startle response. A growth-related increase in the numbers of neurons is reported in several species, including salmonids, and appears to correlate with the increase in numbers of red and white muscle fibers.

Disease. Changes may be associated with space-occupying lesions (Fig. 9.6) such as tumors or with traumatic damage to the vertebral column as seen in lightning strike (see Fig. 12.12). Whirling disease (*Myxosoma cerebralis*) may distort the vertebrae and cause compression lesions. A less dramatic but clinically important result of cord damage may be pigmentary change so that, for example, the posterior part of the fish turns dark. A posterior paralysis has been described in sea bream associated with *Aeromonas hydrophila* infection that tracked along intermyotomal fascia to the spinal cord. Traumatic cord damage is also seen in young rainbow trout as a result of material from intervertebral discs ventrally extruded at the point of maximum flexure, beneath the dorsal fin. The resulting posterior paralysis may be compounded by ischemic damage due to concomitant aortic thrombosis (see Fig. 6.17). Cord lesions include hemorrhage and malacia, with swollen eosinophilic axons and over time, mild perivascular migration of leukocytes and gemastocytic (reactive) astrocytosis.

Myxosporeans have been described within the axons of minnows and lungfish and are also seen in goldfish. In the minnows, prespore stages were present within brainstem, dorsal roots, and even sinoatrial

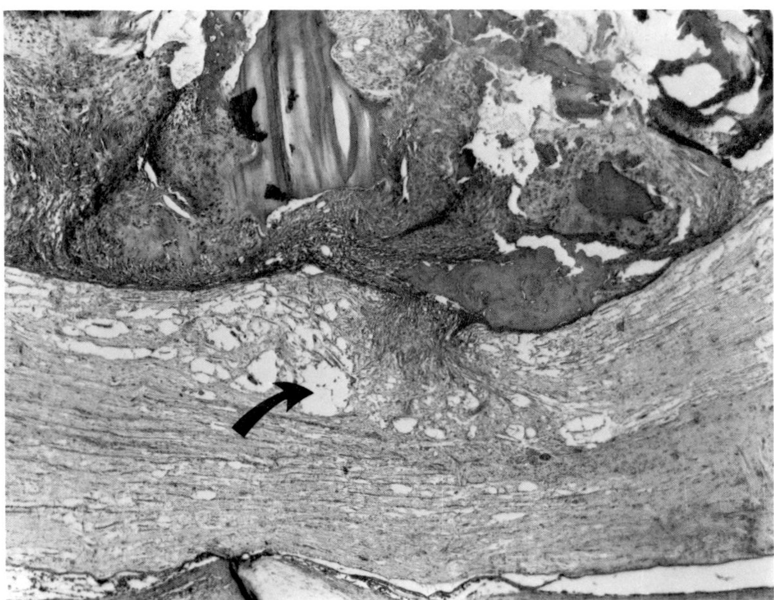

Fig. 9.6. Spinal cord from rainbow trout with pronounced malacia and cavitation (*arrow*) due to compression by a fracture of vertebral column just beneath dorsal fin.

pacemaker, while spores identified as *Myxobolus* were found in the dorsal root ganglia and spinal canal (Fig. 9.7). These locations strongly suggest that the parasites migrated down the axons and out along the dorsal roots into the ganglia where they matured into spores. Despite parasitic occupation of many axons, which greatly increased in diameter and had attenuated myelin sheaths, obvious clinical disease in the fish was lacking, possibly because a thin peripheral rim of axoplasm still remained, even in heavily parasitized axons. A glial response in the minnows was only present in brainstem, along with perivascular cuffing. A response was seen, however, in the cord of the lungfish, possibly because the parasites in this latter species did not restrict themselves to the axons. It should surely come as little surprise to those having some experience with this amazing group of parasites to find that it is the Myxosporea that have been capable of exploiting this highly unusual but very well-protected niche within the fish.

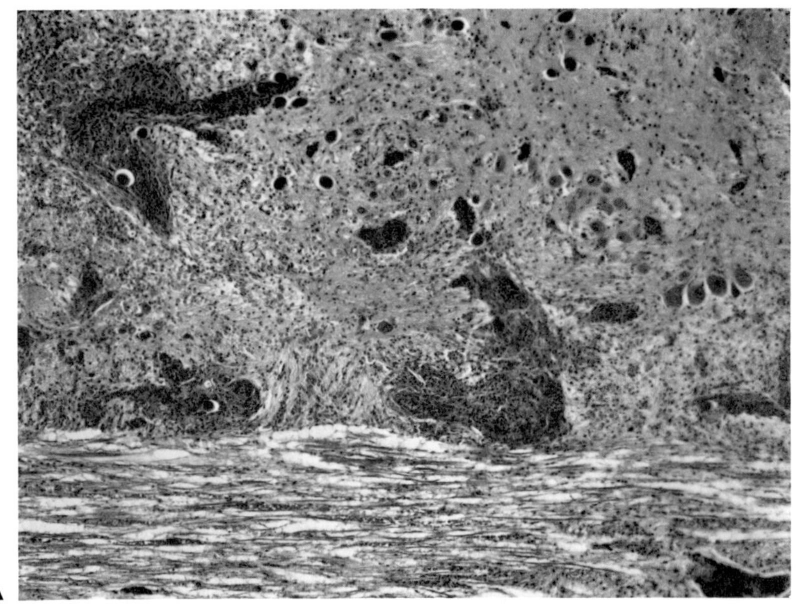
A

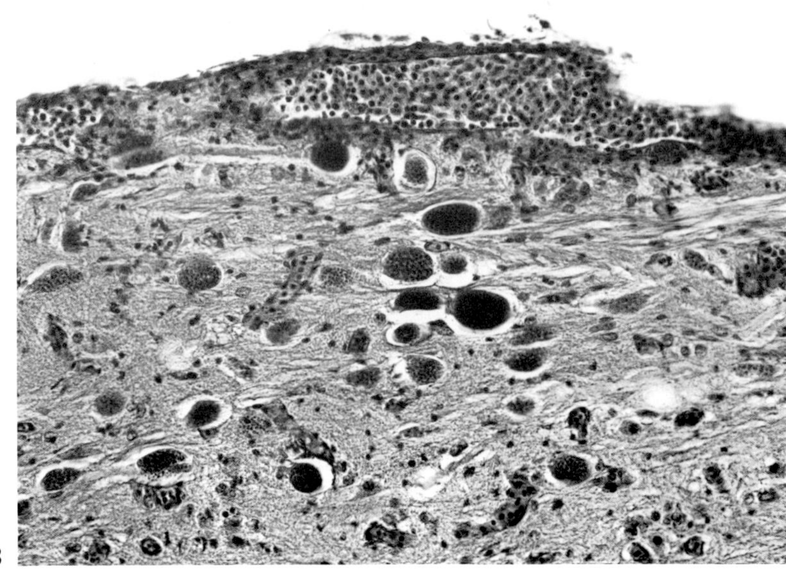
B

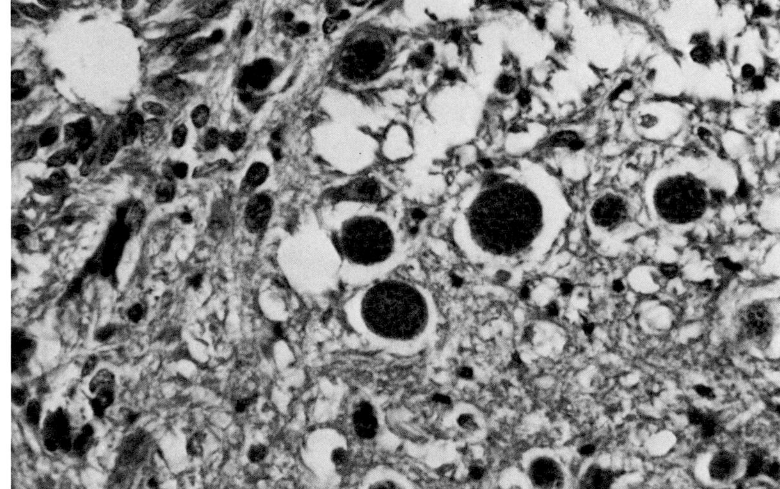

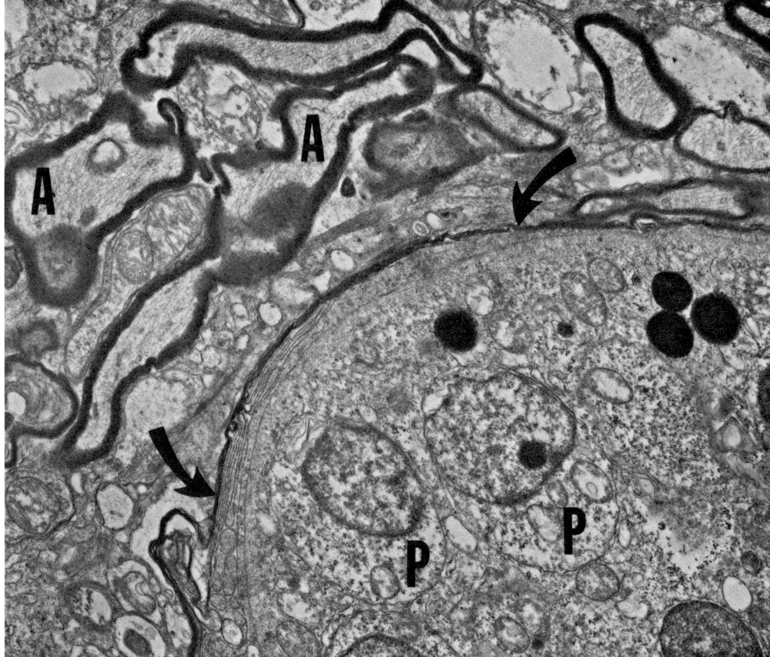

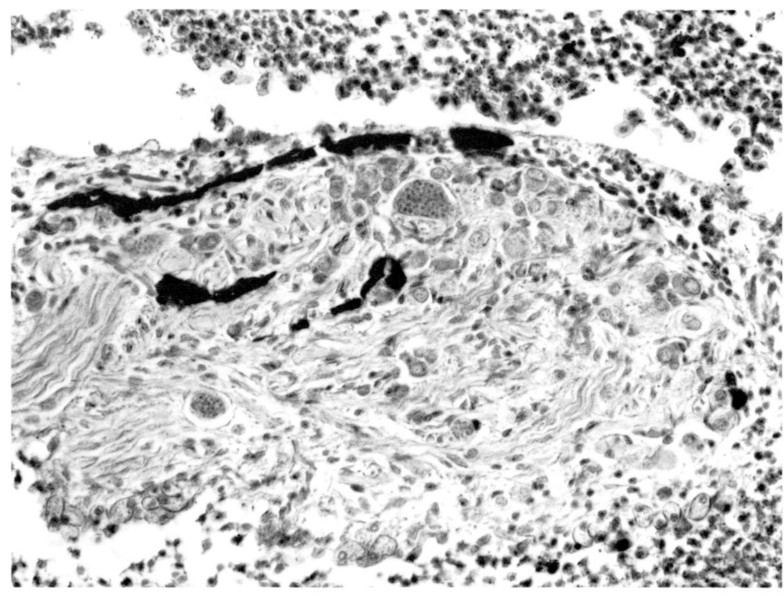

Fig. 9.7. Common shiner with prespore myxosporean infection of nervous system. **A.** Brainstem encephalitis with marked perivascular cuffing and gliosis. **B.** High power of brainstem showing numerous aggregates of prespore stages of parasite distending axons. There is congestion of meningeal vessels. **C.** Spinal cord approximately midway along its length showing axons distended to varying degrees by parasites. **D.** Electron micrograph of spinal cord showing several uninfected myelinated axons (*A*) plus one greatly distended axon containing several nucleated prespore myxosporean parasites (*P*). Note thin but still complete myelin sheath (*arrows*) and axoplasm. **E.** Sinoatrial pacemaker with several prespore myxosporean aggregates. Developing stages of the parasite were also present within myocardial trabeculae, possibly accessed via the conduction system. The clinical impact of either of these locations remains unknown.

References

Alfei, L., and G. Sesti. 1984. Neuronal increase in the trout (*Salmo trutta fario*) spinal cord during development. *Cell. Mol. Biol.* 30:471–78.

Ali, N. M. 1984. Mast cells enter a teleost's brain by Xth cranial nerve in response to *Diplostomum phoxini* (Trematoda). *Experientia* 40:197–98.

Bernstein, J. J. 1970. 1. Anatomy and physiology of the central nervous system. In *Fish Physiology,* ed. W. S. Hoar and D. J. Randall, vol. 4, *The Nervous System, Circulation, and Respiration,* 1–90. New York: Academic Press.

Carmichael, J. W. 1966. Cerebral mycetoma of trout due to a *Phialophora*-like fungus. *Sabouraudia* 6:120–23.

Castejon, O. J., and A. J. Caraballo. 1980. Light and scanning electron microscopic study of cerebellar cortex of teleost fishes. *Cell Tissue Res.* 207:211–26.

De Kinkelin, P. 1980. Occurrence of a microsporidian infection in zebra danio *Brachydanio rerio* (Hamilton-Buchanan). *J. Fish Dis.* 3:71–73.

Egusa, S. 1985. *Myxobolus buri* sp.n. (Myxosporea:Bivalvulida) parasitic in the brain of *Seriola quinqueradiata* Temminck et Schlegel. *Fish Pathol.* 19:239–44.

Ferguson, H. W., J. Lom, and I. Smith. 1985. Intra-axonal parasites in the fish *Notropis cornutus* (Mitchill). *Vet. Pathol.* 22:194–96.

Halliday, M. M. 1974. Studies on *Myxosoma cerebralis,* a parasite of salmonids. III. Some studies on the epidemiology of *Myxosoma cerebralis* in Denmark, Scotland and Ireland. *Nord. Vet. Med.* 26:165–72.

______. 1974. Studies on *Myxosoma cerebralis,* a parasite of salmonids. IV. A preliminary immunofluorescent investigation of the spores of *Myxosoma cerebralis.* Nord. Vet. Med. 26:173–79.

Hoffman, G. L., and J. B. Hoyme. 1958. The experimental histopathology of the "tumor" on the brain of the stickleback caused by *Diplostomum baeri eucaliae* Hoffman and Hundley, 1957 (Trematoda:Strigeoidea). *J. Parasitol.* 44:374–78.

Kent, M. L., and G. L. Hoffman. 1984. Two new species of Myxozoa, *Myxobolus inaeguus* sp. n. and *Henneguya theca* sp. n. from the brain of a South American knife fish, *Eigemannia virescens* (V.). *J. Protozool.* 3:91–94.

Kruger, L., and D. S. Maxwell. 1967. Comparative fine structure of vertebrate neuroglia: teleosts and reptiles. *J. Comp. Neurol.* 129:115–42.

Langdon, J. S. 1987. Spinal curvatures and an encephalotropic myxosporean *Triangula percae* sp. nov. (Myxozoa:Ortholineidae) enzootic in redfin perch, *Perca fluviatilis* L. in Australia. *J. Fish Dis.* 10:425–34.

Langdon, J. S., J. D. Humphrey, L. M. Williams, A. D. Hyatt, and H. A. Westbury. 1986. First virus isolation from Australian fish: an iridovirus-like pathogen from redfin perch, *Perca fluviatilis* L. *J. Fish Dis.* 9:263–68.

Lewis, D. H., and L. R. Udey. 1978. Meningitis in fish caused by an asporogenous anaerobic bacterium. Fish Disease Leaflet no. 56. U.S. Dept. of the Interior. Washington, D.C.

Major, R. D., J. P. McCraren, and C. E. Smith. 1975. Histopathological changes in channel catfish (*Ictalurus punctatus*) experimentally and naturally infected with channel catfish virus disease. *J. Fish. Res. Board Can.* 32:563–67.

Markiw, M. E., and K. Wolf. 1974. *Myxosoma cerebralis*: isolation and concentration from fish skeletal elements—sequential enzymatic digestions and purification by differential centrifugation. *J. Fish. Res. Board Can.* 31:15–20.

Marquet, E., and H. Sobel. 1970. A protozoon in the central nervous system of the lungfish *Polypterus enlicheri. J. Protozool.* 17:71–76.

Meyer, F. P., and T. A. Jorgenson. 1983. Teratological and other effects of malachite green on development of rainbow trout and rabbits. *Trans. Am. Fish. Soc.* 112:818–24.

Mitchell, L. G., C. L. Seymour, and J. M. Gamble. 1985. Light and electron microscopy of *Myxobolus hendricksoni* sp. nov. (Myxozoa:Myxobolidae) infecting the brain of the fathead minnow, *Pimephales promelas* Rafinesque. *J. Fish Dis.* 8:75–89.

Miyazaki, T., and J. A. Plumb. 1985. Histopathology of *Edwardsiella ictaluri* in channel catfish, *Ictalurus punctatus* (Rafinesque). *J. Fish Dis.* 8:389–92.

Roberts, R. J. 1978. The pathophysiology and

systemic pathology of teleosts. In *Fish Pathology,* ed. R. J. Roberts, 55–91. London: Bailliere Tindall.

Smith, C. E. 1984. Hyperplastic lesions of the primitive meninx of fathead minnows, *Pimephales promelas,* induced by ammonia: species potential for carcinogen testing. *Natl. Cancer Inst. Monogr.* 65:119–25.

Smith, J. R. 1984. Fish neurotoxicology. In *Aquatic Toxicology,* vol. 2, ed. L. J. Weber. New York: Raven Press.

Stephens, E. B., M. W. Newman, A. L. Zachary, and F. M. Hetrick. 1980. A viral aetiology for the annual spring epizootics of Atlantic menhaden *Brevoortia tyrannus* (Latrobe) in Chesapeake Bay. *J. Fish Dis.* 3:387–98.

Ventura, M. T., and I. Paperna. 1985. Histopathology of *Ichthyophthirius multifiliis* infections in fishes. *J. Fish Biol.* 27:185–203.

10 The Eye

By B. P. Wilcock and T. W. Dukes

Introduction

The paucity of information on diseases of the fish eye results, not from any lack of ocular disease, but rather from the rarity of people who combine a professional interest in fish with a measure of competence in ophthalmology. Furthermore, since the fish eye represents an evolutionary pathway quite distinct from that of man, it has not benefited, as have domestic mammals, from the spillover of human biomedical research. This evolutionary path has resulted in an eye no less remarkable than that of mammals, but one nevertheless that cannot be treated as a stepping stone to the mammalian eye. Instead, the eyes of fish have adapted to each species' environmental niche with seeming disregard to phylogeny, for the eyes of quite unrelated species inhabiting the same niche are often more similar than the eyes of closely related species inhabiting very different environments. Indeed, one can predict a great deal about the lifestyle of a species by examining its eye, and perhaps the best evidence for the importance of vision for most fish is that the eye has been adapted to extremes of environment far in excess of those experienced by terrestrial vertebrates.

Interested readers should consult the remarkable text by Walls (1967), which despite its age remains the pillar of comparative ophthalmology. The scope of this chapter is restricted to diseases of the eyes of those species of teleost fish that are important as food, game, or aquarium species. Even in this group there is wide variation in morphology, and the unwary investigator will be prone to confuse lesions with normal structures. For this reason it is appropriate to discuss the normal teleost eye in some detail and to consider the reactions of its various parts to injury before beginning a detailed account of specific ocular diseases. It is also appropriate to mention that, as with mammals, autolysis of various ocular structures is rapid. The need for rapid fixation in acid fixatives (Bouin's fluid is the most widely used) is paramount if one wishes to make credible histologic interpretations, particularly of the retina.

Normal Teleost Eye and Responses Injury

The typical teleost eye (Fig. 10.1) is a sphere that has been flattened in its antero-

B. P. Wilcock, B.A., D.V.M., M.S., Ph.D., is an associate professor in the Department of Pathology, Ontario Veterinary College, University of Guelph. T. W. Dukes, D.V.M., M.S., is a pathologist with Animal Disease Research Institute, Agriculture Canada, Ottawa.

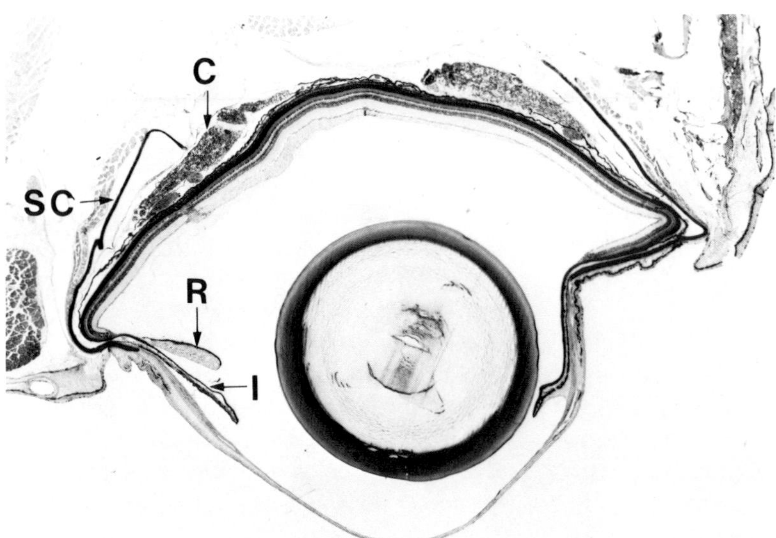

Fig. 10.1. Structure of typical teleost eye. Note large lens, delicate iris, and absence of ciliary apparatus. (*I* denotes iris; *R*, retractor lentis muscle; *C*, choroidal rete; *SC*, scleral cartilage)

posterior (cornea-retina) axis by loss of corneal curvature. In terrestrial vertebrates, the strongly curved cornea is a major factor in refraction, but in an aqueous environment, the refractive index of the cornea is so similar to that of the water that it is irrelevant as an optical surface. Fish corneas are not only flattened but may have contour irregularities that would give rise to incapacitating impairment (astigmatism) in land mammals. The cornea may become streamlined as an ellipse (head-to-tail) in fast-swimming predators. As usual, globe size varies with body size but is disproportionately large in deep-water species where dim light forces species to extremes if they wish to retain useful vision. These species not only have biochemical and histological adaptations of retina to maximize the effect of what light is available, but the large globe permits maximum enlargement of the retinal image for improved acuity. A clever adaptation in species attempting to retain a large globe in a small head is the development of tube-shaped eyes. Among terrestrial vertebrates, the owl has borrowed this adaptation, which, like a core tunneled from the middle of an apple, allows maximum focal length (and hence, image size) within a minimum of overall space. A few cave-dwelling and bottom-dwelling species apparently have given up the evolutionary struggle for vision and have sightless, vestigial eyes.

The globe is held within the orbit by three pairs of striated muscles similar to those of mammals, but fish eyes possess only involuntary ("gyroscopic") movement. Tracking of objects by sight is done by changing body position to "aim" the eyes. In at least a few species (marlins, sailfish, spearfish, butterfly mackerel) the superior rectus muscle is regionally altered to create an "eye heater," presumably to warm the eye and adjacent brain during deep dives. A portion of the muscle is modified by the loss of most myofilaments and the addition of numerous mitochondria and profiles of smooth endoplasmic reticulum. Grossly it resembles liver more than muscle. Its vasculature is also modified into a rete mirabile from the internal carotid artery, and the warmed rete blood draining from the heater tissue then supplies the uvea.

Eyelids. Eyelids and adnexa such as lacrimal apparatus are absent in most fish, simply because they are not needed in an aqueous environment (as Walls [1967] points out, such structures in terrestrial animals reflect efforts to carry the sea onto dry land). The periorbital skin transforms directly into cornea without intervening conjunctival sac, although in some species there is a slight folding of the skin to create thin immovable vertical "lids" and a shallow equivalent of the mammalian conjunctival sac.

Sclera. The sclera is similar to that of other vertebrates and consists of laminated fibrous tissue reinforced by hyalin cartilage. In some species the anterior rim of the sclera is further reinforced by plates of bone, the scleral ossicles. Scleral thickness varies with globe size. Increased thickness in deep-water species is not necessary because, with increasing depth, internal ocular pressure rises in proportion to the external pressure to prevent collapse of the globe.

Cornea. The typical teleost cornea resembles that of birds and mammals. The epithelium is nonpigmented stratified squamous, nonkeratinizing, and rests on a prominent basement membrane. The epithelium represents a larger percentage of overall corneal thickness in teleosts than in mammals, although even in teleosts there is wide variation. In general, it is much thinner in marine species than in freshwater fish (approximately 15% as opposed to 40% of overall corneal thickness). The stroma varies widely in its thickness, but it is lamellar, cell poor, and avascular as are other vertebrate corneas. There is a distinct and often thick, hyalin, eosinophilic zone of subepithelial stroma that may be analogous to Bowman's layer of primates, a layer that is inconspicuous or absent in domestic mammals. Descemet's membrane and corneal endothelium are as in mammals, although neither has received detailed study in fish. In some species the endothelium, Descemet's membrane, and innermost stroma are seen as a more or less distinct layer, the autochthonous ("locally created") layer, so named to denote our ignorance of its origin. In most commercial species (salmonids, tilapia, catfish) the endothelium arises from a multilayered mass of spindle cells external to the ciliary cleft. The same cells spread across the ciliary cleft to line the anterior face of the iris, forming a "bracket" as if to stabilize the iridocorneal angle. This bracket, called the annular ligament, varies considerably in its magnitude but seems to be a variation on the pectinate ligament of domestic mammals. Its function is unknown, as is the developmental relationship between it and the autochthonous layer (Fig. 10.2). The name ligament is a misnomer inasmuch as it does not stain with Masson's trichrome stain for collagen, nor for elastin.

The existence of three distinct corneal layers in fish mirrors corneal embryogenesis in mammals, in which the cornea is formed by clarification of supraoptic epidermis combined with two waves of invasion by perioptic mesenchyme to form the corneal endothelium and stroma (analogous to the scleral and autochthonous stromal layers of fish?).

This dual origin from epidermis and mesenchyme is still obvious in primitive fish such as lampreys in which the corneal stroma (sclera) is only loosely affixed to the corneal epithelium (skin) by a myxoid extracellular matrix. In many teleosts, the fusion of the scleral and cutaneous portions is more or less complete and results in a cornea histologically similar to that of mammals (salmonids, minnows, carp, pike). The imperfect fusion of scleral and cutaneous portions reappears, inexplicably, in many higher species, particularly bottom dwellers such as flounder, hake, halibut, and burbot. The layering is accentuated by artifacts of histological preparation, and in these eyes the cornea is often separated into a superficial half consisting of epithelium and stroma (secondary spectacle or modified skin) and a more deeply staining compact scleral portion (Fig. 10.2C). In mammals the contrast between superficial and deep stromal layers is

Plate 1. Some common eye diseases.

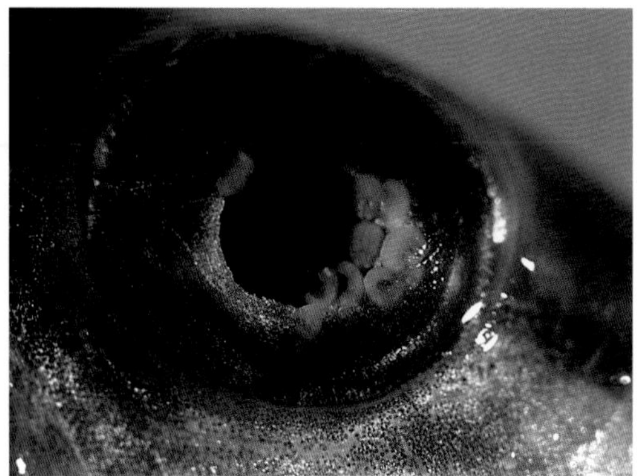

A. Trematode metacercariae in anterior chamber of cichlid.

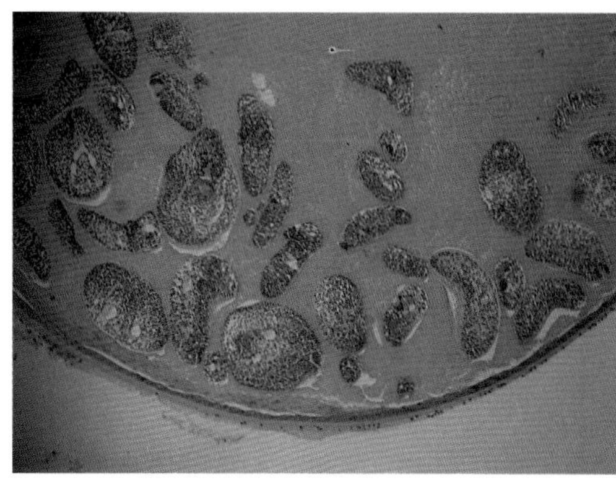

B. Numerous metacercariae in the liquified lens cortex of rainbow trout.

C. Cataract in rainbow trout with zinc deficiency.

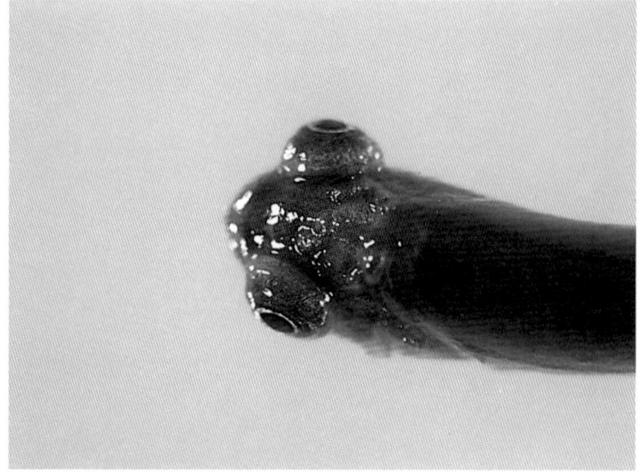

D. Exophthalmos (gas bubble disease) in young rainbow trout.

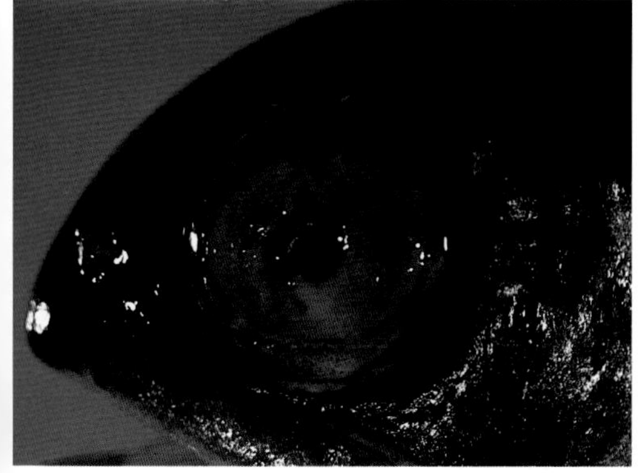

E. Anterior chamber hemorrhage typical of endophthalmitis.

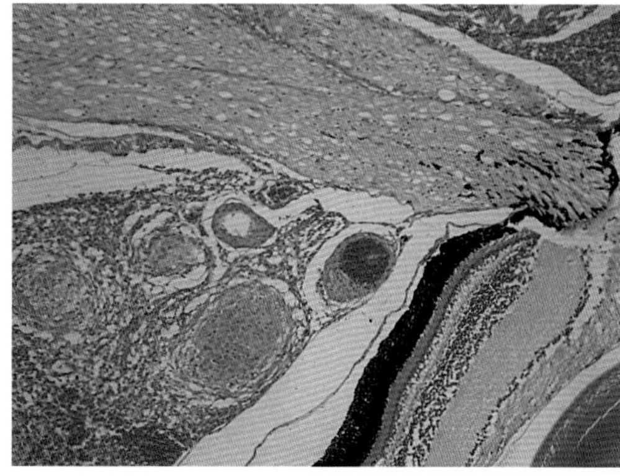

F. Choroidal granulomas (mycobacterial) in Siamese fighting fish.

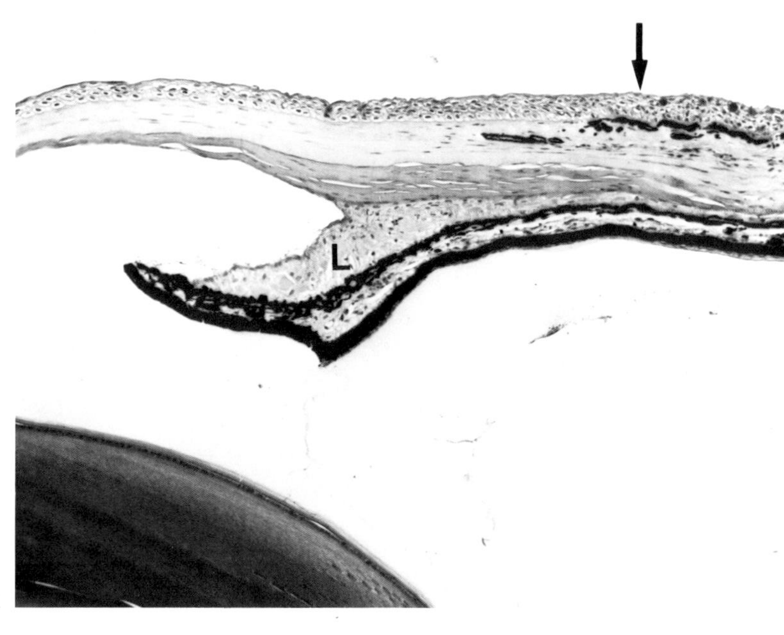

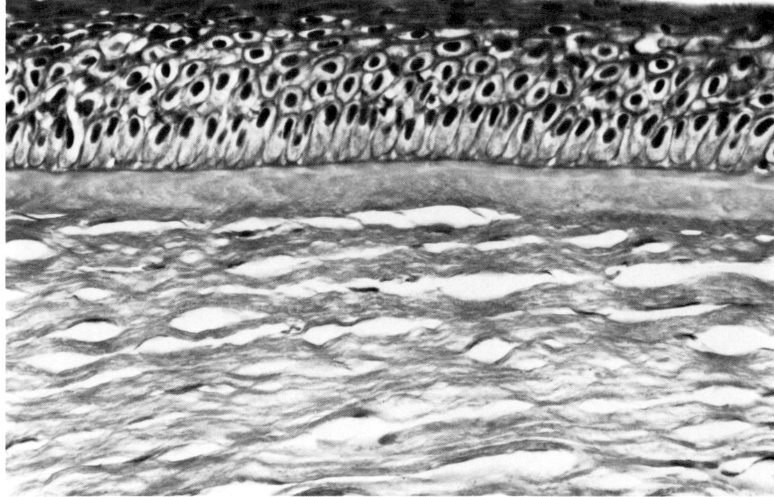

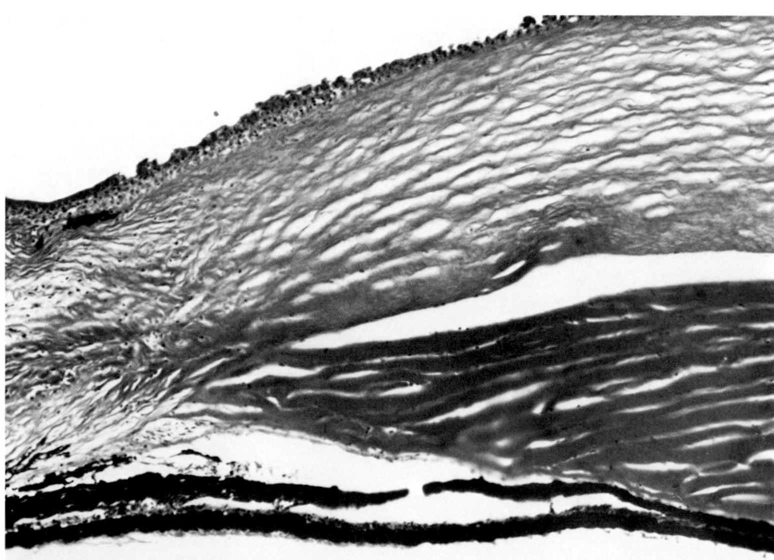

Fig. 10.2. A. Smooth transition from skin to cornea without intervening conjunctiva, typical of teleosts. *Arrow* marks the approximate site of transition from epidermis to corneal epithelium. The annular ligament (*L*) is conspicuous. **B.** Thick stratified squamous epithelium typical of freshwater teleosts. The superficial stroma in this specimen is condensed and resembles Bowman's layer of primates. The stroma resembles that of birds and mammals. **C.** Acute corneal ulceration results in edema of the dermal half of the stroma. The cleft between superficial (dermal) and deep (scleral) stroma is a common artifact.

occasionally made visible by stromal edema. The contrast is thought to result from differences in the water-binding capacity of stromal ground substance between superficial (chondroitin sulfate) and deep (keratan sulfate) stroma, which may, in turn, be vestiges of the cornea's evolutionary origins.

Little is known of corneal physiology in teleosts. In terrestrial vertebrates the cornea is highly organized to passively (hydrophobic epithelium) or actively (corneal endothelium) exclude water in order to maintain the dehydration critical to corneal clarity. In teleosts the epithelium seems to be the major determinant of corneal dehydration, although the endothelium is assumed to play a minor role.

The general pathology of the fish cornea is largely unknown. The mammalian cornea, on the other hand, has been extensively studied as a model for inflammation, edema, wound healing, and immune rejection. Without evidence to the contrary, we must assume that the mechanisms of corneal inflammation, degeneration and repair resemble those of birds and mammals. Loss of corneal epithelium in a variety of marine and freshwater species results in transient corneal edema accompanied by epithelial sliding and, finally, epithelial replication in a manner similar to that seen in mammals. The cornea of freshwater species develops more severe edema than that of marine species, presumably the result of the greater osmotic difference between corneal stroma and the surrounding freshwater environment. Freshwater species also develop cataract that resolves shortly after the disappearance of the corneal edema, the latter requiring the reestablishment of an intact epithelium (4 to 7 days after diffuse ulceration). The cataract is assumed to result from osmotic dilution of the aqueous humor, with lens epithelium unable to maintain lens deturgescence against an unusually steep osmotic gradient. In specimens from naturally occurring disease, changes of epithelial or stromal edema, epithelial ulceration, epithelial repair by sliding and hyperplasia, and chronic irritative changes of epithelial hyperplasia, pigmentation and goblet cell metaplasia (i.e., corneal epidermalization) occur much as they do in mammals (Fig. 10.3). Corneal vascularization seems to occur only in the most chronic and severe cases. Artifacts caused by autolysis, rough handling, or improper histological technique are common. Focal loss of corneal epithelium (especially in marine fish), diffuse loss of corneal endothelium, and splitting of corneal lamellae are especially frequent.

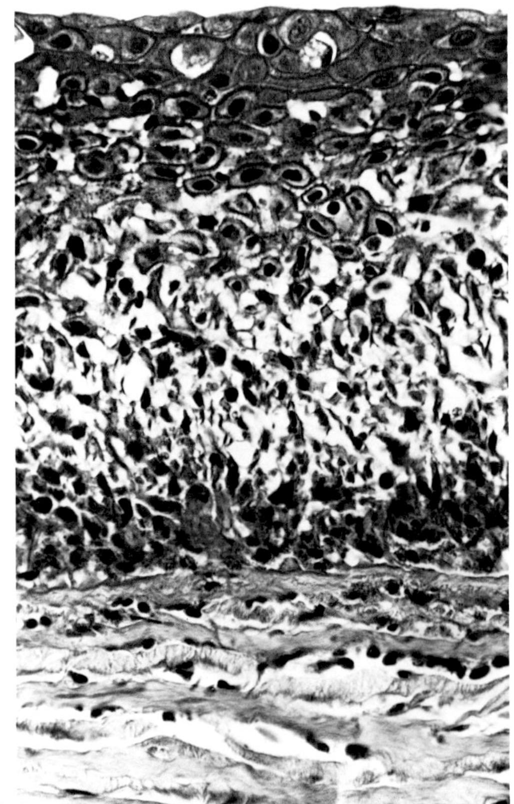
A

Filtration angle. The filtration angle occupies the iridocorneal angle. It is filled with a loose meshwork of mesenchymal-appearing cells and is separated from the anterior chamber by the laminated structure of the annular ligament (Fig. 10.4). The subject of aqueous production and drainage has only recently been investigated. Production cannot be from ciliary epithelium as in mammals because none exists in fish, but the posterior layer of epithelium overlying the root of the iris has ultrastructural features suggesting that it may be the site of aqueous secretion. The route of aqueous drainage is unknown. Glaucoma, if it occurs in teleosts, has not been documented.

Lens. Large and spherical, the teleost lens bulges into the anterior chamber, sometimes touching the cornea. It is minimally elastic compared to the mammalian lens. Be-

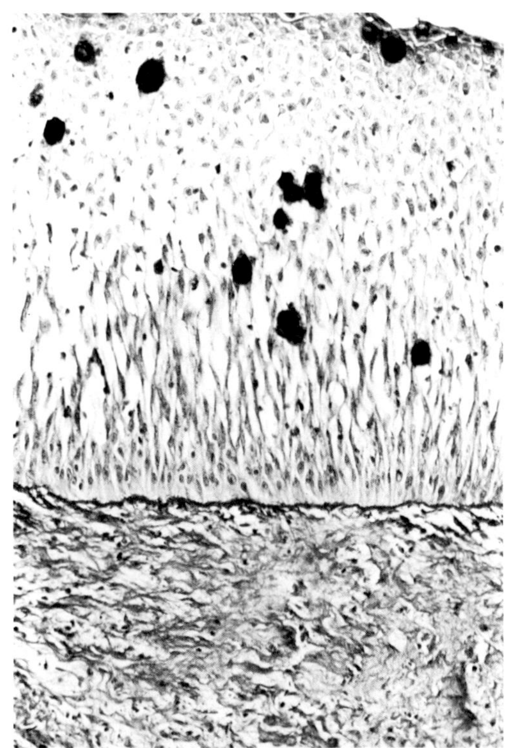

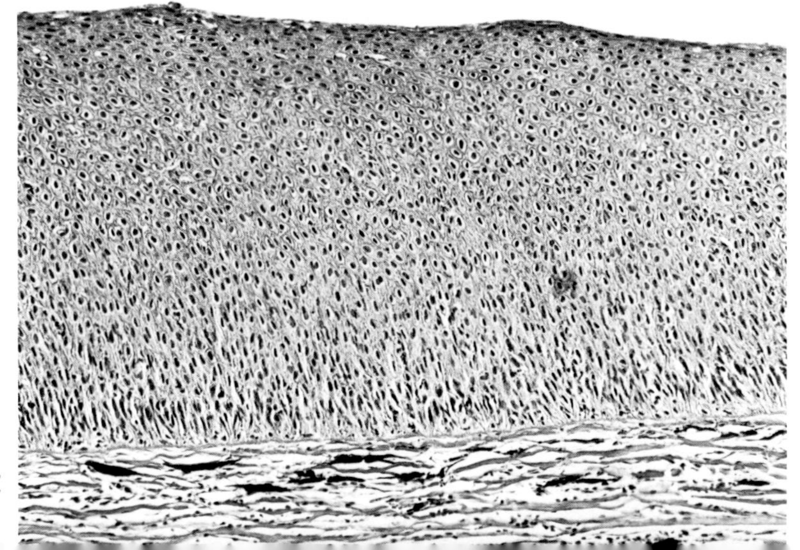

Fig. 10.3. A. Intraepithelial edema in white sucker cornea. Such changes usually progress to ulceration. **B.** Chronic keratitis in white sucker. Epithelium is hyperplastic and has acquired goblet cells, presumably in response to chronic irritation. Underlying stroma is edematous and coagulated. **C.** Epithelial hyperplasia and chronic superficial stromal keratitis in white sucker. Melanin-laden macrophages are numerous.

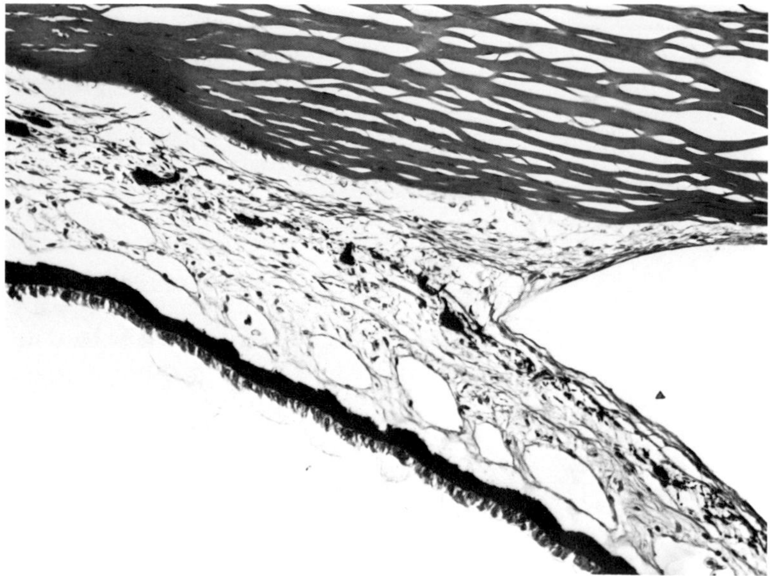

Fig. 10.4. Corneoscleral angle (filtration [?] angle) containing stellate spindle cells that blend with annular ligament as well as corneal endothelium. Drainage route for aqueous humor in fish is unknown. Note the prominent iris vasculature that is susceptible to hemorrhage.

cause of its rigidity, accommodation (focusing) cannot occur by changes in lens shape subsequent to the contraction or relaxation of the ciliary muscle as in mammals; the ciliary muscle itself is absent or vestigial. Instead, teleosts focus by changing the position of the lens along the visual axis. The lens is suspended within the pupil by a dorsal suspensory ligament, and its position is adjusted by the action of a retractor lentis muscle. This muscle originates from the ventral floor of the globe at the anterior end of the falciform process (i.e., near the junction of retina with iris). It has a central vascular stalk and is ensheathed in cuboidal neuroectoderm continuous with the sensory retina. The muscle fibers themselves seem to derive from the retinal pigment epithelium (Fig. 10.5). This same epithelium in mammals is known to retain its ability for pleuripotential metaplasia (fibroblastic, phagocytic, osteocytic) and, in the iris, gives rise to the dilator muscle fibers. The muscle inserts on the ventral aspect of the lens capsule anterior to the equator. This arrangement permits the lens to swing in a hingelike fashion within the visual axis, altering either the lens-retina distance or changing the position of the lens within the pupillary aperture. Even slight movement may be sufficient for useful focusing inasmuch as the fish lens, unlike that of mammals, is not optically uniform, and its refractive power increases significantly from the center to the periphery of the lens.

The lens is histologically similar to that of mammals. It is surrounded by a hyalin, collagenous capsule that is the basement membrane of the underlying layer of simple cuboidal epithelium. The epithelium is continuous over the anterior one-half to two-thirds of the lens, but posterior to the equator the epithelial cells migrate into the lens cortex (nuclear bow). There is thus no epithelium on the posterior pole of the lens. As in other vertebrates, the lens substance is exclusively epithelial and consists of interlocking elongated fibers derived from the nuclear bow.

The general pathology of the lens is similar to that of mammals. Injury via trauma, malnutrition, or infectious agent results in hydropic swelling of lens fibers, lysis of fibers, and attempted fiber regeneration. Regeneration is probably never successful, but nonetheless epithelial hyperplasia, cap-

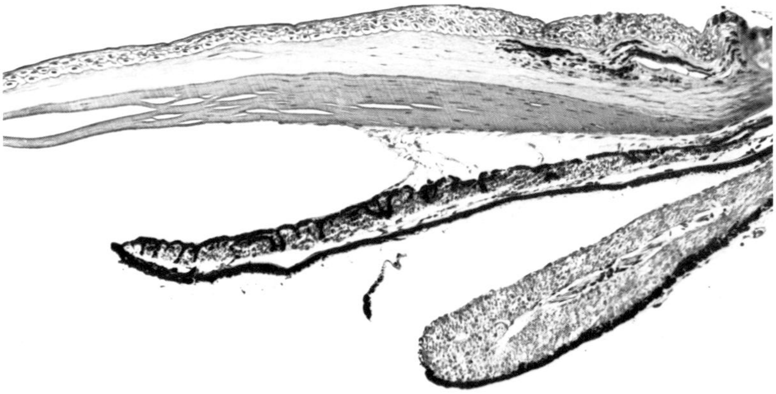

A

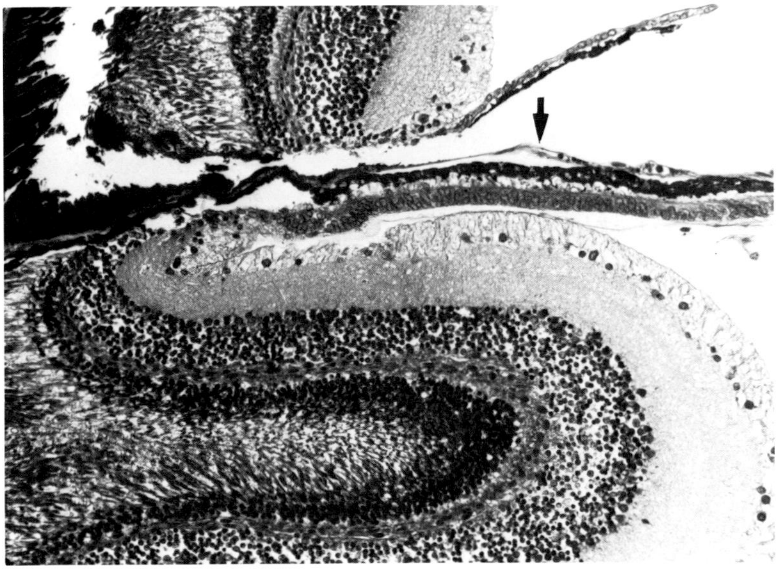

B

Fig. 10.5. A. Anterior segment of rainbow trout showing prominent bilaminated corneal stroma, thin iris, and tangential section of retractor lentis muscle en route to lens insertion. **B.** Retractor lentis merges with retina via stalk composed of vascular core (*arrow*) surrounded by inner pigmented and outer nonpigmented neurectoderm. Pigmented layer is continuous with retinal pigment epithelium, and nonpigmented layer arises from sensory retina. **C.** Pigmented and nonpigmented neurectoderm seem to give rise to muscle of retractor lentis. Continuity of the epithelium with the retina, and its metaplasia to form muscle fibers, is analogous to the epithelium and dilator muscle of the mammalian iris.

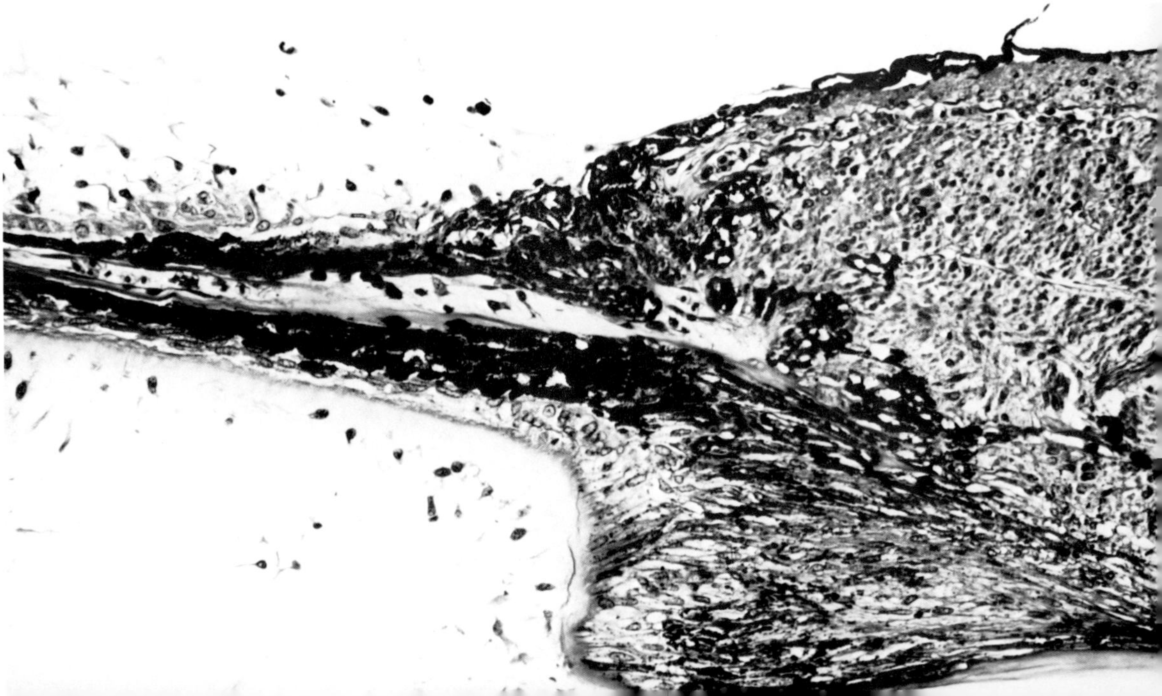

sular reduplication, and sometimes massive intralenticular migration of surface epithelium are all frequent changes (Fig. 10.6). Epithelial proliferation may be particularly extensive following rupture of lens capsule, but the widely migrating epithelium retains its cuboidal character rather than assuming fibroblastic morphology as occurs with rupture of mammalian lenses.

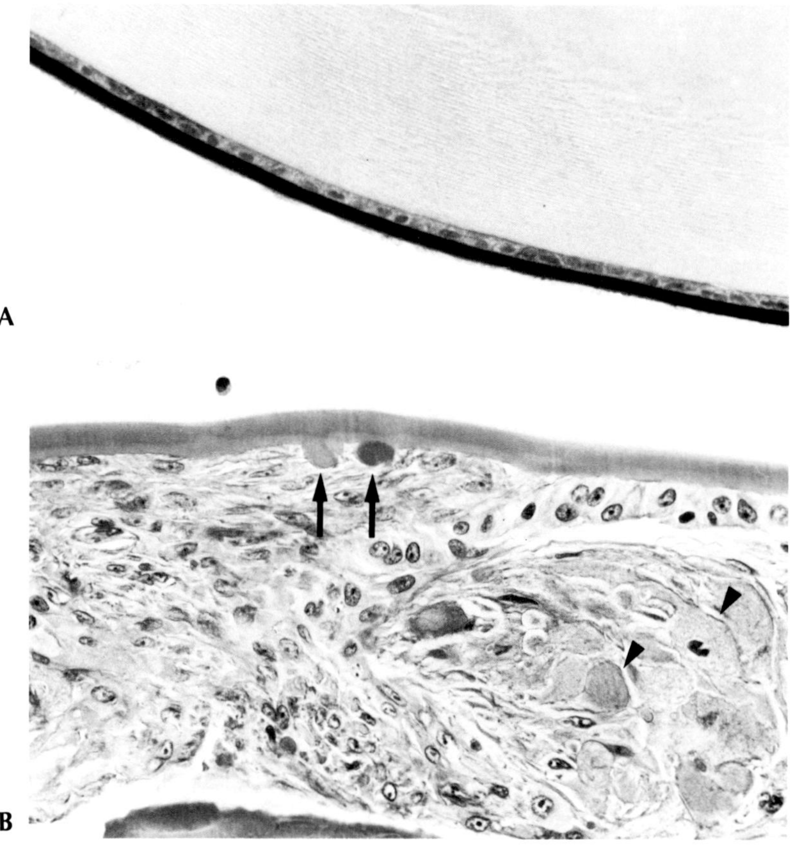

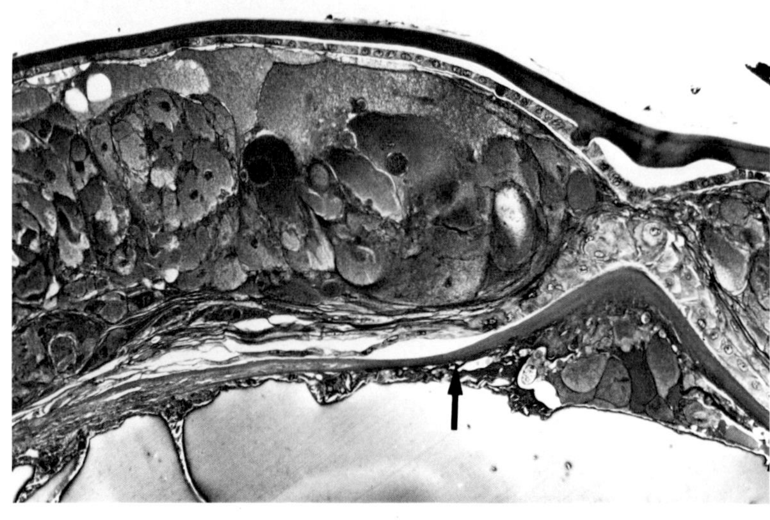

Fig. 10.6. A. Anterior cortex of normal lens with PAS-positive capsule and uniform cuboidal lens epithelium. **B.** Anterior polar cataract in trout with suspected riboflavin deficiency. There is focal capsular reduplication (*arrows*), massive aberrant epithelial proliferation, and formation of bladder cells (*arrowheads*). **C.** Capsular reduplication (*arrows*) in trematode-induced cataract. Original capsule is to the top. Bladder cells are abundant.

Uveal tract. The uveal tract of teleosts contains several unique structures and lacks several important elements of the mammalian uvea.

IRIS. The iris is much more fragile appearing than its mammalian counterpart. It is formed by a posterior bilayer of neuroectoderm similar to that of mammals, but the thin stroma contains neither dilator nor constrictor muscle. The teleost iris is thus immobile. Pupillary mobility is an adaptation to rapid changes in light intensity required by terrestrial animals but not by most fish. Whether species such as reef fish, which seemingly would benefit from a mobile pupil as they dart from brightness to darkness, have developed iris muscles or some other form of rapid light-to-dark adaptation has not been reported. The stroma contains blood vessels, variable numbers of melanin-laden cells, and in some species an anterior stromal layer of guanophores that appear as a nonstaining refractile yellow band that is birefringent with polarized light. This layer is responsible for the silvery reflective iris of these species, although its evolutionary purpose remains obscure. Embryologically, it is probably the anterior persistence of the choroidal argentea (see *choroid*). The anterior border of the iris fuses with the laminations of the annular ligament (Fig. 10.7).

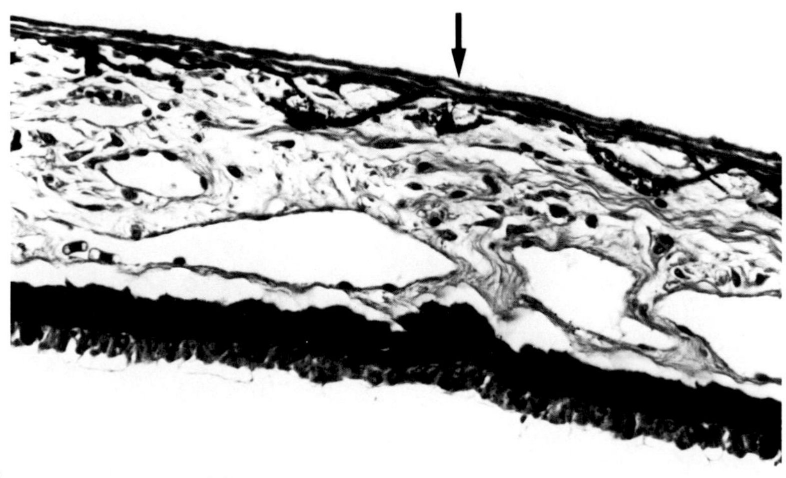

Fig. 10.7. A. High magnification of iris with bilayer of posterior neuroectodermal epithelium, heavily vascularized stroma devoid of muscle, and thin condensation of annular ligament (*arrow*) forming its anterior (corneal) border. (See also Fig. 10.4.) **B.** Corneal reflective guanophores continuous with the choroidal argentea. Both are accentuated by viewing with polarized light.

CILIARY BODY. The ciliary body is, at best, rudimentary, and there are no ciliary processes nor muscle fibers. Iris epithelium merges directly with retina in a manner similar to the junction of ciliary epithelium with retina at the ora ciliaris of mammals. The posterior (nonpigmented) epithelium becomes sensory retina and the anterior (pigmented) epithelium continues as the retinal pigment epithelium.

CHOROID. The stroma of the iris continues caudally as the choroid. As in other vertebrates, the choroid is the vascular tunic sandwiched between the (outer) sclera and (inner) retina. It consists, as in other vertebrates, of blood vessels, connective tissue, and elongated melanocytes. It differs from the choroid of domestic mammals in two major respects: it contains no tapetum but does contain, in many teleost species, a large and highly structured vascular plexus variously termed *choroid gland* or *choroidal rete.* In histological section, it is usually oriented as a horseshoe "hung" vertically over the optic nerve. It is located near the posterior pole of the eye, and it appears as an abrupt transition from dilated, loosely oriented choroidal vessels to become multilaminated, parallel vascular tubules reminiscent of an automobile radiator (Fig. 10.8). This arrangement suggests that the choroidal rete acts as a countercurrent exchanger, and although a supply of oxygen to the retina is proposed, precisely what is exchanged remains in doubt. The rete reportedly exists only in species with a pseudobranch, and among commercial species is therefore absent in catfish.

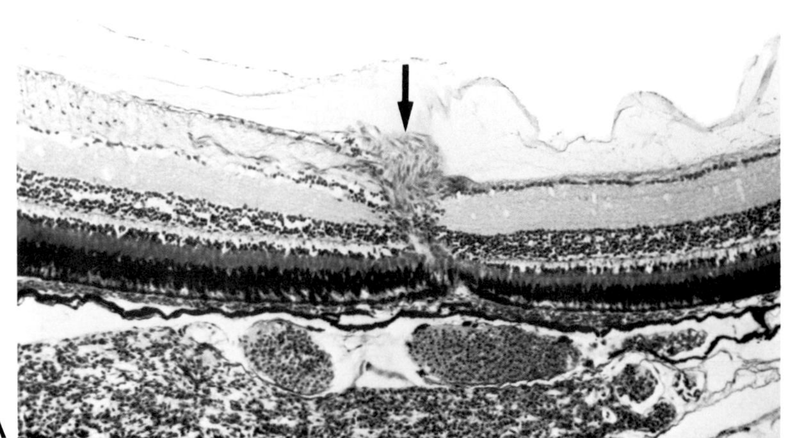

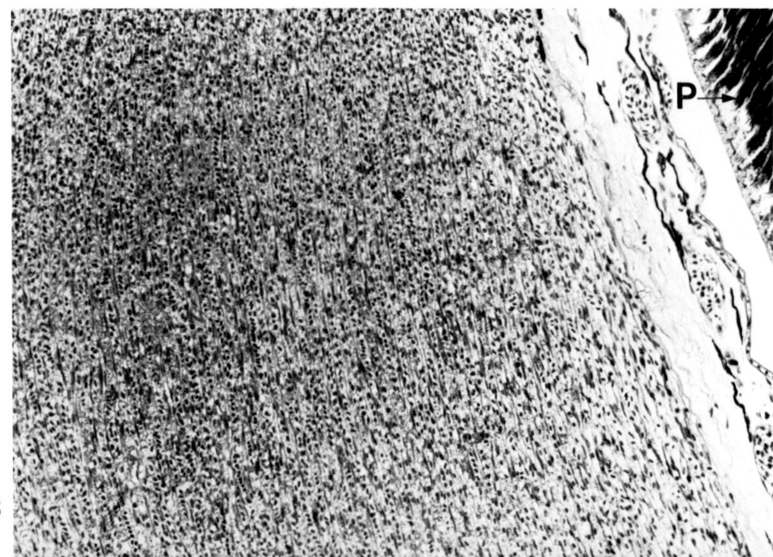

Fig. 10.8. A. Posterior pole of trout eye showing optic papilla (*arrow*) and vascular arcades of choroidal rete. **B.** Higher magnification of choroidal rete. Tightly packed parallel capillaries are typical. Photoreceptors (*P*) are to the *right,* scleral cartilage to the *left.*

ARGENTEA. The argentea is a silvery yellow reflective layer in the outer choroid, adjacent to sclera. It may also continue anteriorly as the reflective layer in the anterior iris stroma, although it is by no means clear whether the two can exist independently or whether, for example, a species can have a reflective iris yet lack a choroidal argentea. The argentea is claimed to be frequently present in fish larvae, but it is not as frequent in adults. In the transparent larvae it may function to hide the choroidal melanin (however scant) and thus improve the larva's camouflage. In some of the species known to have argentea as adults (*Scorpanidae, Clinidae, Batrachoidae*), there is no tapetum and little choroidal melanin. It is the argentea that gives these species "eyeshine." The argentea may have some function in vision enhancement, but the popular theory is that the silvery eyeshine helps to hide the dark pupil and thus reduce the chances of the pupil serving as a target for predators.

TAPETUM LUCIDUM. A tapetum lucidum of some kind exists in most teleosts, but it appears that there are numerous variations in location and structure. Some of the described variation may result from imprecise descriptive terminology related to "retinal" tapeta and "choroidal" tapeta, which on closer inspection of photographs appear to reside in retinal pigment epithelium. Elasmobranchs have a truly choroidal tapetum somewhat similar to the fibrous tapetum of ungulates, but the majority of tapetum-bearing teleosts have the reflecting substance within the cytoplasm of retinal pigment epithelium itself. The presence of a tapetum and the extent of its development seem to be rather sensibly related to each species' visual needs. It is best developed in species feeding in dim light and absent in those inhabiting complete darkness. The fish tapetum, as in its mammalian counterpart, acts to reflect incident light, thereby doubly stimulating the photoreceptors and improving the retinal absorption of light by up to 50%. Because fish lack pupillary mobility, however, species with efficient tapeta may suffer light-induced photoreceptor damage. Species that, in their natural habitat, feed in bright waters have developed one or both of two unique protective mechanisms: occlusible tapeta and photoreceptor motility (see this latter mechanism later under *retina*).

Occlusible tapeta probably come in many designs. The best known is that in which the reflecting particles (triglyceride or guanine) are intermixed with melanin granules within the cytoplasm of the retinal pigment epithelium (RPE). When the melanin granules migrate within the cytoplasm in response to light (basal in dim light, apical in bright), they serve to mask or unmask the tapetal granules.

The structure of tapetal granules has received scant attention. Guanine crystals are probably the most frequent reflector, although a number of common species (snook, sea trout, carp, cusk eels, weakfish) have tapetal granules of almost pure triglyceride. Among terrestrial vertebrates, the opossum has a lipid-rich tapetum, albeit cholesterol rather than triglyceride.

FALCIFORM PROCESS. The falciform process is the final choroidal specialization unique to fish, although the avian pectin is probably a close relative. The process is a ridge of heavily vascularized mesenchyme protruding into the vitreous from an axially oriented cleft in the ventral floor of the globe (the embryonic cleft or fissure). It is thus similar to the mammalian primary vitreous, although the latter regresses during embryogenesis to be replaced by intrinsic retinal vasculature. The vascular network persists in fish to nourish the avascular retina, presumably by diffusion of nutrients through the vitreous. Alternatively, the vasculature of some species resembles that of the mammalian eye during a later stage of its organogenesis and has a preretinal (intravitreal) vascular network resembling the mammalian hyaloid artery system (Fig. 10.9).

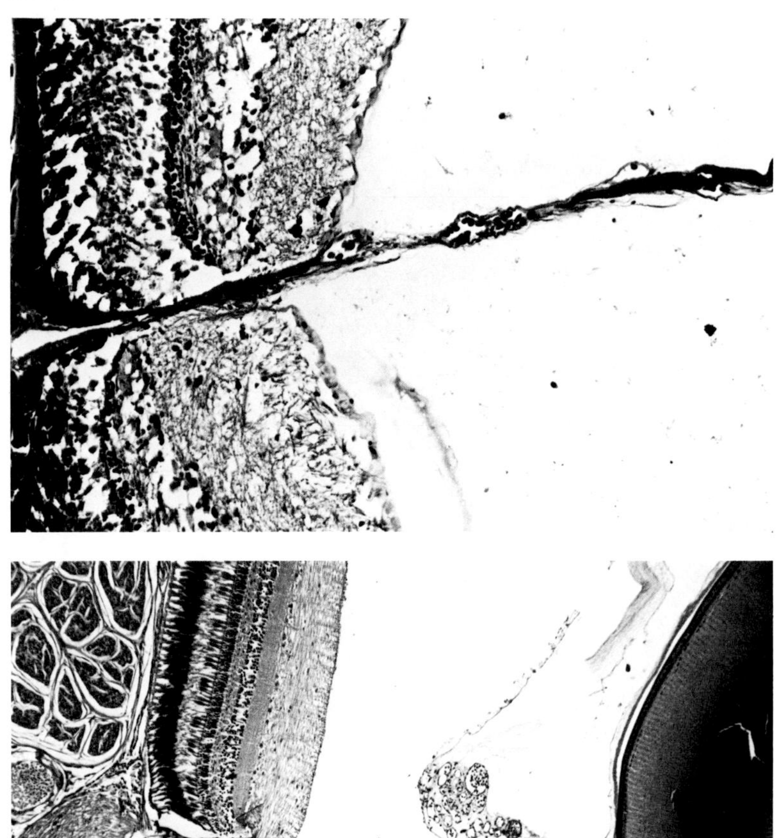

Fig. 10.9. A. One of many variants of falciform process, illustrating its continuity with choroidal vasculature. B. Cross section of falciform process within vitreous cavity between optic nerve and lens. Some species have no discrete falciform process and have, instead, a preretinal vascular plexus (see Fig. 10.10C).

UVEAL PATHOLOGY. Uveal general pathology is largely unexplored. As in mammals it is a frequent site for hematogenous or transscleral lodgment of infectious agents, but in fish these microbial, protozoan, and metazoan agents often produce remarkably little host response. Necrosis and granuloma formation are most frequent, but common mammalian reactions of serous retinal separation or metaplasia of adjacent RPE have not been described in teleosts. Similarly, the dreaded sequels of synechiae and secondary glaucoma have not been described despite ample opportunity for their occurrence. Infiltration by leukocytes, particularly lymphocytes, is quite common, but seldom is the etiology determined (Fig. 10.10). The choroidal rete seems to be a particularly sensitive target. Massive infiltrates are seldom seen in the iris, which instead seems prone to massive vascular engorgement. This engorgement, with presumed subsequent rupture, may explain the observation of hyphema as a more frequent macroscopic observation in fish than in mammals with uveitis.

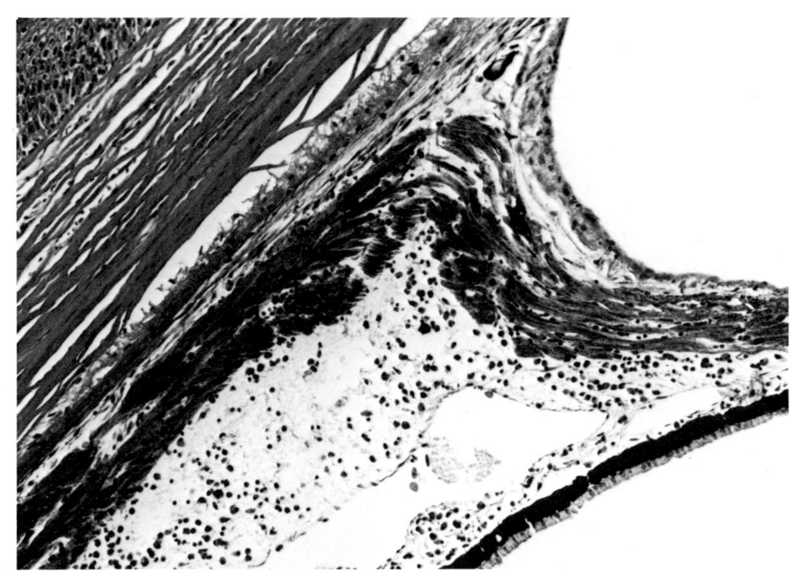

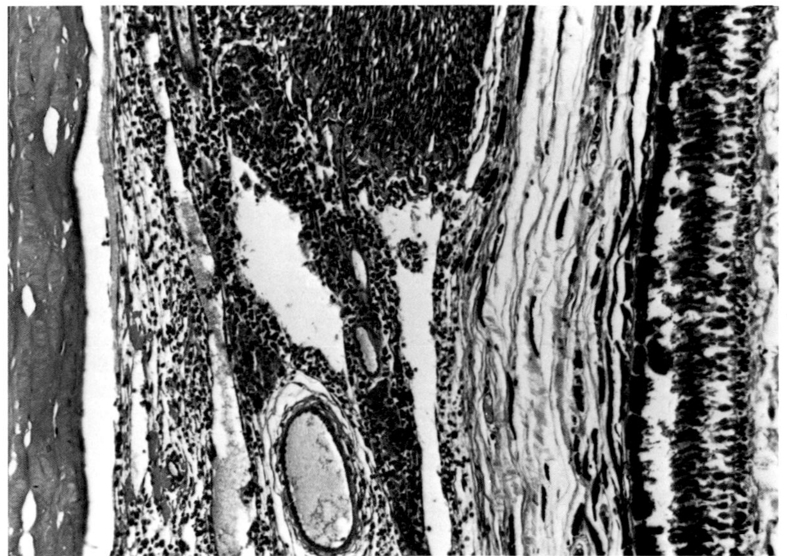

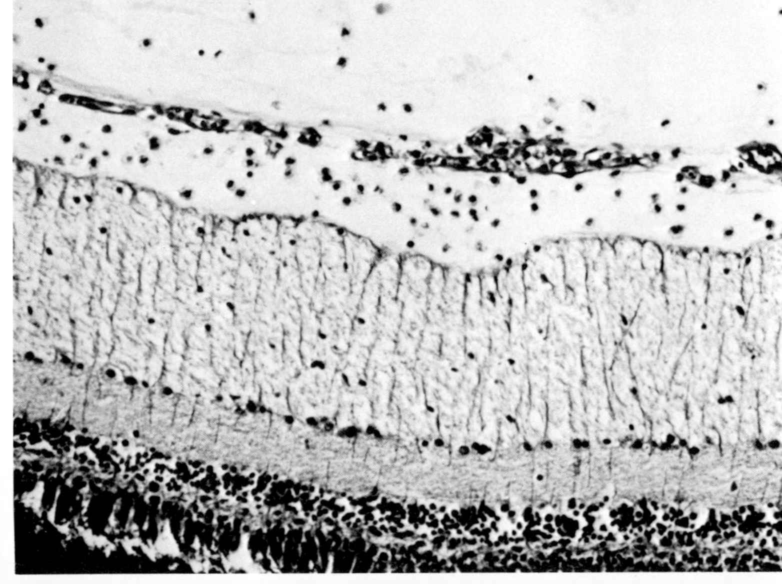

Fig. 10.10. A. Root of iris distended with edema and leukocytes in acute diffuse uveitis in white sucker. **B.** The choroidal rete seems a particular target for embolic lodgment. Acute neutrophilic choroiditis in a white sucker. **C.** Preretinal vascular plexus in an oscar accentuated by inflammation, part of a diffuse uveitis.

Retina. The teleost *retina,* in contrast to all other parts of the eye, has been subjected to extensive structural and physiological investigation. Other than its previously mentioned avascularity, it resembles that of mammals and birds in the number, naming, and location of its various layers. Departures from the avian and mammalian model, however, are several. The retina of at least some teleosts continues to grow by neuronal addition throughout life, in sharp contrast to the postmitotic retina of mature birds or mammals. The well-defined single optic disc of mammals is seen in fish as one or more "discs" that vary in size, shape, number, and location, depending on the species. The most frequent form is a ventral-posterior slitlike "disc" at the posterior termination of the falciform process, but multiple exit points occur in some species. Bullheads reportedly have five such optic papillae, with some African catfish (*Mochokidae*) having twenty-five or more. All apparently converge to form a single retrobulbar optic nerve.

The photoreceptor layer has been extensively studied and is superbly adapted to each species' environment, both in terms of its morphology and in the spectral sensitivity of its photopigments (Ali 1974; Ali and Anctil 1976; Levine and MacNichol 1982). It is in this layer that we see the greatest interspecies variation and the greatest departure from the mammalian model (Fig. 10.11).

A
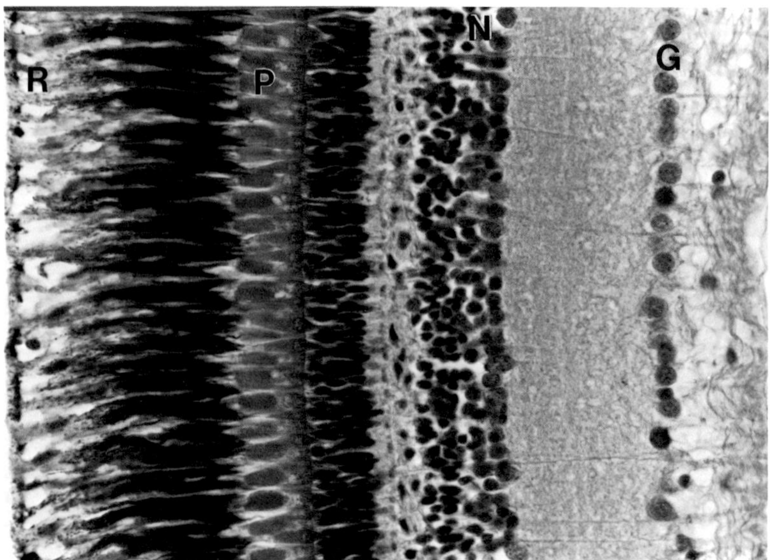

B
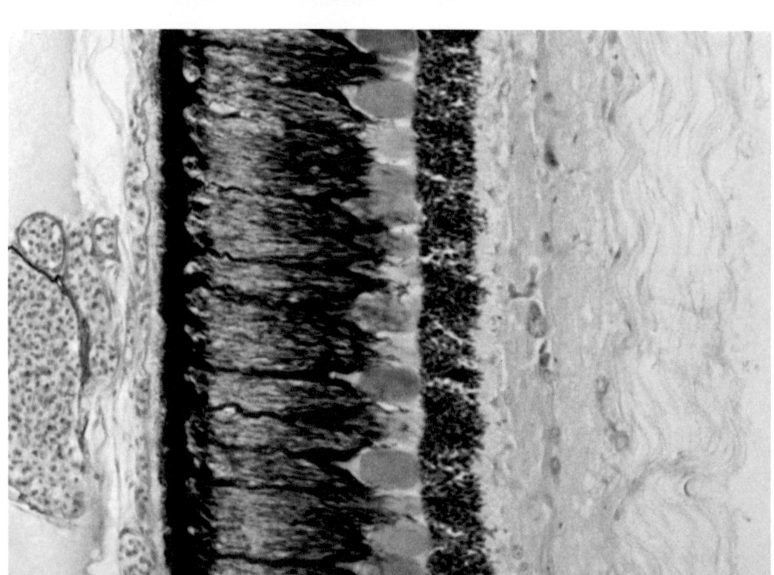

Fig. 10.11. Retinal morphology provides many clues about habitat and behavior of teleost species.
A. Oscar. Numerous photoreceptors (*P*) include both rods and cones. Combined with thick inner nuclear layer (*N*) and ganglion (*G*) cell layer, this retina is built for good color perception and visual acuity in brightly lit waters. Most aquarium and game species have this "typical" teleost retina (*R* denotes retinal pigment epithelium)
B. Pacific cod. Elongated photoreceptors, thick outer nuclear layer, and very sparse inner nuclear and ganglion cell layers are typical adaptations for visual sensitivity in dim waters at the expense of acuity.

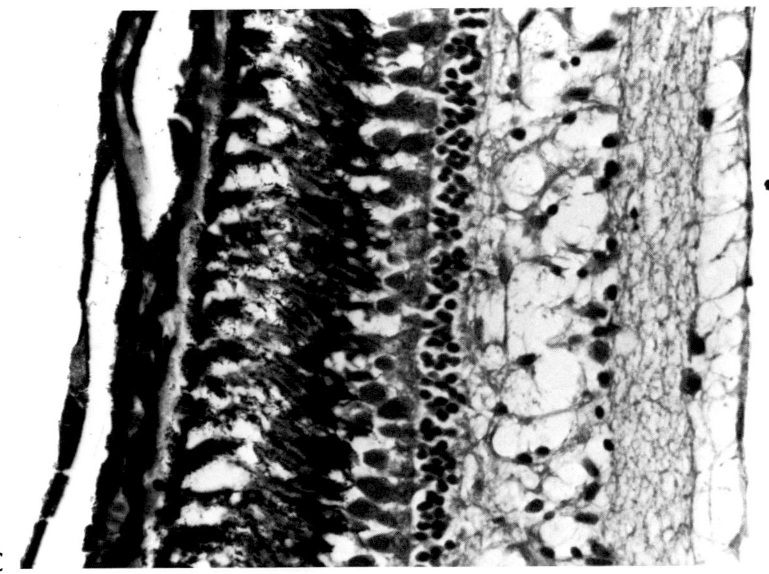

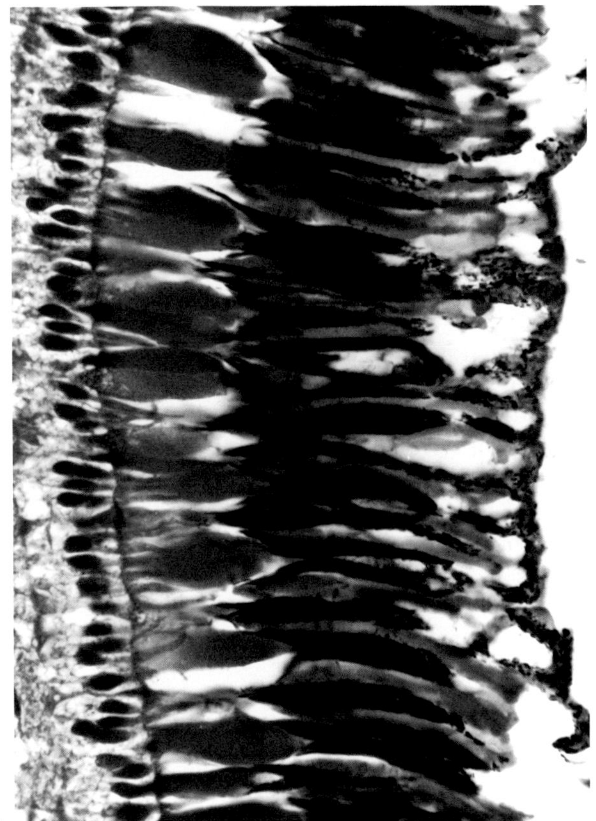

C. White sucker. Poorly developed retina appropriate to this bottom-feeding species that probably relies on tactile and olfactory stimuli rather than on vision. **D.** High magnification of grayling retina showing blunt, fat cones, and long slender rods. The position of the pigmented processes of the retinal pigment epithelium suggests this eye is from a fish caught in daylight.

Deep-sea (bathypelagic) species have disproportionately large eyes to enhance image size, well-developed tapeta and photoreceptors that are exclusively rods. Their golden photopigments, chrysopsins, are most sensitive at the monochromatic wavelengths of light (blues in clear water, greens in slightly turbid water) that dominate at depths greater than 25 meters. The struggle for vision under dim light has also resulted in extraordinarily long photoreceptors or in multiple layers of photoreceptors to maximize the amount of visual pigment in each area of retina. Such retinas often have a markedly attenuated ganglion cell layer in order to maximize the number of rods converging to trigger each ganglion cell. At such depths visual sensitivity is more precious than acuity.

The photoreceptors of species inhabiting shallower, brighter waters have great variation in morphology and density. In general, species that are nocturnal or that inhabit turbid water have rod-rich retinas. Diurnal species, especially if predacious, have cone-rich retinas. Surface-feeding, daytime predators have almost exclusively cones (pike, bass, groupers, barracudas). Cone morphology has some unique teleost features: single cones, twin cones, and double-to-quadruple cones are variations that must serve some, as yet undiscovered, purpose.

Species requiring vision in both bright and dim light have developed a unique retinomotor response in place of the mobile pupil and eyelids of mammals. This response, active in most sport and commercial species, controls the amount of light reaching the photoreceptors by one of two mechanisms. The more common method involves photoreceptor contractility, usually rods but sometimes cones as well. In response to changes in light intensity, the photoreceptor slides into or out of deep recesses among the pigmented processes of the RPE, thereby shielding itself (bright light) or fully exposing itself (dim light) for optimal visual function. A second method involves movement of pigment within RPE processes, so that the stationary photoreceptors can be either exposed or shielded in a manner already described for occlusible tapeta. The RPE pigment is all located basally in the cell body of the dark-adapted eye, while in the light-adapted eye the photoreceptors are enshrouded by pigment in the distal processes of the RPE. These retinomotor responses are slow, unlike the contraction-expansion of the avian or mammalian iris. The two hours or so that are required for the transition from light-to-dark (or vice versa) adaptation in fish is probably not critical in species accustomed to the gradual onset of dawn or dusk, but what additional adaptations must occur in species that move rapidly from shadows to bright light (as in reef fish) are as yet undiscovered.

RETINAL PATHOLOGY. Retinal general pathology has received scant attention. Artifacts are common and include diffuse retinal separation, photoreceptor autolysis, and edema within inner plexiform, ganglion cell, and nerve fiber layers. The retinal separation is usually between choroid and RPE so that the RPE remains attached to the photoreceptor layer. In mammals the RPE remains with the choroid, the separation occurring between the photoreceptors and the RPE. Depending on the state of light adaptation at the time of fixation, the photoreceptor layer will be either as visible as in mammals or shrouded in pigment. The morphology of the retinal layers is so variable among different species that a diagnosis of atrophy or dysplasia should be made cautiously and preferably after examining a healthy specimen of the same species. Most of these variations reflect habitat adaptations and make for fascinating study in evolution.

Retinal pigment epithelium undergoes hypertrophy, hyperplasia, fibrous metaplasia, and phagocytic activity, as it does in mammals, when stimulated by nearby retinal injury. Neural retina itself has been described as undergoing focal necrosis subsequent to metazoan parasite localization, but details of its reaction are scant. Its avascularity should make embolic injury rare but increase its susceptibility to irreversible ischemic injury. The absence of retinomotor response in deep-dwelling species should make light-induced retinopathy a frequent aquarium disease in these species, but there are no published descriptions. The apparent retention of mitotic capability by neuronal elements of retina should create unusual postnecrotic retinal reparative lesions, but again these are as yet undiscovered.

Specific Ocular Diseases

Keratitis. Corneal disease in fish results most often from physical or chemical disruption of the hydrophobic epithelial barrier, the former occurring frequently in crowded, transported fish. The edematous stroma is frequently colonized by opportunistic bacteria or fungi (particularly *Saprolegnia* spp.), which leads to stromal dissolution, panophthalmitis, and phthisis bulbi. The ulcerated cornea often has remarkably little histologic evidence of inflammation, perhaps related to the rapid progression of the lesion, dilution of mediators, or dominating osmotic forces. In freshwater fish, dilution of the aqueous results in cataract, which resolves following corneal healing.

Colonization of the corneal epithelium or stroma by skin or gill parasites such as *Ichthyophthirius* or by microsporidians, digenetic metacercariae, or copepods usually results in little damage other than a thin rim of edema or fibroplasia surrounding the parasite. In one remarkable report (Beck and Mansfield 1969), 84% of Greenland sharks had copepods embedded in both corneas with no apparent ill effects.

Many eyes with panophthalmitis have corneas characterized by edema, granulation tissue, and stromal dissolution. It is usually not possible to determine whether the corneal lesion preceded the ocular infection, or whether infection of anterior chamber eventually involved the adjacent cornea.

The cornea may be affected by lymphocystis, with the classical ballooned giant cells found within corneal stroma and adjacent sclera. Inflammation is usually absent except in late stages. The virally infected cells are usually found elsewhere in the eye as well, particularly in the choroid.

Several outbreaks of corneal disease in lake trout have been associated with undue exposure to sunlight because of errors in hatchery management. This trout species normally lives in deep water and requires protection from sunburn if forced to live in the shallow water of confinement rearing. The earliest corneal lesions involve vacuolation and swelling of corneal endothelium and stroma but no change in corneal epithelium. Later corneal changes are those of a nonspecific chronic keratitis with ulceration and corneal granulation tissue. Affected fish also develop cataracts, and it is unclear whether the keratitis is an inevitable progression of the marked corneal edema or is the result of trauma to the eye of fish blind from one or both of cataract and corneal edema.

Vitamin A deficiency leads to corneal epithelial and stromal changes in hatchery trout. The changes are described as increased stromal thickness (edema?) and epithelial hyperplasia followed by degeneration.

Cataract. Cataract is by far the most frequent ocular condition likely to be submitted for histologic evaluation and is said to be second only to gas-bubble exophthalmos as the most prevalent ocular disorder. Cyprinids and rainbow trout seem particularly susceptible. The lesion has been described under the general reactions of the lens to injury, and usually one gains little insight into the etiology of a specific example of cataract by histologic examination. Because of its simple structure, lens response to a wide variety of insults tends to involve a stereotyped sequence of vacuolation, fiber lysis, and abortive fiber regeneration regardless of the nature of the insult.

The causes of cataract can be broadly grouped into those resulting from dietary deficiency or excess, from intralenticular invasion by trematodes, and from coexistent endophthalmitis. Cataracts caused by excessive exposure to sunlight, to cold temperatures, or to dilution of aqueous humor following corneal ulceration are additional unique categories.

Nutritional cataracts result from dietary deficiency in methionine, riboflavin, vitamin A, or zinc and are thus likely to be seen only in cultured fish. Such dietary imbalances usually result from "bargain-basement" feeding programs utilizing unusual dietary products to formulate the ration. Undoubtedly over time the list will grow longer, to include more dietary deficiencies and environmental pollutants: the carcinogen thioacetamide, for example, causes cataracts in rainbow trout. Sunlight-induced cataract is reported in young lake trout under the same conditions as those leading to corneal opacity, although the two lesions do not necessarily coexist in each affected fish. While raceways are usually protected from sunlight to prevent this well-recognized syndrome, overcrowding may force some fish away from the protection and eventually result in cataract. This cataract, at least in its early stages, is quite characteristic, and consists of anterior epithelial proliferation unaccompanied by the vacuolation, lysis, and bladder cell formation typical of other types of cataract (Fig. 10.12).

A

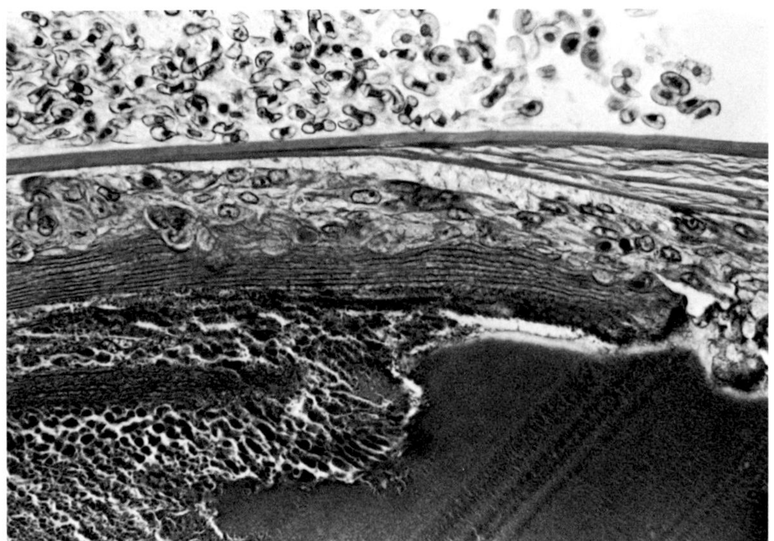

B

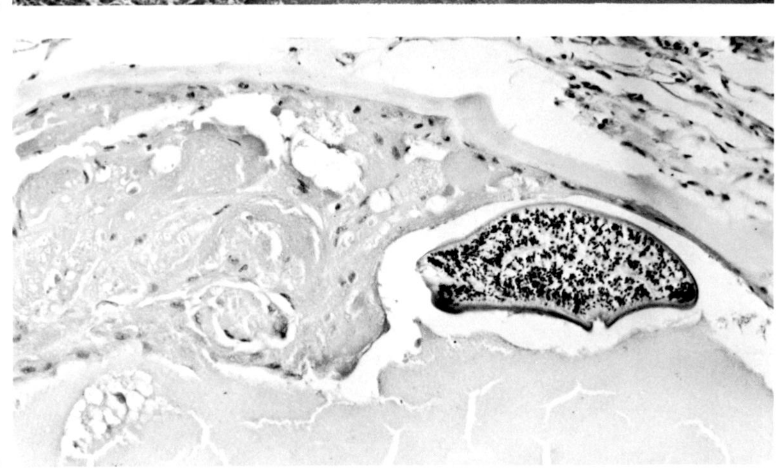

Fig. 10.12. A. Anterior polar subcapsular cataract in juvenile lake trout, consistent with sunlight-induced cataract. Epithelial hyperplasia seems to be the initial lesion. **B.** Cortical liquefaction associated with intralenticular migration of digenetic trematode metacercariae. Perilenticular fibroplasia is probably in response to lens protein that has leaked through sites at which trematodes perforated the lens capsule. **C** and **D.** Metacercariae are occasionally found within extralenticular sites such as cornea (**C**) and retina (**D**), usually with minimal inflammatory reaction.

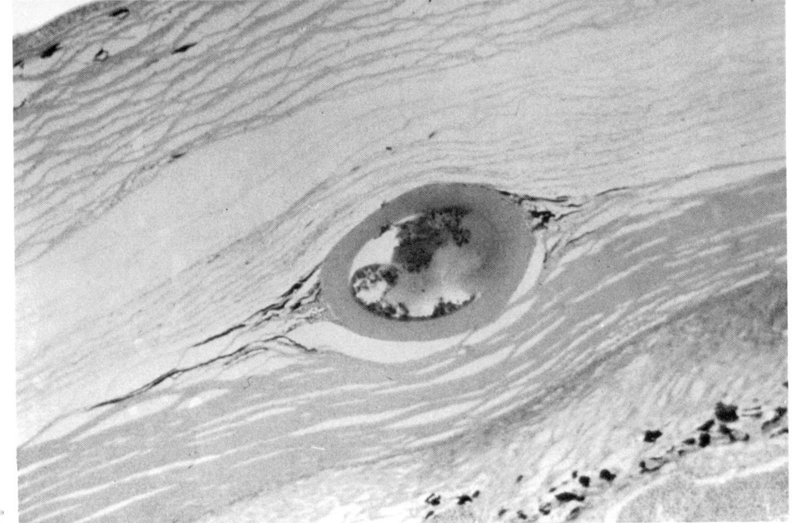

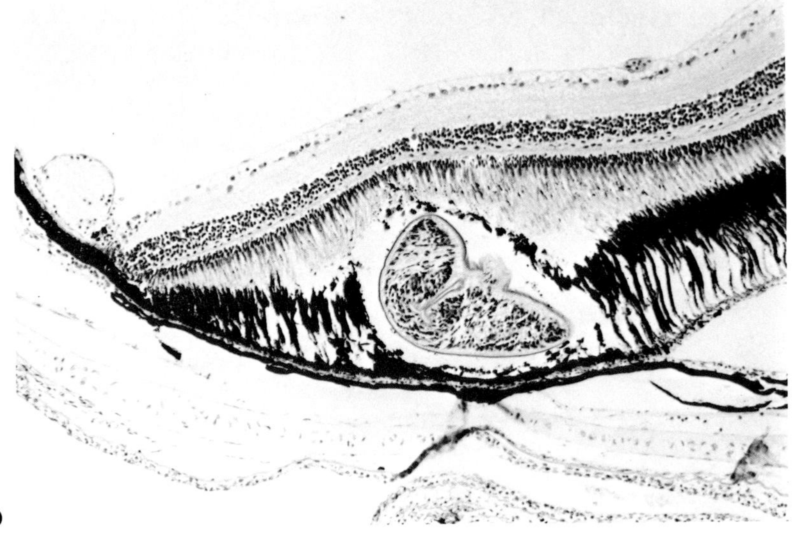

Cold-induced precipitation of lens protein is a relatively frequent cause of reversible lenticular opacity in mammals, particularly neonates born in winter. Fish develop similar opacity that is, however, only partially reversed by warming. The fish lens becomes turbid at temperatures much lower (−10°C to −20°C) than those required for similar results with mammalian lenses, and one wonders if such changes are likely to occur under natural conditions.

Parasitic cataract ("eye fluke") is among the best known of fish diseases, affecting numerous freshwater species in a wide variety of habitats. Trematode metacercariae of the genus *Diplostomum* (syn. *Diplostomulum*) develop within aquatic snails and, following release from the snail, penetrate the skin of nearby fish. Transcutaneous invasion is rapid, and their widespread migration within the fish may be lethal if large numbers are present. Immature flukes can be found in many tissues including all portions of the eye (Fig. 10.12), often causing very little host reaction. It is only within the lens—their preferred niche—that the larval flukes mature. Ingestion of infected fish by carnivorous birds releases the entrapped flukes to complete their life cycle in the avian intestine.

Descriptions of the lenticular pathology are numerous. Among the best are those by

Ashton et al. (1969) and Shariff et al. (1980). The metacercariae pierce the posterior lens capsule as quickly as 15 minutes after transcutaneous penetration. Their site of entry is marked by a tiny capsular perforation through which focally liquified cortical lens protein exudes. Occasionally the perforations lead to lens rupture and a severe endophthalmitis, but in most instances there is no extralenticular sequel to the perforation. The initial focal liquefaction is replaced within days to weeks by generalized cortical liquefaction as flukes migrate to the anterior cortex, accompanied by proliferation of lens epithelium. The proliferation may be extensive, filling the lens and escaping through the posterior capsular perforations to grow within vitreous and posterior chamber. The cells usually retain their epithelial character, in contrast to the fibroblastic metaplasia that predominates in the mammalian reaction to lenticular perforating injury. Fluke larvae, which are numerous within the lens—sometimes several hundred—may also be seen randomly within choroid, retina, sclera, and retrobulbar tissue. Rarely is there significant inflammation, except when larvae die. In that instance, there is a localized granulomatous reaction that, if many flukes die, may coalesce to a massive panophthalmitis that destroys the eye.

Cataract secondary to endophthalmitis occurs in fish as it does in other vertebrates. Interference with lens nutrition or exposure of lens to the hostile environment of a toxin-laden aqueous humor are convenient but unproven theories for pathogenesis. The capsule itself is impermeable to bacterial or

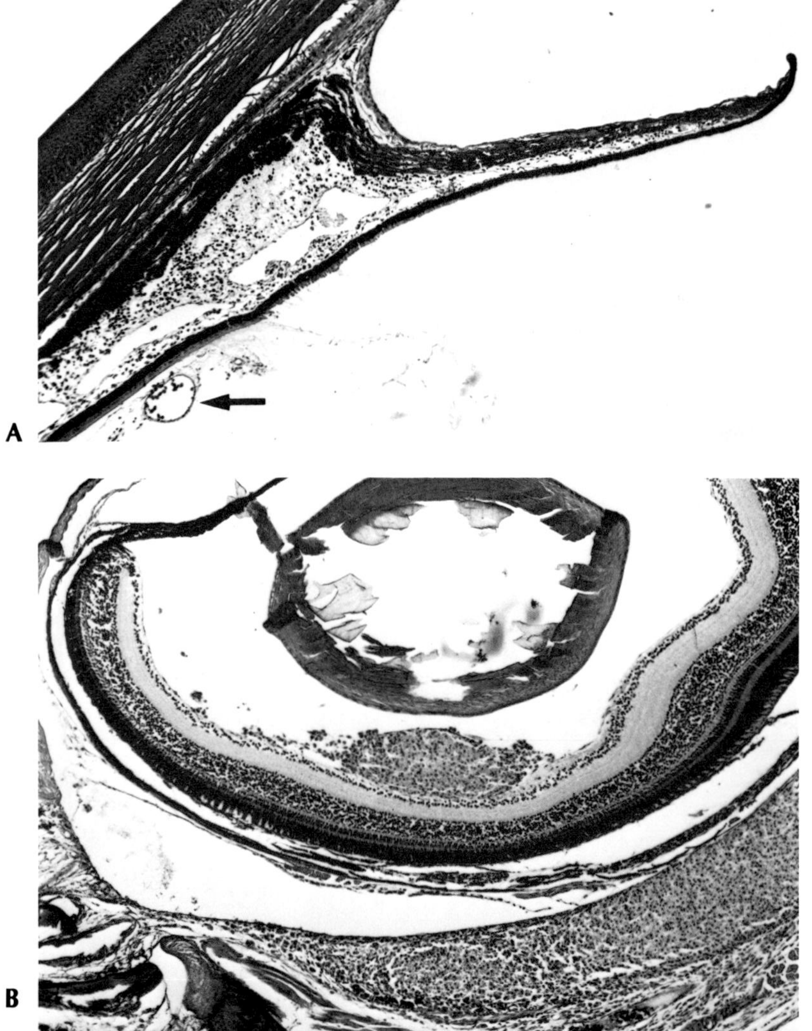

Fig. 10.13. A. Edema and leukocytic infiltration of iris and anterior choroid in suspected gram-negative septicemia. Distended vessels of iris and anterior vitreous (*arrow*) seem prone to hemorrhage and are the probable source of the frequently observed hyphema in fish with uveitis. **B.** Mycobacterial endophthalmitis. Coalescing granulomas have virtually replaced the choroidal rete. The accumulation in the posterior vitreous probably originated in the falciform process or preretinal vascular plexus. **C** and **D.** Mycobacterial granulomas in iris (**C**) and choroid (**D**).

fungal penetration. Observation of such agents, or of leukocytes, within lens is evidence for traumatic capsular rupture.

Endophthalmitis, Panophthalmitis, and Orbital Cellulitis. Endophthalmitis is a convenient term to encompass a variety of septic ocular inflammations that vary from mild localized uveitis to end-stage phthisis bulbi. The usual route of entry is embolic localization within uveal vasculature, particularly the choroidal rete. A similar pathogenesis underlies most cases of orbital cellulitis, with the infection localizing originally within orbit or spreading via panophthalmitis to eventually involve orbital tissue. A few specific examples may result from penetration through cornea or sclera, as with complications of ulcerative keratitis or following corneoscleral perforation by copepod crustaceans.

Hematogenous infection of the uveal tract occurs with a wide range of diseases in fish as it does in birds and mammals, and it serves little purpose to create endless lists of such diseases. Virtually any of the viremic diseases—particularly those with endothelial tropism such as the rhabdoviruses, or the bacteremic diseases such as those due to *Aeromonas* or *Vibrio* spp.—may result in uveal lesions, as may wandering protozoan or metazoan parasites. A few of the most frequent examples are noted here. The lesions vary from mild focal lesions to destructive panophthalmitis with orbital cellulitis, the range presumably dependent upon host resistance and the size of the infective dose (Fig. 10.13).

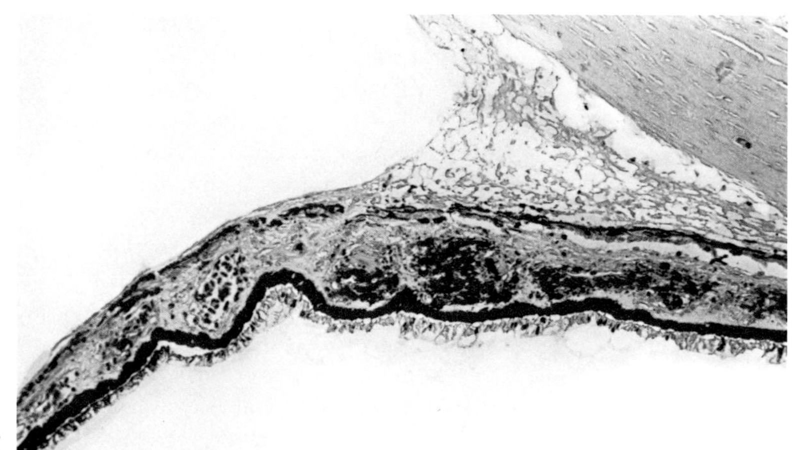

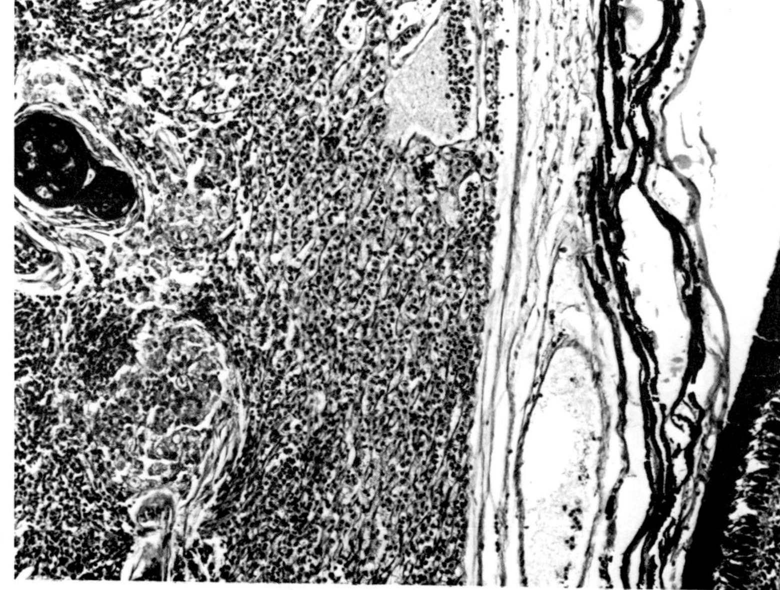

Granulomatous panophthalmitis is associated with mycobacterial infections, as well as infection with *Nocardia* and *Flavobacterium* spp. The lesions are characterized by coalescing choroidal granulomas that may extend to involve any or all parts of the eye (except lens). They resemble lesions caused by similar agents in birds or mammals (except that giant cells are rare), with definitive diagnosis requiring demonstration of the agent by culture or immunofluorescence.

Gram-negative septicemias often cause uveitis, which tends to have a protein-rich and hemorrhagic exudate in its acute stage. Neutrophils predominate later. Iris vasculature becomes particularly engorged and probably accounts for the frequent occurrence of hyphema. Leukocyte exudation may occur from any portion of the uveal vasculature, but choroidal rete seems a favorite site. In those species in which it exists, the preretinal (vitreal) vascular network is another susceptible site. The retina itself is initially spared because it is avascular. Serous detachment from choroidal edema is frequent. Unlike the similar lesion in mammals, the separation virtually always cleaves retinal pigment epithelium from the choroid. In mammals, the split is between photoreceptors and the pigment epithelium, the latter remaining firmly attached to the choroid. A variety of retinal lesions (photoreceptor degeneration, plexiform and ganglion cell layer edema, pigment dispersal) are seen in eyes with severe endophthalmitis, but the mechanisms for such changes and their relationship to the uveal inflammation remain unknown. As previously mentioned, severely affected eyes often have cataract and ulcerative keratitis. Rupture of the globe is frequent, with the resultant *phthisis bulbi* consisting of massive fibrous proliferation, clumps of melanomacrophages, and remnants of the more durable ocular structures such as lens and scleral cartilage.

Lesions associated with *copepod infestation* occur in fresh and marine fish. The copepods vary widely in ecological niche and life cycle, resulting in an equally wide range of ocular lesions. Some are relatively harmless surface browsers that may find the shallow corneoscleral recess ("conjunctival sac") a convenient shelter. It is possible that some cause mild corneal irritation. A few species habitually penetrate or even perforate the cornea. Penetrating species seem relatively harmless, causing only focal epithelial hyperplasia and stromal fibroplasia adjacent to the embedded cephalothorax. A few species perforate the cornea and extend their mouthparts across the globe to embed themselves within the choroid. The best-known example is that seen in Pacific arrowtooth flounder infected with *Phrixocephalus cincinnatus*. One or both eyes may be infected, seldom with more than one parasite per eye. This obligate parasite perforates the cornea and eventually elongates sufficiently to embed its feeding mouthparts in the choroid. Lateral anchoring appendages do further damage to lens or retina, resulting in focal inflammation, hemorrhage, or cataract. As with any of the larger parasites, death of the parasite within the eye usually initiates devastating inflammation that often destroys the eye.

There are a few reports of other parasites within the uvea, sclera, or retrobulbar tissues. These are photogenic but seldom are associated with significant ocular lesions. Myxosporea, microsporidia, nematodes, and flukes are a partial list, and interested readers should consult the reference list for further information.

Exophthalmos (Pop-eye). Exophthalmos results from retrobulbar accumulation of inflammatory exudate or accumulation of gas bubbles. In the former there is usually an associated endophthalmitis or panophthalmitis, although happenstance localization of viral, bacterial, or parasitic agents within the retrobulbar tissue alone is occasionally seen.

Exophthalmos caused by the presence of gas bubbles ("gas bubble disease") within

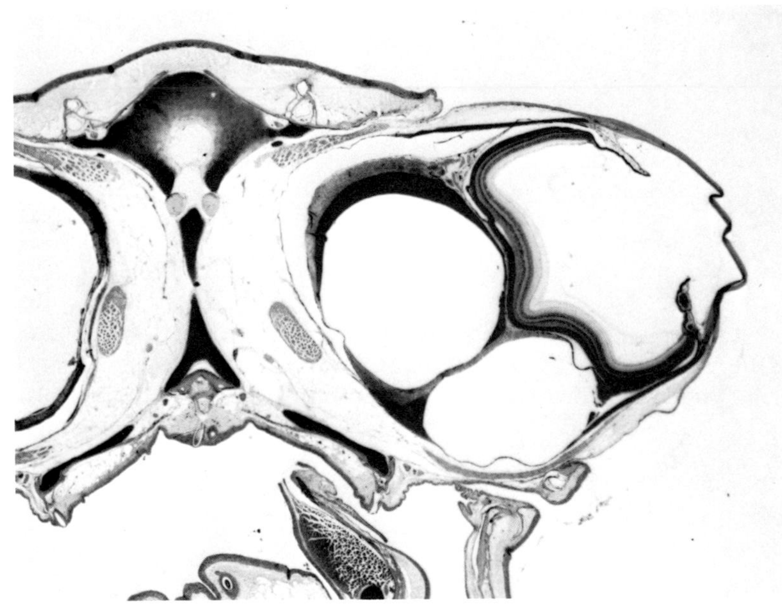

Fig. 10.14. Large retrobulbar gas bubbles causing exophthalmos in trout fingerling.

and behind the eye is common (Fig. 10.14) and is entirely analogous to the "bends" of human divers. It happens under circumstances in which water becomes supersaturated with nitrogen or oxygen and is found naturally, during periods of strong sunlight (causing rapid photosynthesis and hence oxygen production) or artificially, when groundwater or deep lake water is drawn into fish tanks or ponds without adequate opportunity for pressure equalization. When water is pumped, gas may be sucked in under pressure (through leaky manifolds) and forced into solution (Venturi effect). Whatever the reason, the fish equilibriate with the dissolved gas that then comes out of solution in the bloodstream, forming gas emboli in a variety of tissues, especially those with large capillary beds such as eye and gills. These emboli are occasionally fatal, but exophthalmos in otherwise healthy fish is the most frequent clinical observation. The bubbles behind the globe predispose to traumatic corneal injury. Those forming within the globe may physically disrupt delicate choroidal or retinal structure and result in loss of the globe.

References

Ali, M. A., ed. 1974. *Vision in Fishes: New Approaches in Research.* New York: Plenum Press.

Ali, M. A., and M. Anctil. 1976. *Retinas of Fishes: An Atlas.* New York: Springer-Verlag.

______. 1977. Retinal structure and function in the walleye (*Stizostedion vitreum vitreum*) and Sauger (*S. canadense*). *J. Fish. Res. Board Can.* 34:1467–74.

Ali, M. A., and I. Hanyu. 1964. Occurrence of multiple lenses in the eyes of brown bullheads (*Ictalurus nebulosus*). *Copeia* 4:704–5.

Allison, L. N. 1963. Cataract in hatchery lake trout. *Trans. Am. Fish. Soc.* 92:34–38.

Arnott, H. J., N. J. Maciolek, and J. A. C. Nicol. 1970. Retinal tapetum lucidum: a novel reflecting system in the eye of teleosts. *Science* 169:478–80.

Ashton, N., N. Brown, and D. Easty. 1969. Trematode cataract in freshwater fish. *J. Small Anim. Pract.* 10:471–78.

Beck, B., and A. W. Mansfield. 1969. Observations on the Greenland shark, *Somniosus microcephalus,* in Northern Baffin Island. *J. Fish. Res. Board Can.* 26:143–45.

Bennett, P. S. 1964. On *Bomolochus sardinellae* sp. nov. (Copepoda:Cyclopoida) parasitic on *Sardinella albella*. *J. Mar. Biol. Assoc. India* 6:84–88.

Betterton, C. 1974. Studies on the host specificity of the eyefluke, *Diplostomum spathaceum,* in brown and rainbow trout. *Parasitol.* 69:11–29.

Block, B. A. 1986. Structure of the brain and eye heater tissue in marlins, sailfish, and spearfishes. *J. Morphol.* 190:169–89.

Braekevelt, C. R. 1974. Fine structure of the retinal pigment epithelium, Bruch's membrane, and choriocapillaris in the northern pike (*Esox lucius*). *J. Fish. Res. Board Can.* 31:1601–5.

Brandt, T. M., R. M. Jones, Jr., and J. R. Koke. 1986. Corneal cloudiness in transported largemouth bass. *Prog. Fish Cult.* 48:199–201.

Bridges, C. D. B., and C. E. Delisle. 1974. Evolution of visual pigments. *Exp. Eye Res.* 18:323–32.

Burnside, B., and N. Ackland. 1984. Effects of circadian rhythm and cAMP on retinomotor movements in the green sunfish, *Lepomis cyanellus. Invest. Ophthalmol. Vis. Sci.* 25:539–45.

Castric, J., and P. de Kinkelin. 1984. Experimental study of the susceptibility of two marine fish species, sea bass (*Dicentrarchus labrax*) and turbot (*Scophthalmus maximus*), to viral haemorrhagic septicaemia. *Aquaculture* 41:203–12.

Copeland, D. E. 1974. The anatomy and fine structure of the eye of teleost. I. The choroid body in *Fundulus grandis. Exp. Eye Res.* 18:547–61.

______. 1976. The anatomy and fine structure of the eye in teleost. IV. The choriocapillaris and the dual vascularization of the area centralis in *Fundulus grandis. Exp. Eye Res.* 22:169–79.

Copeland, D. E., and D. S. Brown. 1976. The anatomy and fine structure of the eye in teleosts. V. Vascular relations of choriocapillaris, lentiform body and falciform process in rainbow trout (*Salmo gairdneri*). *Exp. Eye Res.* 23:15–27.

Crescitelli, F., M. McFall-Ngai, and J. Horwitz. 1985. The visual pigment sensitivity hypothesis: further evidence from fishes of varying habitats. *J. Comp. Physiol.* 157:323–33.

D'Aoust, B. G., and L. S. Smith. 1974. Bends in fish. *Comp. Biochem. Physiol.* 49A:311–21.

Dechtiar, A. O. 1972. Parasites of fish from Lake of the Woods, Ontario. *J. Fish. Res. Board Can.* 29:275–83.

Dehadrai, P. V. 1966. Mechanism of gaseous exophthalmia in the Atlantic cod, *Gadus morhua* L. *J. Fish. Res. Board Can.* 23:909–14.

Dukes, T. W. 1975. Ophthalmic Pathology of Fishes. In *The Pathology of Fishes,* ed. W. E. Ribelin and G. Migaki, 383–98. Madison: Univ. of Wisconsin Press.

Dukes, T. W., and A. R. Lawler. 1975. The ocular lesions of naturally occurring lymphocystis in fish. *Can. J. Comp. Med.* 39:406–10.

Easter, S. S., Jr. 1971. Spontaneous eye movements in restrained goldfish. *Vision Res.* 11:333–42.

Eckelbarger, K. J., R. Scalan, and J. A. C. Nicol. 1980. The outer retina and tapetum lucidum of the snook *Centropomus undecimalis* (Teleostei). *Can. J. Zool.* 58:1042–51.

Edelhauser, H. F. 1983. Pathogenesis of corneal edema. (Roberts Memorial Lecture) Trans. 14th Annual Prog. *Am. Coll. Vet. Ophthalmol.* 167–92.

Edelhauser, H. F., D. L. Van Horn, and R. O. Schultz. 1969. Corneal opacity associated with eye disease in hatchery-reared lake trout. *Soc. Exp. Biol. Med.* 130:835–38.

Ferguson, M. S., and R. A. Hayford. 1941. The life history and control of an eye fluke. *Prog. Fish Cult.* 54:1–13.

Fournie, J. W., and R. M. Overstreet. 1985. Retinoblastoma in the spring cavefish, *Chologaster agassizi* Putnam. *J. Fish Dis.* 8:377–81.

Grizzle, J. M., and W. A. Rogers. 1976. Anatomy and Histology of the Channel Catfish. Auburn University, Agricultural Experiment Station, 59–60.

Groman, D. B. 1982. Histology of the striped bass. *Am. Fish Soc. Monogr.* 3:84–87.

Herrick, C. J. 1941. The eyes and optic paths of the catfish, Ameiurus. *J. Comp. Neurol.* 75:255–86.

Hoffman, G. L., and J. A. Hutcheson. 1970. Unusual pathogenicity of a common metacercaria of fish. *J. Wildl. Dis.* 6:109.

Hughes, S. G., R. C. Riis, J. G. Nickum, and G. L. Rumsey. 1981. Biomicroscopic and histologic pathology of the eye in riboflavin deficient rainbow trout (*Salmo gairdneri*). *Cornell Vet.* 71:269–79.

Kabata, Z. 1967. Morphology of *Phrixocephalus cincinnatus* Wilson, 1908 (Copepoda:Lernaeoceridae). *J. Fish. Res. Board Can.* 24:515–26.

———. 1969. *Phrixocephalus cincinnatus* Wilson, 1908 (Copepoda:Lernaeoceridae): morphology, metamorphosis, and host-parasite relationship. *J. Fish. Res. Board Can.* 26:921–34.

Kabata, Z., and C. R. Forrester. 1974. *Atheresthes stomias* (Jordan and Gilbert 1880) (Pisces:Pleuronectiformes) and its eye parasite *Phrixocephalus cincinnatus* Wilson 1908 (Copepoda:Lernaeoceridae) in Canadian Pacific waters. *J. Fish. Res. Board Can.* 31:1589–95.

Ketola, H. G. 1979. Influence of dietary zinc on cataracts in rainbow trout (*Salmo gairdneri*). *J. Nutr.* 109:965–69.

Kreuzer, R. O., and J. G. Sivak. 1984. Spherical aberration of the fish lens: interspecies variation and age. *J. Comp. Physiol.* 154:415–22.

Lagler, K. F., J. E. Bardach, R. R. Miller, and D. R. M. Passino. 1977. *Ichthyology.* 2d ed. New York: John Wiley & Sons.

Larson, O. R. 1965. *Diplostomulum* (Trematoda:Strigeoidea) associated with herniations of bullhead lenses. *J. Parasitol.* 51:224–29.

Lee, W. R., R. J. Roberts, and C. J. Shepherd. 1976. Ocular pathology in rainbow trout in Malawi (*Zomba* Disease). *J. Comp. Pathol.* 86:221–33.

Lester, R. J. G., and R. S. Freeman. 1975. Penetration of vertebrate eyes by cercariae of Alaria marcianae. *Can. J. Public Health* 66:384–87.

Lester, R. J. G., and H. W. Huizinga. 1977. *Diplostomum adamsi* sp.n.: description, life cycle, and pathogenesis in the retina of *Perca flavescens. Can. J. Zool.* 55:64–73.

Levine, J. S., and E. F. MacNichol, Jr. 1982. Color Vision in Fishes. *Sci. Am.* 246:140–49.

Lewis, P. 1931. Cataract in the eyes of freshwater fishes, due to the invasion of the larvae of trematoid worms. *Am. Acad. Ophthalmol. Otolaryngol. Trans.* 36:143–58.

Llewellyn, L. C. 1980. A bacterium with similarities to the redmouth bacterium and *Serratia liquefaciens* (Grimes and Hennerty) causing mortalities in hatchery reared salmonids in Australia. *J. Fish Dis.* 3:29–39.

Loewenstein, M. A., and F. A. Bettelheim. 1979. Cold cataract formation in fish lenses. *Exp. Eye Res.* 28:651–63.

Matty, A. J., D. Menzel, and J. E. Bardach. 1958. The production of exophthalmos by androgens in two species of teleost fish. *J. Endocrinol.* 17:314–18.

Molnar, K. 1974. On Diplostomosis of the grass-carp fry. *Acta Vet. Acad. Sci. Hung.* 24:63–71.

Nicol, J. A. C. 1980. Studies on the eyes of toadfishes *Opsanus.* Structure and reflectivity of the stratum argenteum. *Can. J. Zool.* 58:114–21.

Nicol, J. A. C., and H. J. Arnott. 1973. Studies on the eyes of gars (Lepisosteidae) with special reference to the tapetum lucidum. *Can. J. Zool.* 51:501–8.

Nicol, J. A. C., and E. S. Zyznar. 1973. The tapetum lucidum in the eye of the big-eye *Priacanthus arenatus* Cuvier. *J. Fish Biol.* 5:519–22.

Partridge, B. L., and T. J. Pitcher. 1980. The sensory basis of fish schools: relative roles of lateral line and vision. *J. Comp. Physiol.* 135:315–25.

Patt, D. I., and G. R. Patt. 1969. *Comparative Vertebrate Histology.* New York: Harper & Row.

Pauley, G. B., and R. E. Nakatani. 1967. Histopathology of "gas-bubble" disease in salmon fingerlings. *J. Fish. Res. Board Can.* 24:867–71.

Poston, H. A., R. C. Riis, G. L. Rumsey, and H. G. Ketola. 1977. The effect of supplemental dietary amino acids, minerals and vitamins on salmonids fed cataractogenic diets. *Cornell Vet.* 67:472–509.

Richardson, N. L., D. A. Higgs, and R. M. Beames. 1986. The susceptibility of juvenile chinook salmon (*Oncorhynchus tshawytscha*) to cataract formation in relation to dietary changes in early life. *Aquaculture* 52:237–43.

Riis, R. C., M. E. Georgi, L. Leibovitz, and J. S. Smith. 1981. Ocular metacercarial infection of the oyster toadfish, *Opsanus tau* (L.). *J. Fish Dis.* 4:433–35.

Rusoff, A. C., and S. S. Easter, Jr. 1980. Order in the optic nerve of goldfish. *Science* 208:311–12.

Schwanzara, S. A. 1967. The visual pigments of

freshwater fishes. *Vision Res.* 7:121–48.

Shariff, M. 1981. The histopathology of the eye of big head carp, *Aristichthys noblis* (Richardson), infested with *Lernaea piscinae* Harding, 1950. *J. Fish Dis.* 4:161–68.

Shariff, M., R. H. Richards, and C. Sommerville. 1980. The histopathology of acute and chronic infections of rainbow trout *Salmo gairdneri* Richardson with eye flukes, *Diplostomum* spp. *J. Fish Dis.* 3:455–65.

Shino, S. M. 1956. Copepods parasitic on Japanese fishes. II. Genus Phrixocephalus. *Fac. Fish.* 2:242–67.

Silvak, J. G. 1985. The Glenn A. Fry Award Lecture: Optics of the crystalline lens. *Am. J. Optom. Physiol. Opt.* 62:299–308.

Ubels, J. L., and H. F. Edelhauser. 1982–1983. Healing of corneal epithelial wounds in marine and freshwater fish. In *Current Eye Research,* vol. 2, 613–19. Oxford: IRL Press Ltd.

______. 1987. Effects of corneal epithelial abrasion on corneal transparency, aqueous humor composition, and lens of fish. *Prog. Fish. Cult.* 49:219–24.

Von Sallmann, L., J. E. Halver, E. Collins, and P. Grimes. 1966. Thioacetamide-induced cataract with invasive proliferation of the lens epithelium in rainbow trout. *Cancer Res.* 26:1819–25.

Walls, G. L. 1967. *The Vertebrate Eye and its Adaptive Radiation.* New York: Hafner Publishing Co.

Yasutake, W. T., and J. H. Wales. 1983. *Microscopic Anatomy of Salmonids: An Atlas.* Chap. XI. Special Sensory Organs and Tissues, 117–20. U.S. Dept. of the Interior, Fish and Wildlife Service, Resource Publication no. 150.

Zadunaisky, J. A. 1972. The electrolyte content, osmolarity and site of secretion of the aqueous humor in two teleost fishes (*Carassius auratus* and *Diplodus sargus*). *Exp. Eye Res.* 14:99–110.

______. 1973. The hypotonic aqueous humor of teleost fishes. *Exp. Eye Res.* 16:397–401.

11 Endocrine and Reproductive Systems

By J. F. Leatherland and H. W. Ferguson

PATHOLOGICAL CHANGES within these systems are rarely reported, and in routine examinations the tissues are often given only a cursory inspection. Examples of specific lesions are accordingly limited. Nevertheless, for the sake of completeness, a brief description of the normal is given here, even when no specific lesions are mentioned. For comparative interest, more mention is made of the other groups of fish than in other chapters.

Endocrine System

Pituitary. The pituitary gland in fish is a complex neuroepithelial structure that, although somewhat variable in structure between species, is remarkably similar within the major classes. The pituitary gland in fish, as in other vertebrates, is derived embryologically from a downpushing of the infundibulum of the brain (forming the pars nervosa) and an upgrowth of the roof of the mouth, termed Rathke's pouch, which gives rise to the partes distalis and intermedia of the adult (Fig. 11.1).

The pars nervosa (neurohypophysis) comprises glial cells, termed pituicytes, together with neurosecretory axons originating in neurones lying in the hypothalamus. Collectively, the hypothalamic neurones giving rise to the axons of the pars nervosa are called the preoptic nucleus. The role(s) of the pituicytes is not known, but they appear to be phagocytic in the normal pars nervosa. In the hagfishes and lampreys a single pars nervosa octapeptide hormone has been identified, arginine vasotocin (8-arginine oxytocin [AVT]). In teleost fishes two pars nervosa hormones are known, ichthyotocin (sometimes called isotocin) (4-serine, 8-isoleucine oxytocin) and AVT. In the ratfish AVT and oxytocin are present, while in the sharks, valitocin (8-valine oxytocin), aspartocin (4-aspartine, 8-valine oxytocin), and AVT are found, and in the rays, glumitocin (4-serine, 8-glutamine oxytocin) and AVT. In lungfishes, the pars nervosa hormones more closely resemble those found in amphibia, consisting of AVT and mesotocin (8-isoleucine oxytocin). Unfortunately, despite considerable research into the pharmacology and physiology of the neurohypophyseal octapeptides in fishes, relatively little is known

J. F. Leatherland, B.Sc., Ph.D., is a professor in the Department of Zoology, College of Biological Science, University of Guelph.

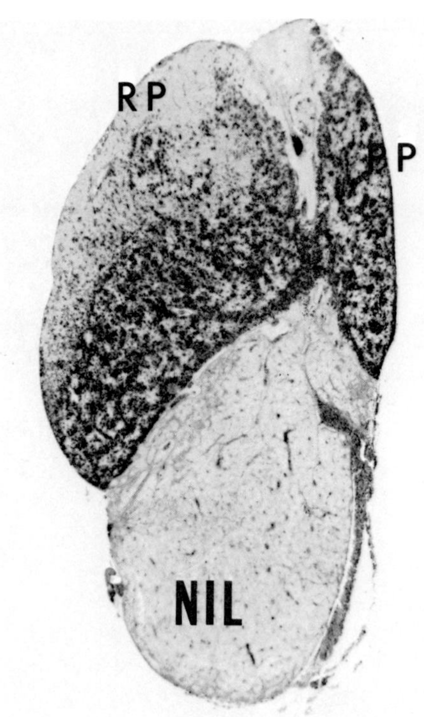

Fig. 11.1. A. Sagittal section through pituitary gland of sexually mature coho salmon. The pars distalis is comprised of rostral (*RP*) and proximal (*PP*) components, and the pars intermedia and pars nervosa are interdigitated to make up the neurointermediate lobe (*NIL*). **B.** Parasagittal section through pituitary gland of adult carp showing similar zonation of gland into rostral (*RP*) and proximal pars distalis (*PP*) and neurointermediate lobe (*NIL*).

about their role. Pressor effects have been reported in a few species after AVT treatment, and spawning behavior has been induced in killifishes by intracranial administration of AVT, but antidiuretic actions, similar to those well established in mammals, are not properly substantiated.

In teleost fish the anterior pituitary or adenohypophysis comprises the rostral and proximal pars distalis and the pars intermedia. In most teleosts the rostral pars distalis secretes prolactin and adrenocorticotropic hormone (ACTH), although some thyrotropic hormone (TSH) is present in some species. The proximal pars distalis contains the somatotropic (growth hormone–secreting) cells, gonadotropic cells, and usually thyrotropic cells. Nongranulated, so-called stellate cells are found throughout the pars distalis. These are thought to play a phagocytic role and may facilitate the transport of secreted hormone from the basilar laminae toward the vascular complex in the pituitary. The pars intermedia secretes melanocyte-stimulating hormone (MSH) and probably endorphin and several other neuropeptide factors that are as yet of unknown function in fish. In addition, some species contain a second pars intermedia cell type that responds cytologically when fish are subjected to altered ambient calcium levels.

In teleosts, prolactin is thought to play a role in freshwater osmoregulation, probably by lowering the permeability of the gill epithelium to water. The hormone may exert its effect by altering calcium binding to the gill membranes, which indirectly changes the permeability of the gill epithelium to water and possibly also to sodium, potassium, and hydrogen ions. Prolactin may also have a metabolic-regulating action in teleosts. ACTH and TSH act to regulate the activity of the interrenal gland and thyroid gland, respectively, in a manner more or less similar to that seen in mammals. Growth hormone (GH), although having a growth-promoting effect in teleosts, also has a metabolic-regulating function. In teleosts there may be two gonadotropic factors, one operating early in gonadal development, the other acting to regulate the synthesis of androgens and estrogens from the gonads during the late maturational stages. The controversy over the existence of one or two gonadotropins has yet to be resolved. Both α- and β-MSH have been isolated from salmon pars intermedia preparations. While MSH has an effect on melanin dispersion in the melanocytes of some fish, the response is usually slow. Recent work shows an increased α-MSH secretion in fish subjected to stressors.

In lampreys and hagfishes, although there is some resemblance in pituitary structure to that of the teleosts, the zonations of the pars distalis cannot be compared. Apart from the MSH content of the pars intermedia, there is little concensus as to the nature or role of the hormones secreted by the pars distalis. In elasmobranchs, the pars distalis has two components, the unique ventral lobe embedded in cartilage ventral to the main pituitary, and the major pars distalis structure in contact with the hypothalamus. The ventral lobe secretes TSH and gonadotropin(s). The other pituitary hormones (prolactin, GH, and MSH) have been found, but their roles are not well understood. Although there have been several histological studies on the pituitaries of crossopterygian, dipnoan, and actinopterygian representatives, little is known about the nature or function of the pituitary hormones in these groups of fish.

There is a simple hypothalamo-hypophyseal portal system in some groups (e.g., the dipnoans and some teleostean representatives), but hypothalamic control of pituitary function may operate by direct innervation or by delivery of the hypothalamic-releasing factors into close proximity of the pars distalis cells by the axons of the hypothalamic neurones. There are close parallels between the nature of hypothalamic control (i.e., stimulatory or inhibitory) of pars distalis cell function in teleosts and mammals. One notable difference is that the TSH cells appear to be under inhibitory control in some teleosts.

Pituitary cysts are common in sticklebacks, and although their significance or pathogenesis is unknown, their prevalence does increase with capture and handling. They have also been reported in salmonids, cyprinids, and in minnows exposed to the herbicide trifluralin.

Basophilic adenomas have been reported in several species of teleost fish, although they tend to be rare and are usually associated with hybrid forms.

Pineal gland. This light-sensitive neuroendocrine structure lies between the eyes just beneath a midline thinning of the cranial cartilage in bony fishes and is covered by a well-vascularized capsule and meninges. As in mammals, the pineal gland in fish appears to secrete melatonin, and elevated plasma melatonin levels have been recorded during the night in lampreys, trout, and eels. The pineal appears to play the role of mediator between photoperiod and the control of daily and perhaps seasonal rhythms, such as the regulation of reproductive cycles, growth, and seasonal migrations.

Thyroid gland. In adult lampreys, hagfishes, and most bony fishes the thyroid gland is diffuse, with follicles concentrated around the ventral aorta (Fig. 11.2). In some species, ectopic thyroid foci are common. These may be present in the eye, kidney, spleen, heart, and other tissues. In some bony fishes

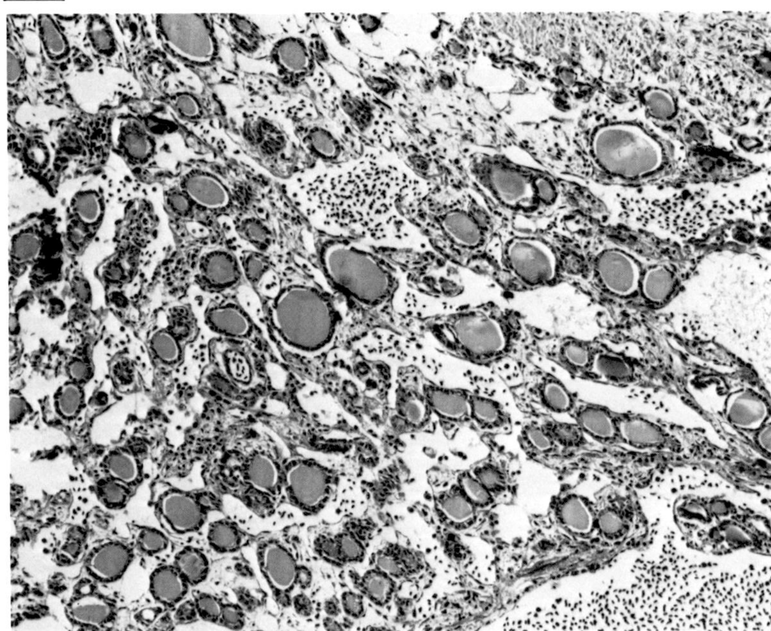

Fig. 11.2. Thyroid follicles in lower jaw of adult, sexually mature coho salmon. Follicles are located in areolar connective tissue surrounding the ventral aorta.

(e.g., the reef parrot fishes) and in most cartilaginous fishes, the thyroid gland is encapsulated.

The microstructure of the thyroid gland in fishes is essentially similar to that of tetrapods. The thyroid follicles comprise epithelial cells, which vary in size depending on the degree of TSH stimulation. The cells form a tight epithelium surrounding the colloid of the follicle lumen and have microvilli and occasional cilia projecting into the follicular lumen.

Thyroxine (T4) is the major product released from the thyroid gland; peripheral monodeiodination of T4 to produce triiodothyronine (T3) takes place in the liver and kidney. The actions of the thyroid hormones at the cellular level in fish are not as well understood as they are in mammals, but specific T3 receptors have been found in hepatic mitochondria and nuclei of several species of bony fishes.

The thyroid hormones have been shown to play roles in growth regulation (most probably acting as a permissive hormone for the action of GH) in bony fishes and cartilaginous fishes but not in lampreys or hagfishes. The hormones appear to act on the growth of muscle, bone, and cartilage as evidenced by retardation of the growth of these tissues in radiothyroidectomized fish. Addition of thyroid hormones to the diet increases skull bone growth and causes hyperplasia of connective tissue in the interorbital area (Fig. 11.3). The thyroid hormones play a part in pigment and guanine deposition, particularly during the smoltification process of salmon, and synergize with gonadotropins in the stimulation of estradiol secretion by the ovarian follicles. Osmoregulatory functions have also been proposed, generally indicating a protective role for the thyroid hormones during seawater adaptation.

Peak thyroid hormone secretion in teleost fishes is most commonly associated with major metabolic events in the life cycle of the fish, such as the phase of rapid growth, gonadal growth, or developmental changes associated with the parr-smolt transformation in salmon. This suggests that the hormones are important regulators of energy-partitioning mechanisms and processes in teleost fish. The concept is sup-

A

B

Fig. 11.3. A. Rainbow trout showing distortion of skull and opercular bones. This was experimentally produced by feeding thyroid hormones (T3). **B.** Normal fish from control group of same experiment.

ported by observations of altered thyroid hormone economy in fasted fish and those subjected to dietary manipulation, particularly of dietary lipid content.

Although low dietary iodide has been shown to produce thyroid hyperplasia in mammals and birds, it is remarkably difficult to induce such iodide-deficiency hyperplasia in fishes because the major source of iodide is obtained from the ambient water across the gill epithelium. Nevertheless, thyroid hyperplasia is not uncommon in wild fish and captive stock, even when captive stock are maintained in seawater that is rich in iodide. Epizootics of thyroid hyperplasia are found in Great Lakes coho, chinook, and pink salmon, and it has also been reported in *Salmo* and *Salvelinus* species from the same regions (Fig. 11.4). Although it was originally considered to have a dietary iodide-deficient etiology, most evidence to date suggests that it is brought about by the presence of goitrogens (pollutants?) in the lake systems; a similar etiology has been proposed to explain the presence of thyroid lesions in both bony and cartilaginous fishes held in aquaria in many parts of the world.

Fish with grossly visible lesions have a

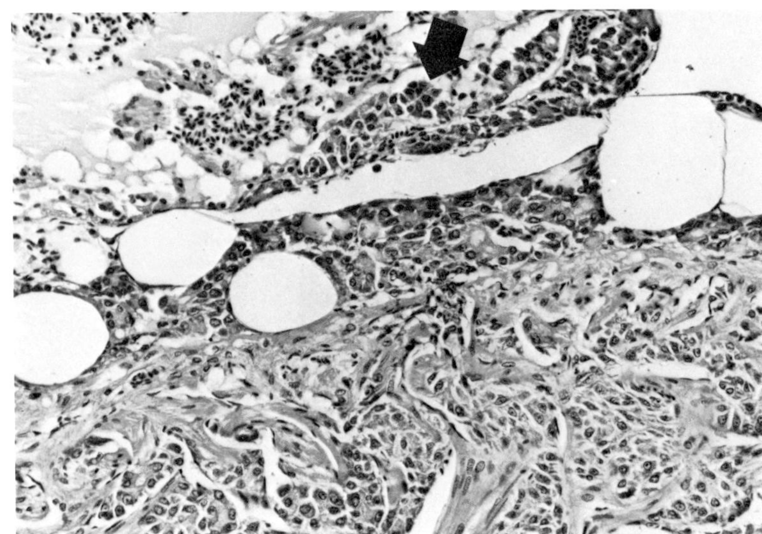

Fig. 11.4. A. Great Lakes coho salmon with first and second gill arches removed to show multinodular enlargement of thyroid in this case of goiter. (*Courtesy of R. D. Moccia*) **B.** Heart from diamond killifish with massively hyperplastic thyroid, showing infiltration of atrial (*arrow*) and ventricular myocardium by thyroid cells. Although probably not neoplastic, it is difficult to imagine that invading cells are not in some way compromising cardiac function and that they are therefore behaving as a malignancy. (Can we term this malignant hyperplasia!)

mass of reddish tissue extending around the ventral aorta, often forming a compact mass in the ventral and lateral pharyngeal areas. In severe cases the lesion invades bone, cartilage, muscles, and blood vessels. This apparently aggressive behavior has led some workers to report the lesions as adenocarcinomas, but the invasive nature of the thyroid tissue is most likely a reflection of the unencapsulated nature of the gland in bony fishes (Fig. 11.4). Histologically, the epithelial cells are generally large, and the colloid is either absent or greatly reduced. Some researchers report that the lesions are alleviated by iodide prophylaxis, but that has not been found to be successful in all instances. The goiter in some species may be accompanied by increased numbers of melanomacrophage centers (MMCs), which are occasionally encountered within the parenchyma of the thyroid in normal fish. Ulcerations and necrotic cysts filled with blood-stained fluid are present in some species.

The thyroid is not infrequently involved in systemic diseases such as mycobacteriosis in which granulomas may cause compression atrophy or necrosis of follicles. Migrating parasites, such as digenetic metacercariae, may result in local inflammation of the thyroid, but the significance of this and other similar lesions is hard to ascertain, and the observations are usually regarded as incidental; certainly, an associated hypothyroidism is not reported, nor is lymphocytic or autoimmune thyroiditis.

Renal Hormones

INTERRENAL CELLS. In elasmobranchs and bony fishes the interrenal cells lying along the major blood vessels (posterior cardinal veins) of the anterior (head) kidney represent the equivalent of the mammalian adrenal cortex. In lampreys, although interrenal cells have been tentatively identified in the pronephros and adjacent to the pericardium, their homology with the interrenal cells of other fishes is not yet fully established.

The interrenal cells have a pale eosinophilic granular cytoplasm with a prominent nucleolus in a rounded nucleus, although their histological appearance varies greatly in response to stress, drugs, salinity changes, and reproduction.

The interrenal cells produce corticosteroids, the most quantitatively important of which in teleost fishes is cortisol, although cortisone and 11-deoxycortisol have also been found in the blood of teleosts. In elasmobranchs, cortisol is not present. The principal corticosteroid in this group is 1α-hydroxycorticosterone, although traces of 11-deoxycorticosterone, 11-dehydrocorticosterone, and corticosterone have also been detected in the plasma. Cortisol, 11-deoxycortisol, corticosterone, and 11-dehydrocorticosterone have been detected in the blood of lampreys and hagfishes. The presence of mineralocorticoids (aldosterone) has been established only in a few fish, notably the dipnoid lungfishes.

The roles of the corticosteroids in normal physiological processes is poorly understood in fish. Cortisol has potent effects on intermediary metabolism, and some evidence suggests that cortisol is important for seawater adaptation in euryhaline species. In addition, cortisol levels are elevated after exposure of teleost fishes to some (but not all) stressors, suggesting that cortisol is a major factor in the stress response. Several stressors have been shown to increase the susceptibility of teleost fish to disease, and treatment of fish with cortisol appears to lower their resistance to fungal, bacterial, and trypanosome infection and disease. It is therefore possible that a final common pathway mediated via cortisol causes depopulation of lymphoid tissues with a concomitant lymphocytopenia and possibly also a neutrophilia. It is also possible that in teleosts the lymphocyte subpopulations show differential sensitivities, just as in mammals where T and B lymphocytes react differently to catecholamines. However, there are several well-established "stressful" situations (such as during extreme debilitating parasitemia) when plasma cortisol levels are not elevated.

Corticosteroid synthesis and release are under the control of pituitary ACTH, and thus hypophysectomy leads to atrophy of the interrenal cells (as does any disease process that involves the pituitary). The cells become smaller with a smaller nucleus and an indistinguishable nucleolus, and they have less active mitochondria and Golgi. By contrast, cold, excess handling, bacterial endotoxins, anesthesia, heavy metals, detergents, and the stress of spawning are just a few situations under which the interrenal cells become hypertrophic and may start to radiate in large numbers into the parenchyma of the kidney (Fig. 11.5).

MSH also has a corticotropic activity in many fish. MSH-regulated production of cortisol is not limited by the negative feed-

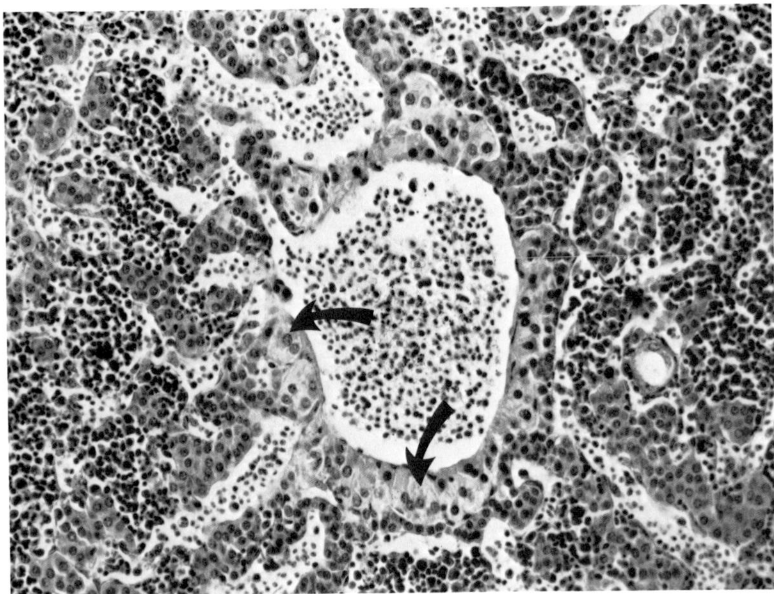

Fig. 11.5. Section through anterior kidney of common shiner to show cords of interrenal cells surrounding posterior cardinal vein radiating out into hemopoietic tissue. Chromaffin cells are also present. These are paler staining and in this case are mostly immediately adjacent to vessel wall (*arrows*).

back control that regulates the ACTH-interrenal relationship, and this may permit the release of extremely large amounts of cortisol from the interrenal cells at certain stages of the life cycle (e.g., during the late stages of reproduction and migration in salmonids).

Interrenal cells are occasionally involved in acute bacteremias and viremias (Fig. 11.6), and compression atrophy may also be found in the severe granulomatous nephritides. Whether these observations see a systemic reflection is unknown.

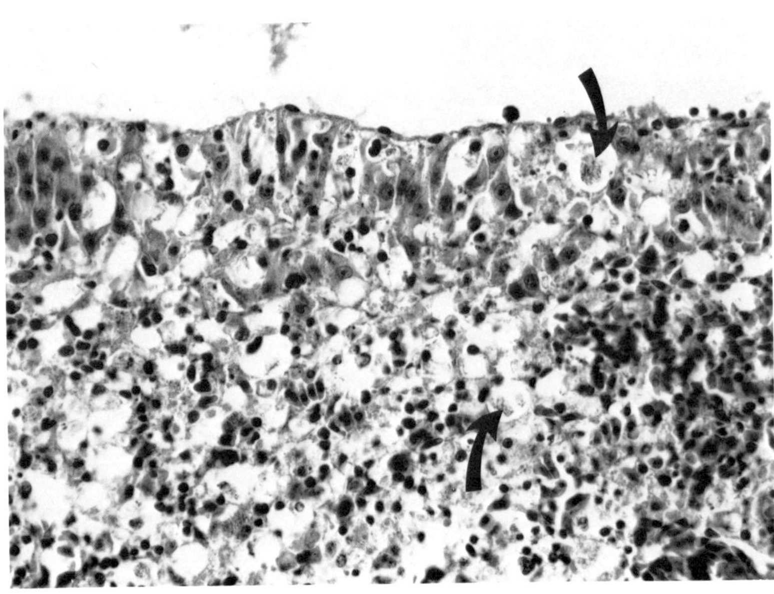

Fig. 11.6. High-power micrograph of anterior kidney from goldfish with severe and widely disseminated bacterial disease. There is marked necrosis of interrenal tissue and hemorrhage into surrounding hemopoietic tissue. Colonies of bacteria are marked with *arrows*.

CHROMAFFIN CELLS. These cells are homologous to the adrenal medulla of mammals. They are mixed among the interrenal cells adjacent to the posterior cardinal veins of the head kidney in teleosts (Fig. 11.5). They are arranged segmentally as masses along the medial border of the kidney in elasmobranchs, while in lampreys and hagfishes they are dispersed in the wall of the sinus venosus along the cardinal veins and between muscle fibers of the heart.

Chromaffin cells tend to be paler than the interrenal cells, and their nucleus is larger and more pleomorphic. However, as with the interrenal cells, their appearance varies considerably in response to physiological, pharmacological, and pathological situations.

Epinephrine and norepinephrine are secreted by the chromaffin tissue of cyclostomes, cartilaginous fishes, and bony fishes. A common function in fishes appears to be the regulation of heart rate and force of cardiac contraction. In addition, catecholamines have been shown to bring about systemic vasoconstriction and a branchial lamellar vasodilation in teleosts. Catecholamines have also been shown to alter respiratory functions of blood and to modify the activity of the chloride cells in the gills.

The catecholamines are released into the blood following stressful situations. Such release usually precedes the release of cortisol and is generally considered to represent the so-called primary stress response. Almost any unusual disturbance of the fish, such as the sampling method itself, will elicit a release into the blood of the catecholamines; thus details of the roles of the hormones in normal physiology are difficult to establish.

RENIN-ANGIOTENSIN SYSTEM. The renin-angiotensin system is found throughout vertebrates except in cyclostomes and elasmobranchs. The enzyme renin is produced by the juxtaglomerular cells, modified smooth muscle cells of the afferent glomerular arterioles. The enzyme appears as cytoplasmic granules in the cells and is released into the blood where it catalyses the conversion of angiotensinogen to form angiotensin I, which in turn is cleaved to form angiotensin II.

It is likely that angiotensin II acts in fish in a manner similar to mammals, in which there is an increase in blood pressure by smooth muscle contraction and sodium retention by the kidney. However, the sodium-retention effect, which in mammals is regulated by aldosterone secreted from the adrenal cortex, is probably regulated by a different mechanism in fish.

In eels, plasma renin levels increase when the fish are transferred from fresh water to seawater, and they decrease in eels transferred from seawater to fresh water, providing some evidence of an osmoregulatory role for the renin-angiotensin system.

In trout, hemorrhage severe enough to cause a reduction in renal perfusion pressure will activate a baroreceptor response and renin release. Of interest, renin levels in trout also increase with increasing levels of environmental un-ionized ammonia, thereby causing a rise in aortic pressure, and possibly an associated rise in glomerular filtration rate (could this help explain the eosinophilic droplets seen in increasing numbers in the cytoplasm of renal tubular epithelial cells of fish exposed to high levels of ammonia?).

No specific pathological changes of the juxtaglomerular cells have been described, but the involvement of the kidney in so many acute and chronic inflammatory diseases suggests that it is a distinct possibility. In particular, one wonders if the edema sometimes associated with the severe granulomatous nephritides may in part be the result of reduced renin release due to compression atrophy of juxtaglomerular cells. This would have the effect, among other things, of reducing glomerular filtration rate (due to reduced blood pressure) and hence the efficiency of urine production from a kidney that may be already compromised from loss of tubules.

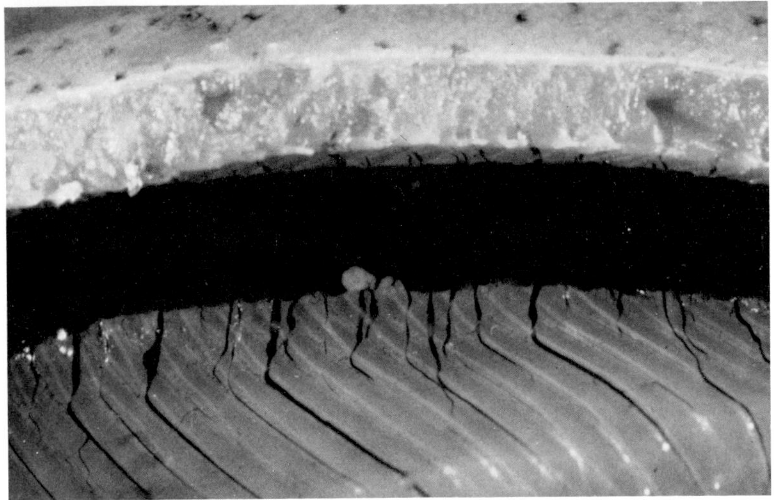

Fig. 11.7. Gross view of kidney from brook trout showing corpuscles of Stannius located along margin.

CORPUSCLES OF STANNIUS. These discrete encapsulated structures are probably unique to holostean and teleostean fishes. They are situated in the renal parenchyma, usually at the lateral borders and approximately midway along the length (Fig. 11.7). In some species, such as salmonids, they lie just beneath the capsule and appear as pale white nodules. In other species they lie deeply embedded within the parenchyma and are not visible except on cut surfaces. They should not be mistaken for granulomas!

The glands are divided into cords or lobules by connective tissue septae that are well supplied with blood vessels and nerves. They have a portal system with blood derived from the caudal vein that then drains back into the renal parenchyma. In some species the lobules are composed of two cell types, granular (type I) and nongranular (type II). Although the functions of the corpuscles of Stannius are incompletely understood, they are known to synthesize a protein or glycoprotein hormone called teleocalcin or hypocalcin, which is involved in calcium homeostasis. The hormone blocks branchial uptake of calcium. Removal of the gland from teleosts (termed either Stanniectomy or Stanniosomatiectomy) causes hypercalcemia, and thus the active principal has an action similar to that of parathyroid hormone. In addition to the calcium-regulating factors, the corpuscles of Stannius of some fish produce pressor substances that are similar in physiology to renin.

As with the other renal endocrine elements, the corpuscles of Stannius are occasionally seen to be involved in acute viremias and bacteremias and may be compressed in the severe granulomatous diseases. They are also occasionally involved in nephrocalcinosis (Fig. 11.8).

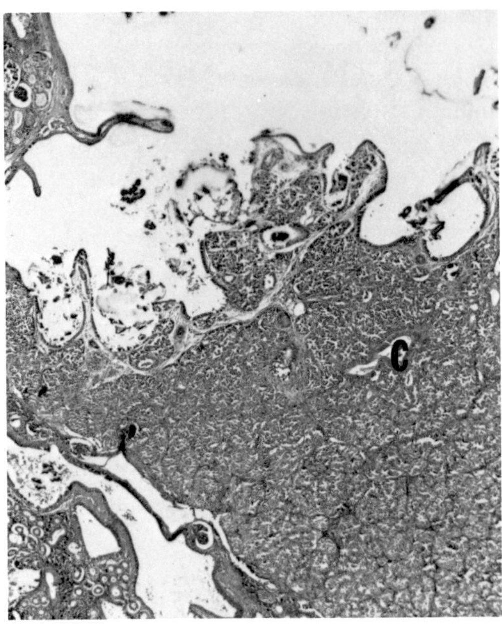

Fig. 11.8. Nephrocalcinosis in Atlantic salmon. A large cyst within the renal parenchyma is compressing adjacent corpuscle (*C*), and associated granulomatous response is extending into endocrine gland.

Ultimobranchial gland. The gland is derived from the last branchial pouch in embryonic fish, and in the adult lies just ventral to the esophagus in the transverse septum separating the heart from the abdominal cavity (Fig. 11.9). In some species it is a single organ whereas in most others, including salmonids, it is paired. It comprises small follicles of columnar cells supplied by a capillary network that drains directly into the sinus venosus. The cells produce calcitonin, a protein hormone that decreases plasma calcium levels in mammals. In fact, salmon calcitonin has a greater biological potency in mammals than any of the known mammalian calcitonins because of its longer half-life in the circulation. However, to date, there has been little conclusive evidence of a direct involvement of calcitonin in calcium homeostasis in fishes. Nevertheless, the general opinion is that calcitonin works in combination with hypocalcin from the corpuscles of Stannius and prolactin from the pituitary to regulate calcium homeostasis. In particular, calcitonin may be involved in calcium regulation at specific parts of a life cycle, such as during egg production and ovulation when the ultimobranchial epithelium decreases in height and cytoplasmic granularity, and when calcitonin levels in females are higher than in males.

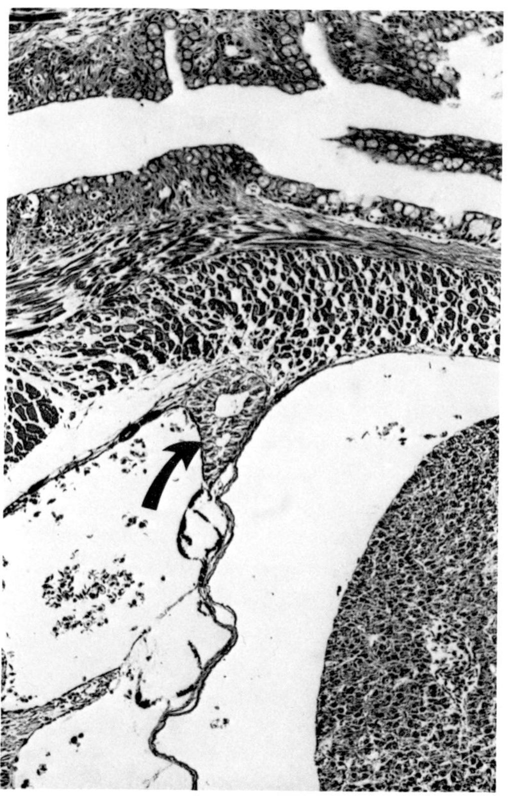

Fig. 11.9. Young rainbow trout; ultimobranchial gland (*arrow*) situated beneath esophagus in septum separating heart from abdominal cavity.

Urophysis. This neurosecretory organ is found in elasmobranchs and teleosts. In most teleosts and some elasmobranchs it lies on the ventral aspect of the distal end of the spinal cord, whereas in other elasmobranchs the urophysis consists of paired neuro-hemal organs lying on the latero-ventral surface of the nerve cord.

The neurosecretory cell bodies, which in elasmobranchs can be very large (Dahlgren cells), have polymorphic nuclei and lie in the spinal cord. Their unmyelinated axons terminate on capillary walls in the neuro-hemal complex. Two peptides have been isolated, urotensin I and urotensin II, although other peptides are probably also produced by the neurosecretory cells. Both urotensins are vasoactive, causing an increase in blood pressure, but their major role may well be their effect on short-circuit currents across epithelial cells and a consequent control over salt and water movement across epithelia such as the urinary bladder found in some species of teleost fish.

Pancreas and Gastrointestinal Hormones. The endocrine portion of the pancreas is represented by the islets of Langerhans. In hagfishes and lampreys the cells are embedded in the wall of the intestinal tract, whereas in most bony fishes and elasmobranchs the islets, as in mammals, are scattered throughout the exocrine pancreas (Fig. 11.10). In some species the islets are very

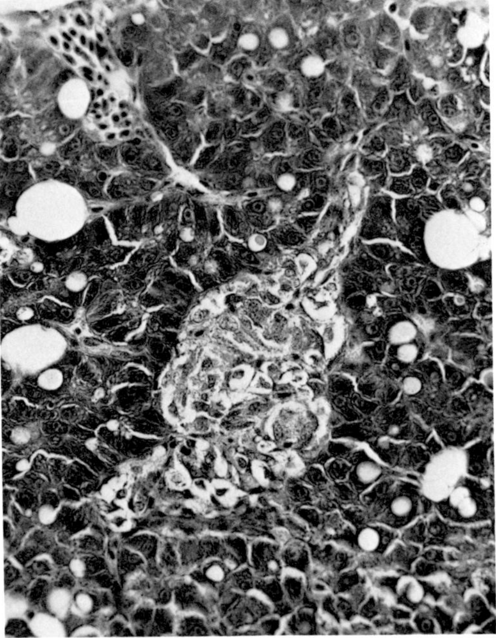

Fig. 11.10. Islet of Langerhans in rainbow trout, surrounded by exocrine pancreatic tissue.

large, often visible grossly, and form Brockman bodies or principal islets. The islet cells of cyclostomes appear to be of one type, the insulin-secreting beta cells, although in elasmobranchs and bony fishes the fine capsule of the islets encloses several cell types including glucagon-secreting alpha cells, insulin-secreting beta cells, and somatostatin-secreting delta cells.

Alterations in blood insulin levels in teleost fishes appear to be much more closely related to changes in amino acid metabolism, and there is some debate as to whether insulin is involved to any significant degree in carbohydrate metabolism of carnivorous fish, which usually have a relatively low dietary intake of carbohydrates. Carnivorous fish, including the salmonids, are generally considered to be essentially diabetic with regard to their ability to cope with dietary glucose. However, recent work using homologous radioimmunoassays (Hilton et al. 1987) has shown that insulin levels are markedly elevated in trout shortly after feeding them experimental diets containing high carbohydrate levels.

The role of glucagon in fishes is poorly understood. Indeed, fish glucagon has yet to be isolated and chemically characterized. Much of the information available regarding the effect of glucagon on fishes has been obtained using mammalian glucagon preparations. Mammalian glucagon generally elicits hyperglycemia and results in metabolic changes in the liver that are indicative of a role in gluconeogenesis.

As with glucagon, the role(s) of islet somatostatin in fishes is poorly understood. Somatostatin is secreted from the hypothalamus in most vertebrates, where it plays a regulatory role by inhibiting the secretion of growth hormone from the pars distalis. The somatostatin secreted in the islets of Langerhans appears to act at the level of the islet cells by inhibiting the release of both insulin and glucagon from the alpha and beta islet cells. In this way somatostatin has a modifying effect on the glucagon and insulin response to changes in blood metabolites. In addition, somatostatin levels in the major blood vessels around the gastrointestinal tract alter depending on the type of food being digested within the gut, suggesting that somatostatin also plays an endocrine role at sites other than those of the islets themselves. It is not known whether similar interactions occur in fishes, although the morphological similarities between fishes and tetrapods might be indicative of a similar relationship between the islet cells in the two groups.

The endocrine pancreas is frequently involved in conditions that affect the exocrine portion, including bacteremias and viremias; parasitic diseases such as coccidiosis, which may almost totally obliterate the entire organ (see Fig. 7.14); and pancreas disease, a condition of unknown etiology affecting farmed Atlantic salmon in which associated lesions can include a rhabdomyopathy and which can be associated with low tissue and plasma levels of vitamin E (see Fig. 7.11). Rancid silkworm pupae were incriminated as the cause of Sekoke disease of farmed carp, a condition reported to result in a diabeteslike syndrome with degranulation of islet beta cells and degeneration of capillaries in retina and glomerulus as just some of the lesions. A severe degenerative myopathy of skeletal muscle was also described in this disease and may in fact be the most significant lesion.

The gastrointestinal hormones in fish are not well understood. Gastrin is produced by the stomach wall (in those species having one) and leads to acid production by the mucosa. It also stimulates the growth of gastric and intestinal mucosa, and as in other vertebrates, histamine leads to a dose-dependent release of acid. Secretin, which in mammals leads to pancreatic secretion, is probably present in teleosts.

Reproductive System

Teleost fish exhibit tremendous variety in breeding patterns, and although the presence of both males and females is the commonest pattern, self-fertilizing hermaphroditism, parthenogenesis (development from an unfertilized egg), and gynogenesis (spermatozoa stimulates egg to develop but does not contribute genetic material) are all found. Moreover, sex reversal in adults is not uncommon. For example, in the coral-reef fish *Anthias squamipinnus,* individuals initially mature as females, and then following changes within social groups (such as removal of a male), an adult female may change to a male. Such changes are associated with increased levels of H-Y antigen, suggesting a feature in common with early gonadal development in mammals (Pechan et al. 1986). The number of eggs produced by a particular species has a rough parallel with the amount of care accorded them. Thus a pelagic species such as the cod, which broadcasts the eggs in open water, may produce many millions per season, whereas sticklebacks, which build complex nests and protect the developing eggs, may produce only 30 to 100. Compare too with fish such as the guppy, which incubates the eggs internally and bears live young.

Testes. These are paired organs situated in the dorsal abdomen below the swimbladder. They vary greatly in size depending on the age of the fish and the season. Mature spermatozoa are conducted to the genital orifice by two ducti deferens that merge caudally. The testis itself lies within a thin tunica albuginea and comprises tubules or lobules that contain spermatogonia either along their entire length (salmonids and cyprinids among others) or at their distal end (antheriniforms). The caudal portion of the testis in some species, such as ictalurids and some cyprinids, is comprised of nongerminal secretory epithelial cells, which may be involved in sperm storage, or in supplying nutrients to the sperm or in contributing to the fluid component of the ejaculate.

The tubules or lobules of the spermatogenic component of the testis (Fig. 11.11) comprise Sertoli cells that, as in mammals, form an uninterupted epithelial barrier between the lumen and the extracellular compartment (the blood-testis barrier). In the interstitium are found small numbers of Leydig cells and incomplete layers of boundary cells around each tubule or lobule. A lymphatic system akin to that seen in mammals is not present (Grier et al. 1980).

Progesterone, 17α-OH progesterone, 17α-OH-20β-dihyroprogesterone, androstenedione, 11β-OH androstenedione, adrenosterone, testosterone, 11β-OH testosterone and 11-ketotestosterone, and 17β-estradiol have been isolated from the testes of several teleost fishes, although the details of the interactions between the Sertoli and Leydig cells in the steroidogenic pathways are not yet understood.

Inflammatory changes including hemorrhage may accompany bacteremias and viremias, or trauma from migrating parasites such as the nematode *Philometra,* which may encyst within the testes, or in the

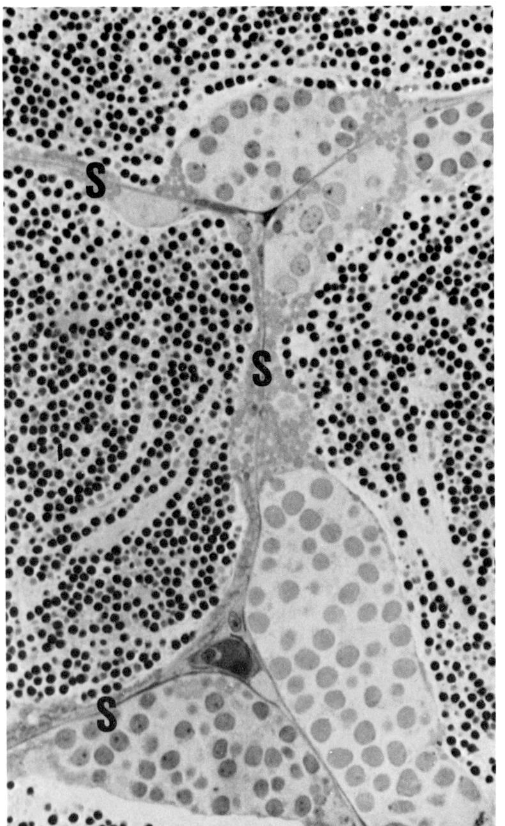

Fig. 11.11. Testicular tissue of goldfish showing lobular nature of seminiferous tubules. Lipid-rich Sertoli cells (*S*) form epithelial lining of tubules, within which are contained cysts of developing germ cells (spermatogonia, spermatocytes, and spermatids). Heads of free spermatozoa fill the lumen of the tubules.

case of the New Zealand snapper, cause total atrophy. Granulomas are also encountered in diseases such as bacterial kidney disease and mycobacteriosis. As in mammals, release of tubular antigens results in an autoimmune response and orchitis, and this has formed the basis for an immunization process aimed at sterilizing intensively cultured fish. Nevertheless, even in widely disseminated bacterial infections, the testes often remain remarkable oases of histological relative normality, a feature of probably no small comfort to the fish! A few protozoa are found in the testes, examples being the myxosporean *Agarella gracilia* Kudo in the South American lungfish *Lepidosiren paradoxa;* the microsporidian *Pleistophora longifilis* Schuberg, which is recorded from the testis of *Barbus flaviatilis* in Germany; and the coccidian *Eimeria brevoortiana,* sporocysts of which are present in the testis of menhaden and are shed with the milt. Parasitic castration is occasionally seen due to the space-occupying effects of parasites such as the larval tapeworm *Ligula* present within the peritoneal cavity, although indirect effects of a heavy ectoparasitic load may also cause failure of proper maturation of gonads.

Ovaries. Ovarian organization varies greatly in teleosts as might be expected in a group with egg-layers as well as live-bearers. Typically, ovaries are paired or bilobed elongated structures suspended from the dorsal abdominal wall by mesovaria and enclosed by a tunica albuginea containing fibrous tissue and smooth muscle. A short oviduct conducts eggs to the outside via an exit between the anus and urinary pore. Salmonids do not have a complete oviduct, and eggs are released dorsally to empty into the peritoneal cavity, gaining access to the outside via a pore behind the anus. The parenchyma of the ovary comprises lamellae containing germinal and follicular epithelium in a well-vascularized connective tissue stroma (Fig. 11.12). Development of eggs from oogonia starts within the lamellae, but they are shed into the ovarian lumen with an investment of thecal cells. Thus a follicle comprises oocyte, a hyaline zona radiata (zona pellucida in some texts), columnar follicular cells (zona granulosa), and theca on the outside. If an oocyte degenerates before ovulation, granulosa cells invade prior to macrophage invasion. Thus examination of some ovaries may reveal eggs in all stages of development and degeneration.

Failure of ovulation leads to atresia and sometimes excessive fibrosis and adhesions

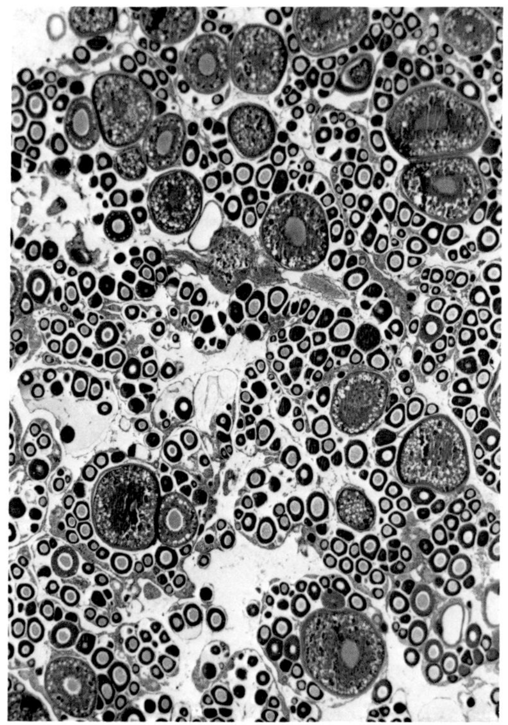

Fig. 11.12. Ovarian tissue of carp showing germ cells in various stages of oogenesis.

to the peritoneal cavity or viscera. Large numbers of melanomacrophages within the stroma are a frequent accompaniment to these changes. Excessively rough handling at manual stripping can cause similar lesions. Postspawning peritonitis due to enteric and other bacteria is another possible consequence of such procedures.

Bacterial infections of the ovary include *Renibacterium salmoninarum,* the organism being present within the eggs at ovulation. Colonies of bacteria are seen within the parenchyma in diseases such as furunculosis and vibriosis and may lead to heavily infected spawning fluids. Protozoan infections include the myxosporean *Wardia ovinocua* Kudo in *Lepomis humilis* and the microsporidians *Glugea* (Fig. 11.13) and *Pleistophora ovariae,* a specific infection of the golden shiner *Notemigonus crysoleucas* in which the ovaries are the prime target. The large cysts are considered to reduce fecundity, cause pressure atrophy of adjacent uninfected ova and parenchyma, and lead to sterility.

Philometra is found in the ovaries as well as testes and although there may be little effect on egg production, in some cases there is hemorrhage, fibrosis, and an increased number of melanomacrophages. Visceral parasites such as the larval cestode *Ligula* will cause pressure atrophy of ovaries. A heavy ectoparasitic load may inhibit proper maturation of eggs, probably due to competition for nutrients; malnutrition can of course have a similar effect. Ovarian abnormalities, including numbers of atretic follicles, are higher in fish from heavily polluted sites, and there may be impaired egg production.

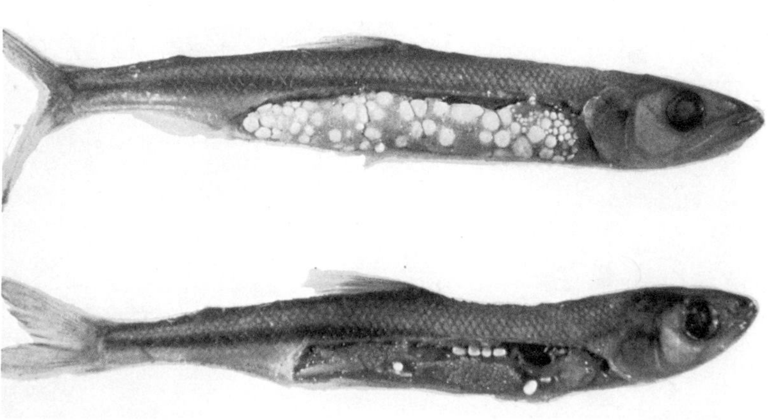

Fig. 11.13. Smelt with microsporidian *Glugea* infection of ovary. Infected cells are creamy and markedly hypertrophic.

References

Antila, E. 1984. Steroid conversion by oocytes and early embryos of *Salmo gairdneri. Ann. Zool. Fenn.* 21:465–71.

Arillo, A., B. Uva, and M. Vallarino. 1981. Renin activity in rainbow trout (*Salmo gairdneri* Rich.) and effects of environmental ammonia. *Comp. Biochem. Physiol.* 68:307–11.

Bailey, J. R., and D. J. Randall. 1981. Renal perfusion pressure and renin secretion in the rainbow trout, *Salmo gairdneri. Can. J. Zool.* 59:1220–26.

Ball, J. N., and B. I. Baker. 1969. The pituitary gland: anatomy and histophysiology. In *Fish Physiology,* ed. W. S. Hoar and D. J. Randall, vol. 2:1–110. New York: Academic Press.

Barrington, E. J. W., and G. J. Dockray. 1972. Cholecystokinin-pancreozymin-like activity in the eel, (*Anguilla anguilla*). *Gen. Comp. Endocrinol.* 19:80.

Benjamin, M. 1985. Experimental studies on cysts in the prolactin zone of the pituitary in the nine-spined stickleback, *Pungitius pungitius* L. *J. Comp. Path.* 95:57–64.

Billard, R., and P. Roubaud. 1985. The effect of metals and cyanide on fertilization in rainbow trout (*Salmo gairdneri*). *Water Res.* 19:209–14.

Billard, R., B. Breton, and M. Richard. 1981. On the inhibitory effect of some steroids on spermatogenesis in adult rainbow trout (*Salmo gairdneri*). *Can. J. Zool.* 59:1479–87.

Billard, R., R. Christen, M. P. Cosson, J. L. Gatty, L. Letellier, P. Renard, and A. Saad. 1986. Biology of the gametes of some teleost species. *Fish Physiol. Biochem.* 2:115–20.

Brinn, J. E. 1973. The pancreatic islets of bony fishes. *Am. Zool.* 13:653–65.

Burke, M. G., and J. F. Leatherland. 1984. Seasonal changes in testicular histology of brown bullheads, *Ictalurus nebulosus* Lesueur. *Can. J. Zool.* 62:1185–94.

Butler, D. G. 1973. Structure and function of the adrenal gland of fishes. *Am. Zool.* 13:839–79.

Chakraborti, P. K., M. Weisbart, and A. Chakraborti. 1987. The presence of corticosteroid receptor activity in the gills of the brook trout, *Salvelinus fontinalis. Gen. Comp. Endocrinol.* 66:323–32.

Chan, D. K. O. 1975. Cardiovascular and renal effects of urotensins I and II in the eel, *Anguilla rostrata. Gen. Comp. Endocrinol.* 27:52–61.

______. 1977. Comparative physiology of the vasomotor effects of neurohypophysial peptides in the vertebrates. *Am. Zool.* 17:751–61.

Chan, D. K. O., and H. A. Bern. 1976. The caudal neurosecretory system: a critical evaluation of the two hormone hypothesis. *Cell Tissue Res.* 174:339–54.

Chan, D. K. O., and N. Y. S. Woo. 1978. Effect of cortisol on the metabolism of the eel, *Anguilla japonica. Gen. Comp. Endocrinol.* 35(3):205–15.

______. 1978. Effect of glucagon on the metabolism of the eel, *Anguilla japonica. Gen. Comp. Endocrinol.* 35(3):216–25.

Chan, D. K. O., J. C. Rankin, and I. Chester-Jones. 1969. Influences of the adrenal and the corpuscles of Stannius on osmoregulation in the European eel (*Anguilla anguilla* L.) adapted to freshwater. *Gen. Comp. Endocrinol.* supp. 2, 342–53.

Clarke, W. C., and Y. Nagahama. 1977. Effect of premature transfer to sea water on growth and morphology of the pituitary, thyroid, pancreas, and interrenal in juvenile coho salmon (*Oncorhynchus kisutch*). *Can. J. Zool.* 55:1620–30.

Couch, J. A. 1984. Histopathology and enlargement of a teleost exposed to the herbicide trifluralin. *J. Fish Dis.* 7:157–63.

Coupland, R. E. 1979. Catecholamines. In *Hormones and Evolution,* ed. E. J. W. Barrington, vol. 1, 309–40, London: Academic Press.

Crim, J. W., W. W. Dickhoff, and A. Gorbman. 1978. Comparative endocrinology of piscine hypothalamic hypophysiotropic peptides: distribution and activity. *Am. Zool.* 18:411–24.

Cristy, M. 1974. Effects of prolactin and thyroxine on the visual pigments of trout, *Salmo gairdneri. Gen. Comp. Endocrinol.* 23:58–62.

Davis, M. S., and T. J. Shuttleworth. 1985. Pancreatic peptides regulate Cl-secretion in the marine teleost gill. *Peptides* 6:379–82.

Dockray, G. J. 1978. Comparative biochemistry and physiology of gut hormones. *Ann. Rev. Physiol.* 41:83.

Donaldson, E. M., and J. R. McBride. 1967. The effects of hypophysectomy in the rainbow trout *Salmo gairdneri* (Rich.) with special reference to the pituitary-interrenal axis. *Gen. Comp. Endocrinol.* 9:93–101.

Duston, J., and N. Bromage. 1986. Photoperiodic mechanisms and rhythms of reproduction in the female rainbow trout. *Fish Physiol. Biochem.* 2:35–51.

Eales, J. G. 1979. Thyroid functions in cyclostomes and fishes. In *Hormones and Evolution,* ed. E. J. W. Barrington, vol. 1, 341–436, New York: Academic Press.

Epple, A., T. L. Lewis. 1973. Comparative histophysiology of the pancreatic islets. *Am. Zool.* 13:567–90.

Farbridge, K. J., and J. F. Leatherland. 1986. A comparative immunohistochemical study of the pars distalis in six species of teleost fishes. *Fish Physiol. Biochem.* 1:63–74.

Farbridge, K. J., M. G. Burke, and J. F. Leatherland. 1985. Seasonal changes in the structure of the adenophypophysis of the brown bullhead (*Ictalurus nebulosus* Lesueur). *Cytobios* 44:49–66.

Fenwick, J. C., and Y. P. So. 1981. Effect of an angiotensin on the net influx of calcium across an isolated perfused eel gill. *Can. J. Zool.* 59:199–221.

Foster, G. D., and T. W. Moon. 1986. Cortisol and liver metabolism of immature American eels, *Anguilla rostrata* (LeSeur). *Fish Physiol. Biochem.* 1:113–24.

———. 1987. Metabolism in sea raven (*Hemitripterus americanus*) hepatocytes: the effects of insulin and glucagon. *Gen. Comp. Endocrinol.* 66:102–15.

Fostier, A., and B. Jalabert. 1986. Steroidogenesis in rainbow trout (*Salmo gairdneri*) at various preovulatory stages: changes in plasma hormone levels and *in vivo* and *in vitro* responses of the ovary to salmon gonadotropin. *Fish Physiol. Biochem.* 2:87–99.

Gervai, J., T. Marian, Z. Krasznai, A. Nagy, and V. Csanyi. 1980. Occurrence of aneuploidy in radiation gynogenesis of carp, *Cyprinus carpio* L. *J. Fish Biol.* 16:435–39.

Gorbman, A., W. W. Dickhoff, S. R. Vigna, N. B. Clark, and C. L. Ralph. 1983. *Comparative Endocrinology.* New York: John Wiley & Sons.

Grier, H. J., J. R. Linton, J. F. Leatherland, and V. I. De Vlaming. 1980. Structural evidence for two different testicular types in teleost fishes. *Am. J. Anat.* 159:331–45.

Hane, S., and O. H. Robertson. 1959. Changes in plasma 17-hydroxycorticosteroids accompanying sexual maturation and spawning of the Pacific salmon (*Oncorhynchus tschawytscha*) and rainbow trout (*Salmo gairdneri*). *Proc. N.Y. Acad. Sci.* 45:886–93.

Hilton, J. W., E. M. Plisetskaya, and J. F. Leatherland. 1987. Does oral 3, 5, 3′-triiodo-L-thyronine affect dietary glucose utilization and plasma insulin levels in rainbow trout (*Salmo gairdneri*)? *Fish Physiol. Biochem.* 4:113–20.

Holmgren, S., C. Vaillant, and R. Dimaline. 1982. VIP-, substance P-, Gastrin/CCK-, Bombesin-, Somatostatin- and Glucagon-like immunoreactivities in the gut of the rainbow trout, *Salmo gairdneri. Cell Tissue Res.* 223:141.

Ichikawa, T., H. Kobayashi, P. Zimmermann, and U. Muller. 1973. Pituitary response to environmental osmotic changes in the larval guppy, *Lebistes reticulatus* (Peters). *Z. Zellforsch.* 141:161–79.

Ishihara, A., and Y. Mugiya. 1987. Ultrastructural evidence of calcium uptake by chloride cells in the gills of goldfish, *Carassius auratus. J. Exp. Zool.* 242:121–29.

Kah, O. 1986. Central regulation of reproduction in teleost. *Fish Physiol. Biochem.* 2:25–34.

Kenyon, C. J., I. Chester Jones, and R. N. B. Dixon. 1980. Acute responses of the freshwater eel (*Anguilla anguilla*) to extracts of the corpuscles of Stannius opposing the effects of stanniosomatiectomy. *Gen. Comp. Endocrinol.* 41:531–38.

Kime, D. E. 1980. Comparative aspects of testicular androgen biosynthesis in nonmammalian vertebrates. In *Steroids and Their Mechanism of Action in Nonmammalian Vertebrates,* ed. G. Delrio and J. Brachet. New York: Raven Press.

Lam, T. J., S. Pandey, and W. S. Hoar. 1975. Induction of ovulation in goldfish by synthetic luteinizing hormone-releasing hormone (LH-RH). *Can. J. Zool.* 53:1189–92.

Lambert, J. G. D. 1970. The ovary of the guppy *Poecilia reticulata.* The granulosa cell as sites of steroid biosynthesis. *Gen. Comp. Endocrinol.* 15:464–76.

Leatherland, J. F. 1982. Environmental physiology of the teleostean thyroid gland: a review. *Environ. Biol. Fish* 7:83–110.

Leatherland, J. F., and C. Y. Cho. 1985. Effect of rearing density on thyroid and interrenal gland activity and plasma and hepatic metabolite levels in rainbow trout, *Salmo gairdneri* Richardson. *J. Fish Biol.* 27:583–92.

Leatherland, J. F., and R. A. Sonstegard. 1978. Lowering of serum thyroxine and triiodothyronine levels in yearling coho salmon, *Oncorhynchus kisutch,* by dietary mirex and PCBs. *J. Fish. Res. Board Can.* 35:1285–89.

______. 1980. Seasonal changes in thyroid hyperplasia, serum thyroid hormone and lipid concentrations, and pituitary gland structure in Lake Ontario coho salmon, *Oncorhynchus kisutch* Walbaum and a comparison with coho salmon from Lakes Michigan and Erie. *J. Fish Biol.* 16:539–62.

______. 1981. Thyroid function, pituitary structure and serum lipids in Great Lakes coho salmon, *Oncorhynchus kisutch* Walbaum, Jacks compared with sexually immature spring salmon. *J. Fish Biol.* 18:643–53.

Leatherland, J. F., R. Moccia, and R. Sonstegard. 1978. Ultrastructure of the thyroid gland in goitered coho salmon (*Oncorhynchus kisutch*). *Cancer Res.* 38:149–58.

Leatherland, J. F., R. A. Sonstegard, and R. D. Moccia. 1981. Interlake differences in body and gonad weights and serum constituents of Great Lakes coho salmon (*Oncorhynchus kisutch*). *J. Biochem. Physiol.* 69:701–64.

Lederis, K. 1977. Chemical properties and the pharmacological actions of urophysial peptides. *Am. Zool.* 17:823–32.

Leung, E., and J. C. Fenwick. 1978. Hypocalcemic action of eel Stannius corpuscle in rats. *Can. J. Zool.* 56:2333–35.

McBride, J. R., and A. P. van Overbeeke. 1969. Hypertrophy of the interrenal tissue in sexually maturing sockeye salmon (*Oncorhynchus nerka*) and the effect of gonadectomy. *J. Fish. Res. Board Can.* 26:2975–85.

______. 1971. Effects of androgens, estrogens, and cortisol on the skin, stomach, liver, pancreas, and kidney in gonadectomized adult sockeye salmon (*Oncorhynchus nerka*). *J. Fish. Res. Board Can.* 28:485–90.

MacGregor, R., III, J. J. Dindo, and J. H. Finucane. 1981. Changes in serum androgens and estrogens during spawning in bluefish, *Pomatomus saltator,* and king mackerel, *Scomberomorus cavalla. Can. J. Zool.* 59:1749–54.

McMillan, P. J., W. M. Hooker, B. A. Roos, and L. J. Deftos. 1976. Ultimobranchial gland of the trout (*Salmo gairdneri*). 1. Immunohistochemistry and radioimmunoassay of calcitonin. *Gen. Comp. Endocrinol.* 28:313–33.

McNulty, J. A. 1981. A quantitative morphological study of the pineal organ in the goldfish, *Carassius auratus. Can. J. Zool.* 59:1312–25.

Marine, D. 1914. Further observations and experiments on goitre (so-called thyroid carcinoma) in brook trout (*Salvelinus fontinalis*). *J. Exp. Med.* 19:70–88.

Marshall, W. S., and H. A. Bern. 1981. Active chloride transport by the skin of a marine teleost is stimulated by urotensin I and inhibited by urotensin II. *Gen. Comp. Endocrinol.* 43:484–91.

Matty, A. J. 1985. *Fish Endocrinology.* Portland, Ore.: Timber Press.

______. 1986. Nutrition, hormones and growth. *Fish Physiol. Biochem.* 2:141–50.

Mazeaud, M. M., and F. Mazeaud. 1981. Adrenergic responses to stress in fish. In *Stress in Fish,* ed. A. D. Pickering. London: Academic Press.

Meissl, H., T. Nakamura, and G. Thiele. 1986. Neural response mechanisms in the photoreceptive pineal organ of goldfish. *Comp. Biochem. Physiol.* 84:467–73.

Milne, R. S., J. F. Leatherland, and B. J. Holub. 1979. Changes in plasma thyroxine, triiodothyronine and cortisol associated with starvation in rainbow trout (*Salmo gairdneri*). *Environ. Biol. Fish.* 4:185–90.

Moccia, R. D., J. F. Leatherland, and R. A. Sonstegard. 1981. Quantitative interlake comparison of thyroid pathology in Great Lakes coho (*Oncorhynchus kisutch*) and chinook (*Oncorhynchus tschawytscha*) salmon. *Cancer Res.* 41:2200–2210.

Murat, J. C., E. M. Plisetskaya, and N. Y. S. Woo. 1981. Review. Endocrine control of nutrition in cyclostomes and fish. *Comp. Biochem. Physiol.* 68:149–58.

Nagahama, Y., H. Kagawa, and G. Young. 1982. Cellular sources of sex steroids in teleost gonads. *Can. J. Fish. Aquat. Sci.* 39:56–64.

Nishimura, H., M. Ogwana, and W. H. Sawyer. 1973. Renin-angiotensin system in primitive bony fishes and a holocephalan. *Am. J. Physiol.* 224:950.

Noakes, D. L. G., and J. F. Leatherland. 1977. Social dominance and interrenal cell activity in rainbow trout, *Salmo gairdneri* (Pisces, Salmonidae). *Environ. Biol. Fish.* 2:131–36.

Oguri, M. 1971. A histological study on the ACTH cells in the pituitary glands of freshwater teleosts. *Bull. Jpn. Soc. Sci. Fish.* 37:577–84.

Pang, P. K. 1973. Endocrine control of calcium metabolism in teleosts. *Am. Zool.* 13:775–92.

Pang, P. K. T., R. K. Pang, V. K. Y. Liu, and H. Sokabe. 1981. Effect of fish angiotensins and angiotensin-like substances on killifish calcium regulation. *Gen. Comp. Endocrinol.* 43:292.

Pechan, P., D. Y. Shapiro, and M. Tracey. 1986. Increased H-Y antigen levels associated with behaviorally induced, female-to-male sex reversal in a coral-reef fish. *Differentiation* 31:106–10.

Peter, R. E. 1982. Neuroendocrine control of reproduction in teleosts. *Can. J. Fish Aquat. Sci.* 39:48.

Peter, R. E., M. Sokolowska, B. Truscott, J. Walsh, and D. R. Idler. 1984. Secretion of progestogens during induced ovulation in goldfish. *Can. J. Zool.* 62:1946–49.

Pickering, A. D., and T. G. Pottinger. 1983. Seasonal and diel changes in plasma cortisol levels of the brown trout, *Salmo trutta* L. *Gen. Comp. Endocrinol.* 49:232–39.

Plisetskaya, E. M., B. N. Leibush, and V. M. Bondareva. 1976. The secretion of insulin and its role in cyclostomes and fishes. In *The Evolution of Pancreatic Islets,* ed. T. A. I. Grillo, L. G. Leibson and A. Epple, 251–69. London: Permagon Press.

Plisetskaya, E. M., N. Y. S. Woo, and J-C. Murat. 1983. Review: thyroid hormones in cyclostomes and fish and their role in regulation of intermediary metabolism. *Comp. Biochem. Physiol.* 74:179–87.

Reinboth, R. 1972. Hormonal control of the teleost ovary. *Am. Zool.* 12:307–24.

Schreibman, M. P., J. F. Leatherland, and B. A. McKeown. 1973. Functional morphology of the teleost pituitary gland. *Am. Zool.* 13:719–42.

Scott, A. P., and S. M. Baynes. 1980. A review of the biology, handling and storage of salmonid spermatozoa. *J. Fish Biol.* 17:707–39.

Scott, A. P., J. P. Sumpter, and P. A. Hardiman. 1983. Hormone changes during ovulation in the rainbow trout (*Salmo gairdneri* Richardson). *Gen. Comp. Endocrinol.* 49:128–34.

Secombes, C. J., A. E. Lewis, L. M. Laird, E. A. Needham, and I. G. Priede. 1984. Agglutination of spermatozoa by autoantibodies in the rainbow trout, *Salmo gairdneri. J. Fish Biol.* 25:691–96.

Slooff, W., and E. Klootwijk-Van Dijk. 1982. Hermaphroditism in the bream, *Abramis brama* (L.). *J. Fish Dis.* 5:79–81.

Sonstegard, R., J. F. Leatherland, and R. Moccia. 1977. Thyroid activity of alewives (*Alosa pseudoharengus*) and rainbow smelts (*Osmerus mordax*) in the Great Lakes. *J. Fish. Res. Board Can.* 34:1242–44.

Spieler, R. E., and T. A. Noeske. 1979. Diel variations in circulating levels of triiodothyronine and thyroxine in goldfish, *Carassius auratus. Can. J. Zool.* 57:665–69.

Stacey, N. E. 1981. Hormonal regulation of female reproductive behavior in fish. *Am. Zool.* 21:305–16.

Stefan, Y., and S. Falkmer. 1980. Identification of four endocrine cell types in the pancreas of *Cottus scorpius* (Teleostei) by immunofluorescence and electron microscopy. *Gen. Comp. Endocrinol.* 42:171–78.

Stott, G. G., W. E. Haensly, J. M. Neff, and J. R. Sharp. 1983. Histopathologic survey of ovaries of plaice, *Pleuronectes platessa* L., from Aber Wrac'h and Aber Benoit, Brittany, France: long-term effects of the *Amoco Cadiz* crude oil spill. *J. Fish Dis.* 6:429–37.

Summerfelt, R. C. 1964. A new microsporidian parasite from the golden shiner, *Notemigonus crysoleucas. Trans. Am. Fish. Soc.* 93:6–10.

Triplett, E., and J. R. Calaprice. 1974. Changes in plasma constituents during spawning migration of Pacific salmons. *J. Fish. Res. Board Can.* 31:11–14.

Villars, T. A., and M. Burdick. 1986. Rapid decline in the behavioral response of paradise fish (*Macropodus opercularis*) to prostaglandin F_2 treatment. *Gen. Comp. Endocrinol.* 63:157–61.

Wagner, G. F., M. Hampong, C. M. Park, and D. H. Copp. 1986. Purification, characterization, and bioassay of teleocalcin, a glycoprotein from salmon corpuscles of Stannius. *Gen. Comp. Endocrinol.* 63:481–91.

Wendelaar Bonga, S. E. 1976. The effect of prolactin on kidney structure of the euryhaline teleost *Gasterosteus aculeatus* during adaptation to fresh water. *Cell Tissue Res.* 166:319–38.

______. 1980. Effect of synthetic salmon calcitonin and low ambient calcium on plasma calcium, ultimobranchial cells, stannius bodies, and prolactin cells in the teleost, *Gasterosteus aculeatus. Gen. Comp. Endocrinol.* 40:99.

______. 1981. Effect of synthetic salmon calcitonin on protein-bound and free plasma calcium in the teleost *Gasterosteus aculeatus. Gen. Comp. Endocrinol.* 43:123–26.

White, A., and T. C. Fletcher. 1986. Serum cortisol, glucose and lipids in plaice (*Pleuronectes platessa* L.) exposed to starvation and aquarium stress. *Comp. Biochem. Physiol.* 84:649–53.

Wilson, J. X. 1984. Coevolution of the renin-angiotensin system and the nervous control of blood circulation. *Can. J. Zool.* 62:137–47.

Yaron, Z., M. Cocos, and H. Salzer. 1980. Effects of temperature and photoperiod on ovarian recrudescence in the cyprinid fish *Mirogrex terraesanctae. J. Fish Biol.* 16:371–82.

Yokote, M. 1970. Sekoke disease, spontaneous diabetes in carp, *Cyprinus carpio,* found in fish farms. I. Pathological study. *Bull. Freshwater Fish. Res. Lab.* 20:39–72.

______. 1970. Sekoke disease, spontaneous diabetes in carp, *Cyprinus carpio,* found in fish farms. V. Hydrocortisone diabetes in carp. *Bull. Freshwater Fish. Res. Lab.* 20:161–74.

Youngson, A. F., H. A. McLay, and T. C. Olsen. 1986. The responsiveness of the thyroid system of Atlantic salmon (*Salmo salar* L.) smolts to increased water velocity. *Aquaculture* 56:243–55.

Zucker, A., and H. Nishimura. 1981. Renal responses to vasoactive hormones in the aglomerular toadfish, *Opsanus tau. Gen. Comp. Endocrinol.* 43:1–9.

12 Musculoskeletal System

DESPITE its size and importance from many standpoints, not least of which are the economic aspects of carcase quality, the musculoskeletal system is often overlooked at a routine diagnostic level, and samples are taken only if there are grossly evident lesions.

General Considerations

Muscle. There are two main and several minor types of muscle fibers. Red "slow" fibers are well vascularized and contain numerous mitochondria. These are often as numerous as in mammalian myocardium. In many species, red muscle is located as a wedge along the lateral line (Fig. 12.1).

Separating the red from the white "fast" fibers are the intermediate or "pink" fibers. In some species, such as salmonids, these fast aerobic fibers comprise a thin scattering situated between the red and white muscle, whereas in others, such as carp, they are much more extensive.

White fibers have a wide range of diameters, especially in growing fish; this becomes less obvious in adults. They are not as well vascularized as red fibers and have a lower number of mitochondria with less well-developed cristae, although these latter

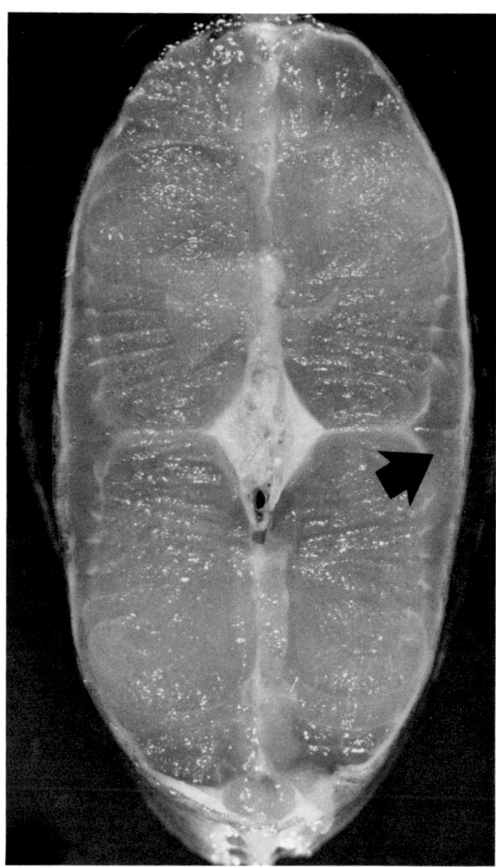

Fig. 12.1. Transverse section through body of trout at level of adipose fin. Wedge-shaped strips of red muscle running down each side are shown (*arrow*).

features are both influenced by nutritional and environmental factors. For example, Johnston (1982) has shown that the white fibers of starved plaice (*Pleuronectes platessa*) contain higher than normal densities of mitochondria. Whereas red fibers have multiple terminal innervation, white fibers have in addition, polyneuronal innervation, a pattern unique to higher teleosts. (Presumably, therefore, the likelihood of denervation atrophy as seen in mammals is reduced as a consequence of this arrangement.)

In many species, muscle fiber numbers increase throughout a large part of the life of the fish (in mammals, of course, numbers are fixed at or just after birth). This increase, which originates from myosatellite cells, must have a significant influence on the reparative potential of fish muscle following even severe disease.

Inflammatory, degenerative, and reparative responses in fish muscle mimic their mammalian counterparts, with myophagia, regeneration, and fibrosis all occurring but at slower rates, proportional to the lower temperature. Thus myosatellite cell proliferation with central rowing of nuclei in regenerating fibers is commonly seen.

Bone. Teleost bone is of two major types: cellular and acellular. The latter is found only in higher teleosts and is another feature unique to this group of vertebrates. Osteogenesis in other respects is similar to mammals, with endochondral and intramembranous ossification both occurring. Spongy bone may be found in the gill arches and cranial bones, but despite the presence of these spaces, hemopoietic elements are not found here. Dermal elements of the skeletal system include the teeth, spines, and fin rays.

Repair and remodeling of damaged skeletal elements involve osteocytes and osteoclasts, although in acellular bone the process is hindered.

Disease

Muscle. Degeneration of muscle has been experimentally produced in salmonids fed a vitamin E/selenium deficient diet, particularly at low water temperatures where requirements would appear to be higher. This "nutritional muscular dystrophy" (NMD) appears to be entirely analagous to similar syndromes seen in other intensively reared animals, especially pigs, cattle, and poultry. Lesions include hyaline and granular degeneration of fibers with loss of striation (Fig. 12.2). Repair sees multiplication of sarcolemmal nuclei, myophagia, and central rowing of nuclei (Fig. 12.3). Mineralization

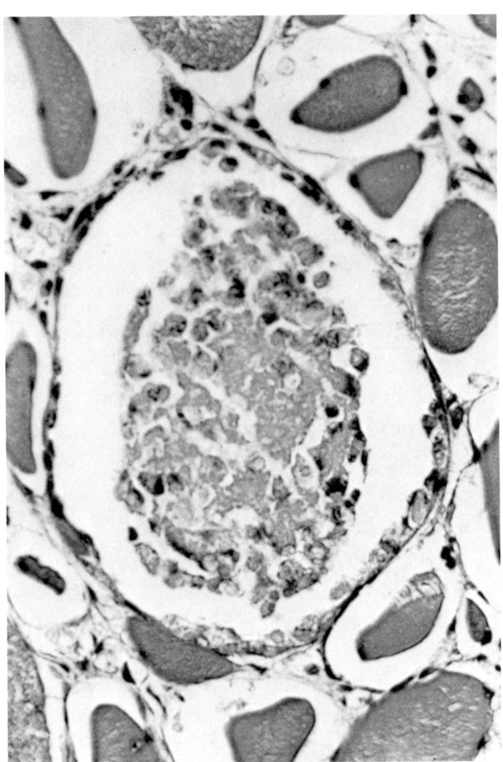

Fig. 12.2. Granular degeneration and necrosis in this section of white muscle from Atlantic salmon with degenerative rhabdomyopathy and pancreas disease, followed by macrophage invasion to remove necrotic tissue and multiplication of sarcolemmal nuclei.

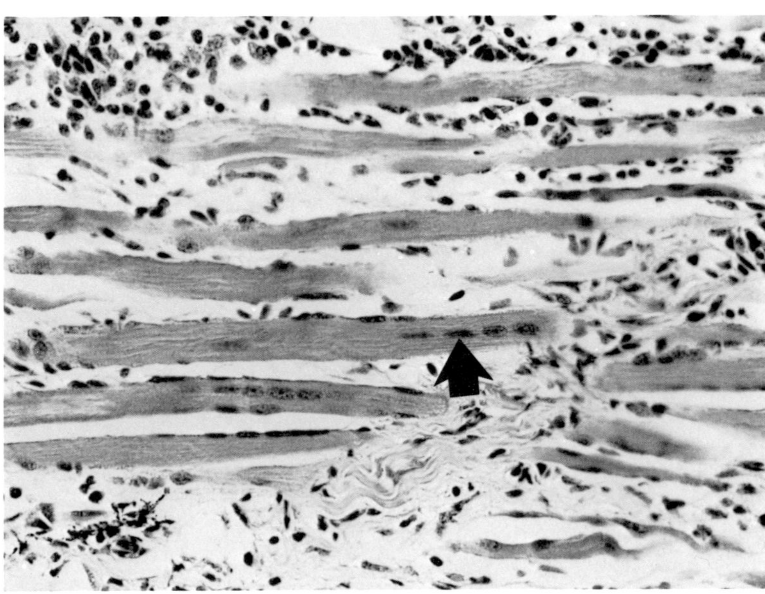

Fig. 12.3. Repairing red muscle in Atlantic salmon with pancreas disease. Note central rowing of nuclei (*arrow*) indicating regeneration.

and inflammation are usually minimal. There are many reports of NMD in intensively cultured fish; in salmonids they are described in the syndromes of pancreas disease and Hitra disease as well as pansteatitis. Similar syndromes occur in channel catfish, puffers, and ayu, as well as in Sekoke disease of carp. This latter disease is associated with rancid oil in the silkworm pupae that are incorporated into the carp diet. In pancreas disease, and probably the other syndromes too, red muscle fibers are more affected than white, but if the lesions are severe and the fish becomes sluggish as a consequence of impaired swimming, feeding responses may be diminished or eliminated. The resulting starvation probably complicates the histological picture due to catabolism of the muscle. Starvation can also be compounded if the striated esophageal muscles are affected to any extent, resulting in the fish being unable to swallow any food it does manage to reach. In addition, severely affected fish may be unable to maintain their position in the water column, with traumatic damage to skin and muscle from the net or tank sides a probability, especially if water flow rates are high. Under these conditions, secondary bacterial or fungal invasion is a distinct probability, thereby complicating the picture even more. A similar degenerative myopathy is seen in coho salmon exposed to the pesticide carbaryl.

Myopathies are seen in wild walleye (Fig. 12.4) and in the Siamese fighting fish under aquarium conditions. Of unknown etiology, these conditions present as a severe necrotizing and granulomatous myositis, the degenerating muscle undergoing caseonecrotic change and becoming the focus for a pronounced inflammatory response. The reason for this very different response (compared to NMD) is unknown, but it suggests much more severe damage with possible loss of myosatellite cells as well as myocytes, thereby inhibiting regeneration.

Muscle hemorrhage is seen in many acute bacteremias and viremias, while in some bacterial diseases, such as furunculosis (Fig. 12.5) and vibriosis (Fig. 12.6), multifocal liquefactive necrosis of muscle can be pronounced, especially in the subacute or chronic stages of these diseases. The affinity of such organisms for vascular endothelium suggests that at least some of these lesions may be infarcts. Certainly the lesions are

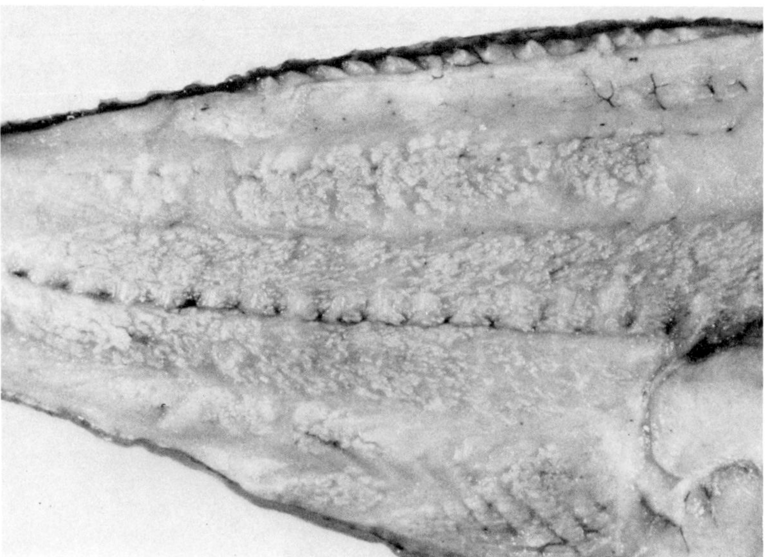
A

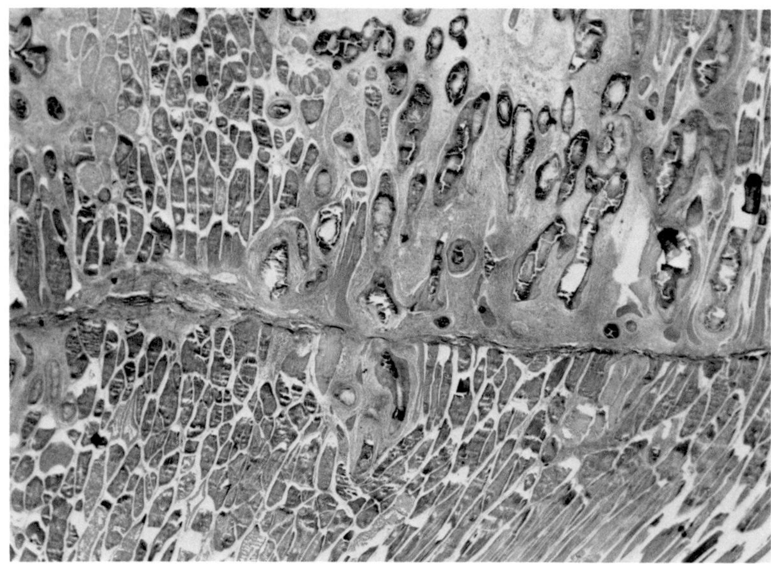
B

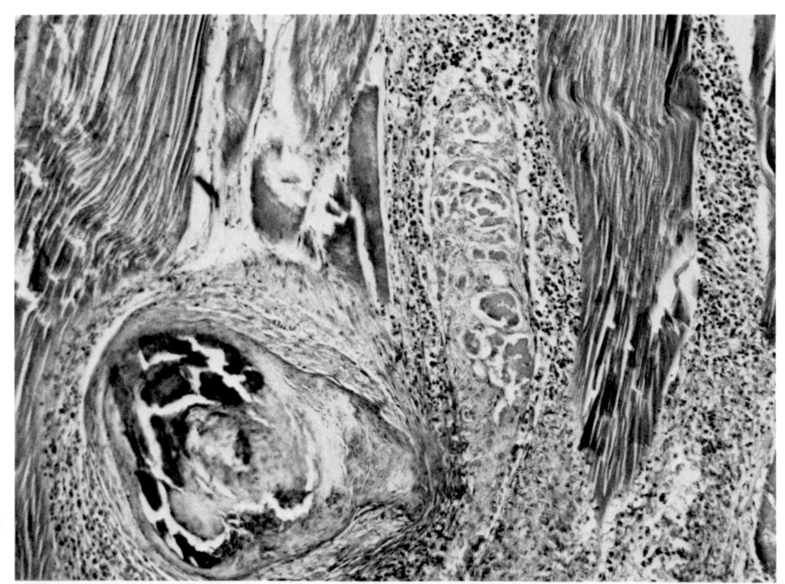
C

Fig. 12.4. Myopathy of walleye. **A.** Fillet showing multifocal to focally extensive areas of fiber necrosis. Such fillets are obviously unacceptable for human consumption. **B** and **C.** Low- and high-power views of **A** showing extent and severity of caseonecrotic reaction and the associated granulomatous response.

A

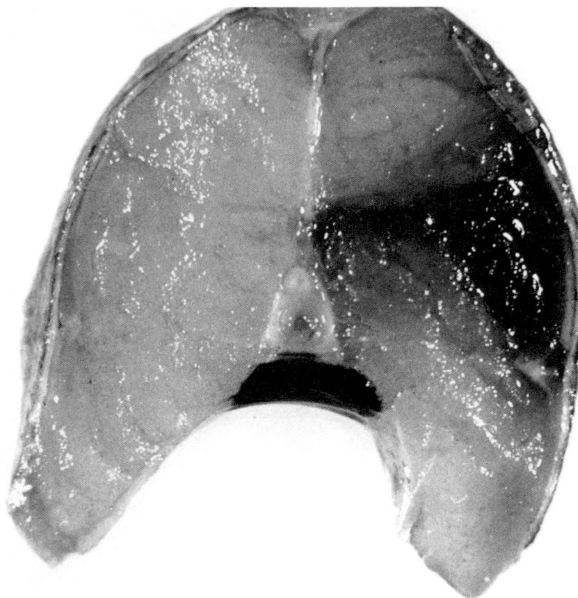

B

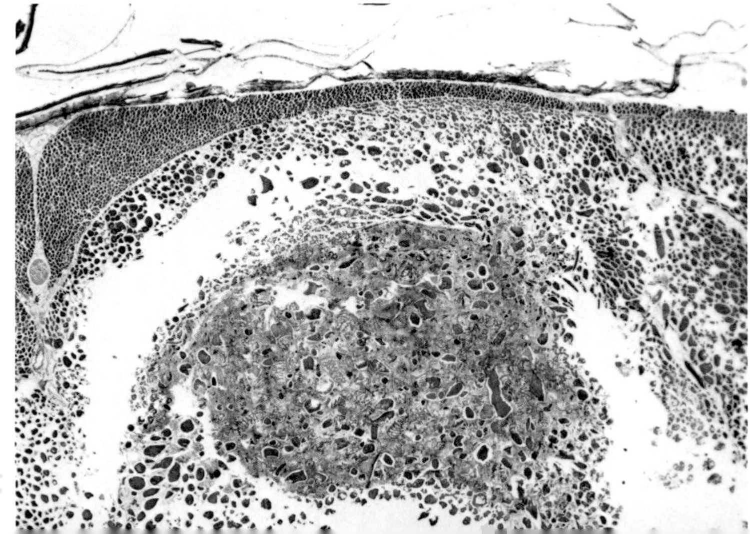

C

Fig. 12.5. Furunculosis (*Aeromonas salmonicida*). **A.** Market-sized rainbow trout with skin partly reflected to show large necrotic and hemorrhagic focal myositis. (*Courtesy of D. H. McCarthy*) **B.** Transverse section through early lesion in brook trout to show extent and severity of hemorrhagic reaction. **C.** Creek chub with large necrotic focus (almost a sequestrum) in dorsal musculature.

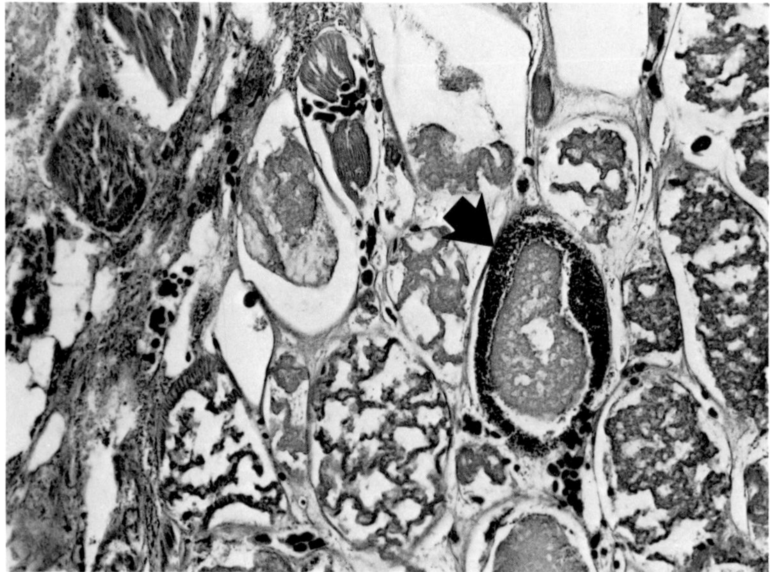

Fig. 12.6. Acute necrotizing myositis in Atlantic salmon with vibriosis (*Vibrio anguillarum*). A large number of bacteria may be seen surrounding one fiber (*arrow*).

usually hemorrhagic and pulpy and often rupture through the overlying skin. Cavitation of muscle is seen in bacterial kidney disease, and if the lesions are present in the subcutis, the usually straw-colored flocculent contents create large, externally visible blisters (see Fig. 3.19). Alternatively, the lesions may be so large as to grossly distort the whole fish (Fig. 12.7). The chronic myositis that accompanies these lesions is usually very pronounced, the cellular infiltrate comprising neutrophils and macrophages. Many of the latter are stuffed with bacteria al-

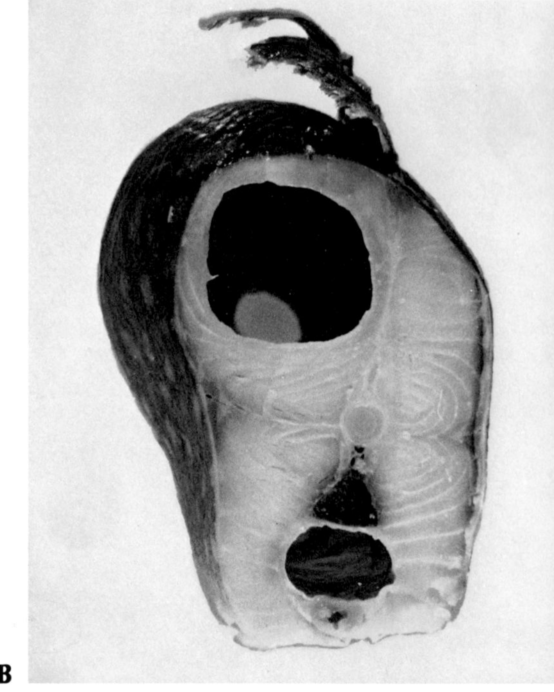

B

Fig. 12.7. A. Bacterial kidney disease in young coho salmon with severe myositis of peduncle region causing gross distortion. **B.** Large cavitating abscess in wild-caught speckled trout. Although bacterial kidney disease was suspected, no organisms were found.

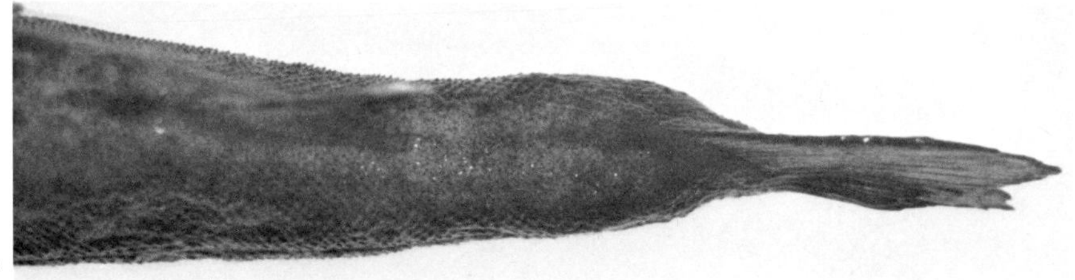

A

though special stains are needed to see this. Similar lesions are seen in salmonids with *Lactobacillus* (recently renamed *Carnobacterium*) *piscicola* infection, which occurs frequently as a postspawning event. Cavitation of the muscles dorsal to the kidney is also seen in very severe cases of nephrocalcinosis (see Fig. 4.3).

Any disease with dermal lesions pronounced enough (such as ulceration) to interfere with osmotic integrity will, in fresh water, result in water-logging of the underlying muscle; inflammation under these circumstances is usually minimal. This is particularly obvious where *Saprolegnia* is involved. By contrast, the disseminated fungal diseases cause a granulomatous myositis. A good example of this occurs with *Ichthyophonus*.

Some of the most economically important diseases of fish include the microsporidian (phylum Microspora) and myxosporean (subphylum Myxosporea) infestations of muscle. These can be so extensive as to cause grossly visible opaque milky white blotches or streaks, thereby ruining the esthetic appearance of the fish (Fig. 12.8). Examples include the myxosporean *Henneguya*

Fig. 12.8. A. Commercially caught yellow perch with creamy white nodular myositis due to myxosporean parasites (*arrow*). A high percentage of such fish may be discarded as unfit for human consumption although this varies from year to year. **B.** Section through typical myxosporean "cyst" to show appearance of spores. These appear refractile under a normal H and E stain but are well demonstrated, as in this case, by toluidine blue. **C.** Microsporidian "cyst" at same magnification as **B** showing very much smaller size of spores.

A

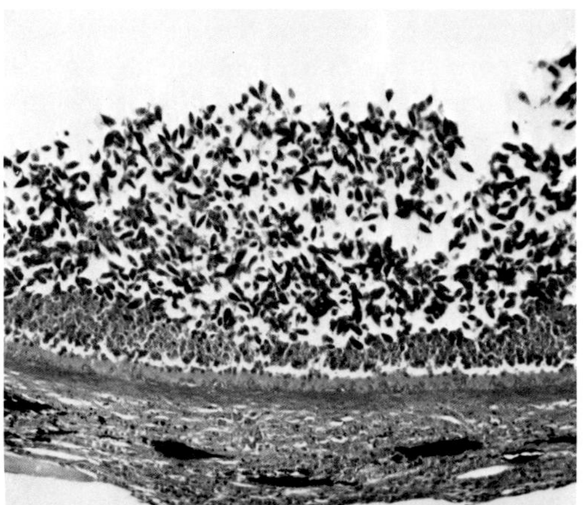

B

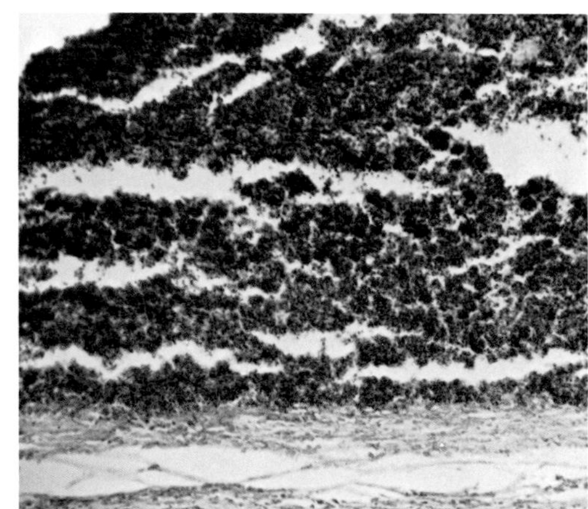

C

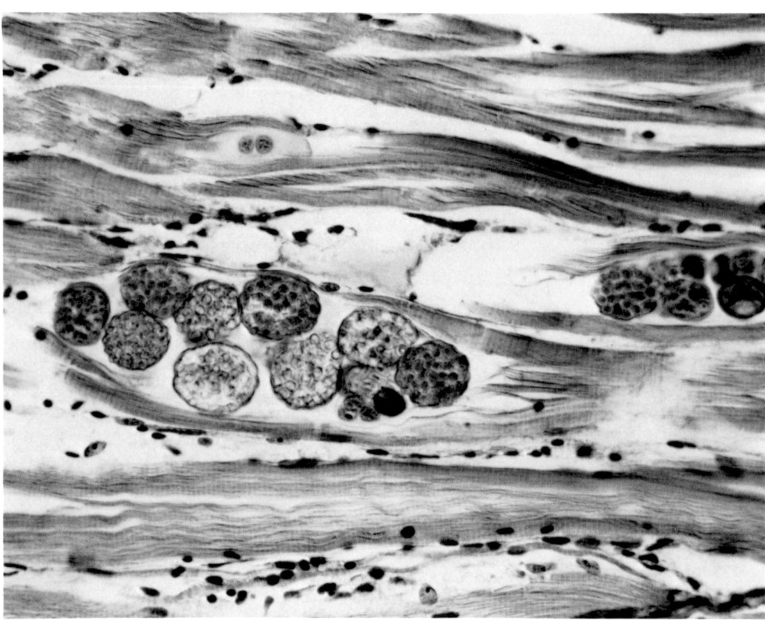

Fig. 12.9. Cardinal tetra with neon tetra disease caused by the microsporidian *Pleistophora hyphessobryconis.* Various developmental stages of the parasite may be seen within several muscle fibers, enlarging them but eliciting no inflammatory response.

salminicola affecting Pacific salmon and the microsporidian *Pleistophora macrozorcidis* affecting ocean pouts *Macrozoarces americanus.* Individual or multiple "cysts" or xenomas within the muscle give the gross appearance of neoplasms. These may cause pressure necrosis of the overlying skin with release of spores into the water as a consequence. Neon tetra disease is an example of a microsporidian (*Pleistophora hyphessobryconis*) muscle infection of an aquarium species (Fig. 12.9). Another microsporidian genus *Glugea* causes serious pathological changes in a variety of tissues. It infects macrophages and other mesenchymal cells that subsequently undergo massive hypertrophy, causing space-occupying compression and deformations. The peritoneal cavity, especially gut and gonads are target sites. Examples include *Glugea stephani* infection of marine flatfish affecting mainly the gut, but ovaries and liver too.

Tissue responses to microsporidian infections are beautifully described by Dykova and Lom (1980). These vary with the species of parasite involved. With *Pleistophora* infections, the parasite does not induce hypertrophy of the myocyte but rather replicates within the sarcoplasm, eventually occupying and destroying it. A significant inflammatory response only develops when the cell ruptures and comprises an infiltrate of macrophages that phagocytose the liberated spores. By contrast, in genera such as *Glugea, Spraguea* (or *Nosema*), and *Ichthyosporidium,* the intracellular parasites replicate and cause cytomegaly, which in turn leads to pressure atrophy and encapsulating fibroplasia. A large xenoma with a refractile wall and filled with refractile spores may elicit the formation of granulation tissue. This often signals the start of degenerative changes in the xenoma wall, its invasion by the surrounding fibroblasts, and eventual disappearance. The result is a granuloma with spores at the center. Macrophages gradually remove the spores and the lesion resolves.

While Microspora are intracellular and affect a wide variety of vertebrates and invertebrates, Myxosporea are largely intercellular and infect mainly fish. (A few myxosporean genera develop intracellularly in skeletal muscle, while others have some development stages that are intracellular.) Classification is based on spore morphology of which there is a wonderful diversity. In section, the spores have a refractile wall and are approximately 10 μm to 12 μm in diameter (microsporidian spores, which also have a refractile wall, are roughly 1 μm to 2 μm). They contain 1 to 6 polar capsules, which stain intensely with toluidine or methylene blue. Many species inhabit organ cavities (coelozoic), but in general the serious pathogens are those that develop within tissues (histozoic), of which skeletal muscle is a prime target (see Fig. 3.23). Even so, initial host inflammatory response to the developing stages is usually minimal and only later is there any appreciable reaction. In some species the skin overlying the "cysts" ulcerates with the consequent release of spores, while in others the host response completely eliminates the parasites. Various species of the genus *Kudoa* are notable for their economic impact on fish muscle. Some authors suggest that the parasites release myolytic enzymes after the death of the fish thereby resulting in rapid liquefaction.

Granulomatous myositis is also seen in severe cases of proliferative kidney disease (presumptively caused by prespore myxosporeans), where the parasite is so widely disseminated as to involve muscle (Fig. 12.10). Another uncommon disease is acute myodegeneration and inflammation associated with the ciliates *Tetrahymena* (Fig. 12.11) and *Uronema,* opportunistic pathogens in guppies, salmonids, and marine aquarium fish.

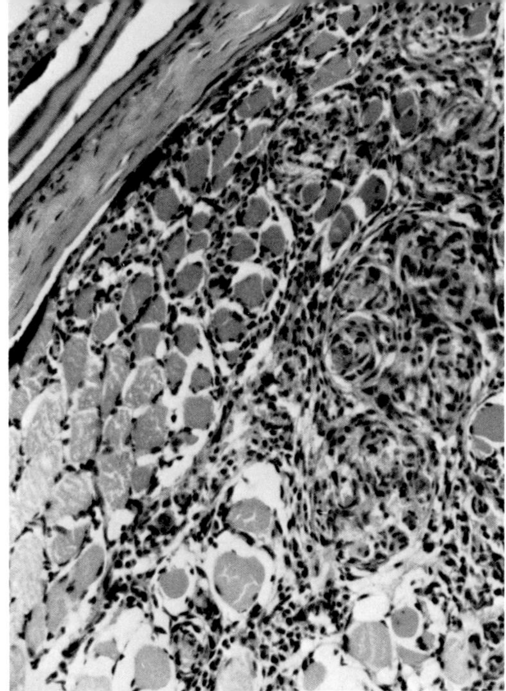

Fig. 12.10. Proliferative kidney disease in rainbow trout showing granulomatous myositis of subdermal muscle. Such lesions may rarely ulcerate.

Metazoan diseases of muscle include encysted metacercariae of digeneans that may or may not evoke a melanin response; if they do, the parasites become a grossly visible black spot (see Fig. 3.24). The host response is otherwise usually minimal, with mild fibrosis, although in unusual cases parasites may be present in numbers large enough to cause ulceration of the body wall (Hoffman

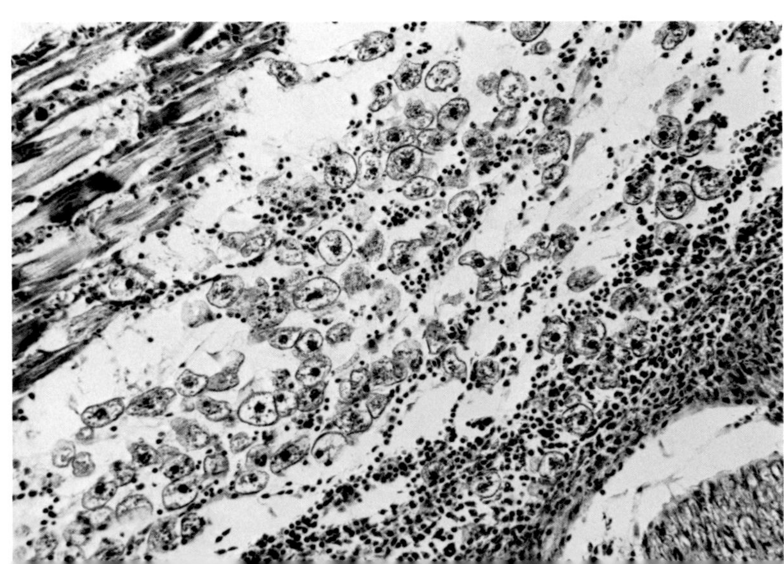

Fig. 12.11. Acute necrotizing myositis of head region in Atlantic salmon parr due to *Tetrahymena* sp., probably *T. corlissi.* Large numbers of ciliates are here accompanied by hemorrhage (see also Fig. 12.16).

and Hutcheson 1970). A much more dramatic response is shown to *Lernea* spp., which inserts part of its crustacean body into the muscle and evokes severe acute inflammation. Larval nematodes such as the anisakids *Phocanema* and *Contracaecum* are found within the muscle, although it is known that some species can migrate there from the peritoneal cavity as a postmortem event. Although the effect on the fish is usually minimal, some of these parasites have zoonotic importance.

Finally, consideration must be given to the possibility that unusual muscle lesions may be the result of trauma from the activities of two- or four-legged hunters and poachers, particularly deep puncture wounds from beaks or gaffs.

Skeleton. Any disease that does cause severe damage to muscle will eventually distort the skeleton. Whenever malformations are encountered therefore, it is important to appreciate that the underlying lesion may lie in muscle, its innervation, or in the skeleton. A dramatic example of the interrelationships between these components is found in lightning strike; severe contraction of the body muscle may break the spinal column at the flexure point (Fig. 12.12). Sublethal organophosphates also lead to hypercontraction of muscle (and eventual spinal deformation) due to cholinesterase inhibition. Similar changes are seen with organochlorines. Vitamin C deficiency in channel catfish leads to "broken back syndrome" due to osteoporosis, dysplasia, and impaired collagen formation. Scoliosis is also seen in salmonids with tryptophan and vitamin C deficiency or vitamin A toxicity (Fig. 12.13). The pathogenesis of this latter lesion is unknown although it may be due to differential development of the vertebral growth plates. Vertebral hyperostosis and dysplasia are seen in salmonids and other fish exposed to the herbicide trifluralin. Asymmetrical mandibular hyperostosis has been occasionally seen in wild pike (Fig. 12.14), although the cause is unknown. Ankylosing spondylosis is reported in the giant perch. Salmonids are prone to vertebral neoplasms, and these will also cause deformation.

Scoliosis, lordosis, and other skeletal abnormalities, particularly of cranial, mandibular, opercular, and branchial bones, are common in young hatchery-reared and wild populations of fish. Many that appear early

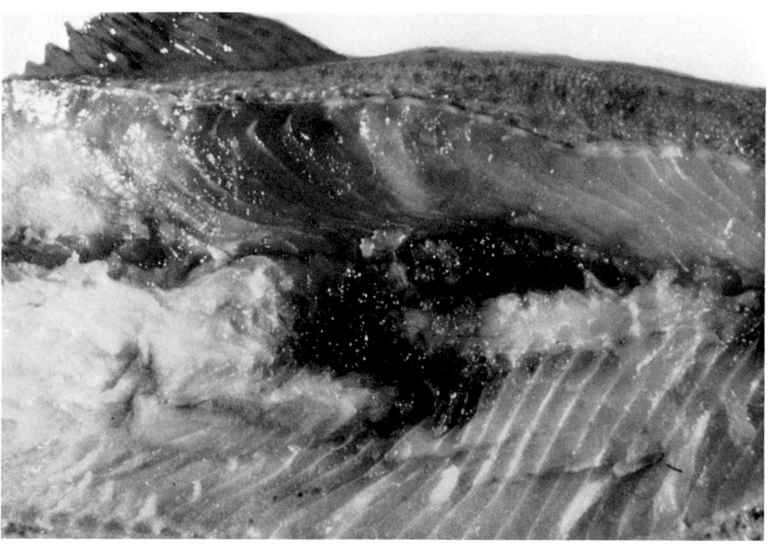

Fig. 12.12. Lightning strike in brood stock rainbow trout. Severe flexure of the fish has snapped spinal column at the fulcrum, just beneath the dorsal fin, resulting also in hemorrhage. (*Courtesy of R. D. Moccia*)

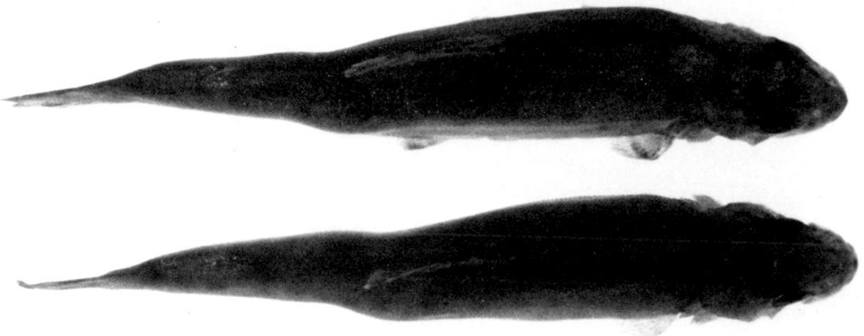

A

B

Fig. 12.13. A and **B.** Scoliosis in rainbow trout due to vitamin A toxicity. (*Material courtesy of J. Hilton*)

in life are lethal, while later ones interfere less with survival. Heavy metals, especially cadmium and lead, are often implicated (lead causes neuronal degeneration), as is zinc, low pH from "acid rain," and harsh incubation conditions including overzealous treatment with disinfecting agents such as malachite green, which is known to have teratogenic potential.

Several species of myxosporean parasites develop within cartilage, causing lysis, and may lead to skeletal abnormalities (Fig. 12.15). The best-known example is *Myxosoma cerebralis,* the cause of the economically important condition of trout and salmon whirling disease. As with other myxosporeans, there is very little known about the life cycle of this parasite. A major but

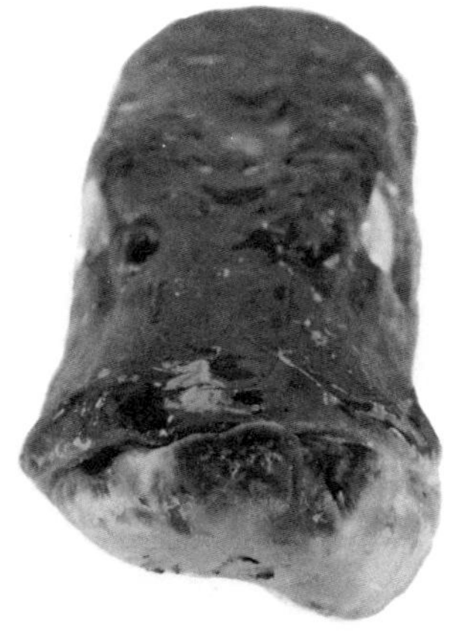

Fig. 12.14. Unilateral hyperostosis of mandible of pike. Cause of the condition is unknown.

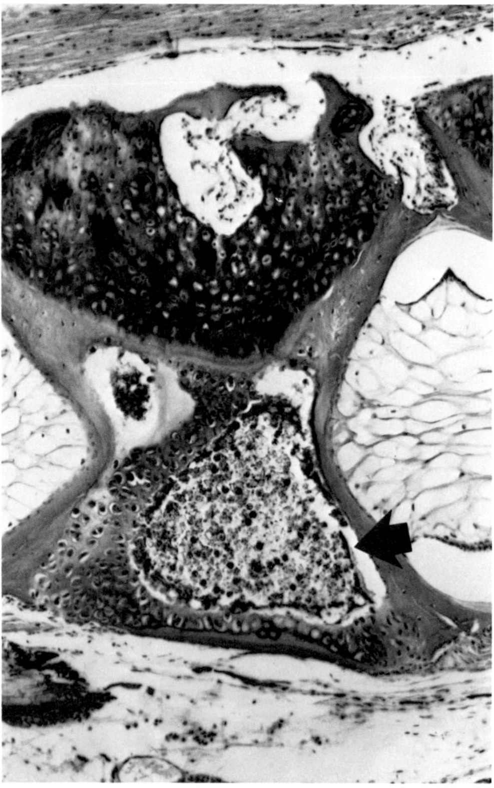

Fig. 12.15. Myxosporean parasites (*arrow*) within vertebral cartilage of wild pumpkinseed. There appeared to be little inflammation or degeneration of surrounding tissues.

controversial advance recently showed that an aquatic oligochaete worm (*Tubifex*) was required for the development of infectivity, thereby conveniently explaining the success of the practical method used for years to control the disease. This method entailed keeping susceptible fish in concrete ponds, or at least away from the mud in which the worms were found (Markiw and Wolf 1983). Regardless of life cycle, once they are within the fish, the myxosporeans develop and form spores in the cartilages of the head and vertebrae of young fish, causing lysis and subsequent deformation. Involvement of the labyrinth may also be seen, doubtless contributing to the whirling behavior from which the disease derives its name.

Other parasites with an affinity for cranial skeletal tissues include *Spironucleus,* the presumptive cause of hole-in-the-head disease of oscars, discus, and other aquarium cichlids. (An alternative hypothesis implicates bacterial infection of the lateral line and cranial sensory canals system as the cause.) This hexamitid is associated with large erosions in the cranial cartilages, which may ulcerate, resulting often in bilaterally symmetrical lesions. Erosion of the cranium has also been associated with *Tetrahymena* sp. in young Atlantic salmon (Fig. 12.16).

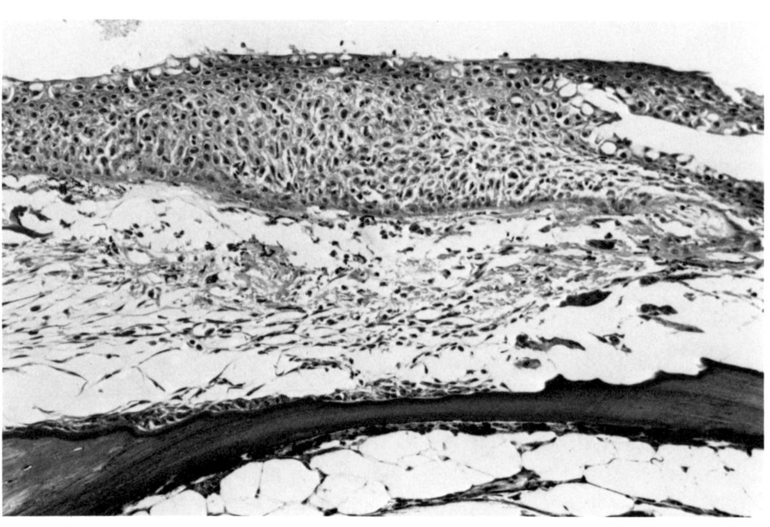

Fig. 12.16. Lysis and osteoclasis of cranium in young Atlantic salmon. These lesions were associated with the ciliate *Tetrahymena* but whether causally related or not is unknown.

Proliferation of branchial cartilage can occur in response to encysted digenetic metacercariae (Fig. 12.17). The pathogenesis of this lesion is unclear. Possibly it is related to altered oxygen tensions affecting the more usual fibroplasia response.

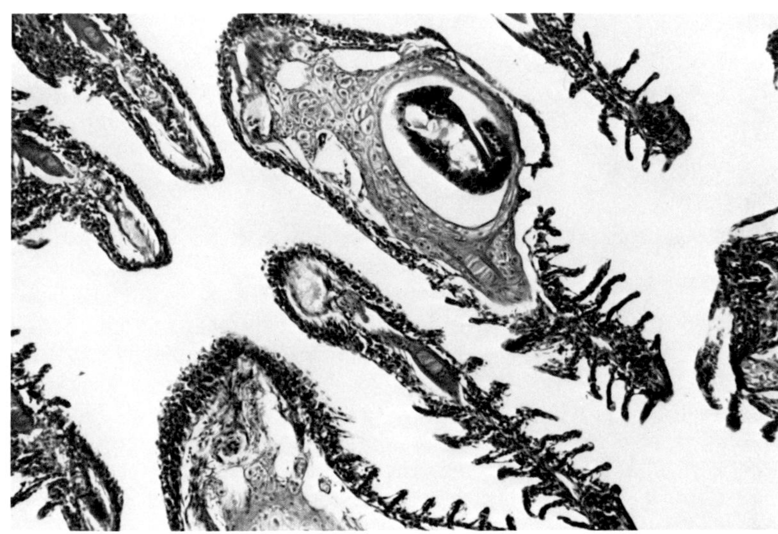

Fig. 12.17. Goldfish gill showing metacercariae encysted within filamental cartilage. The hyperplastic response has distorted the normal architecture. (*Material courtesy of V. Blazer*)

References

Barahona-Fernandes, M. H. 1982. Body deformation in hatchery reared European sea bass *Dicentrarchus labrax* (L). Types, prevalence and effect on fish survival. *J. Fish Biol.* 21:239–49.

Canning, E. U. 1977. Microsporida. In *Parasitic Protozoa: Babesia, Theileria, Myxosporida, Microsporida, Bartonellaceae, Anaplasmataceae, Ehrlichia, and Pneumocystis,* ed. J. P. Kreier, vol. 4, 155–96.

Canning, E. U., E. I. Hazard, and J. P. Nicholas. 1979. Light and electron-microscopy of *Pleistophora* sp. from skeletal muscle of *Blennius pholis. Protistol.* 25:317–32.

Ching, H. L. 1984. Comparative resistance of Oregon (Big Creek) and British Columbia (Capilano) juvenile chinook salmon to the myxozoan pathogen, *Ceratomyxa shasta,* after laboratory exposure to Fraser River water. *Can. J. Zool.* 62:1423–24.

Cone, D. K., and R. C. Anderson. 1977. Myxosporidan parasites of pumpkinseed (*Lepomis gibbosus* L.) from Ontario. *J. Parasitol.* 63:657–66.

Couch, J. A., J. T. Winstead, D. J. Hansen, and L. R. Goodman. 1979. Vertebral dysplasia in young fish exposed to the herbicide trifluralin. *J. Fish Dis.* 2:35–42.

Cowey, C. B., E. Degener, A. G. J. Tacon, A. Youngson, and J. G. Bell. 1984. The effect of vitamin E and oxidized fish oil on the nutrition of rainbow trout (*Salmo gairdneri*) grown at natural, varying water temperatures. *Br. J. Nutr.* 51:443–51.

Davison, W., and G. Goldspink. 1977. The effect of prolonged exercise on the lateral musculature of the brown trout (*Salmo trutta*). *J. Exp. Biol.* 70:1–12.

Dykova, I., and J. Lom. 1980. Tissue reactions to microsporidian infections in fish. *J. Fish Dis.* 3:265–83.

Ferguson, H. W., D. A. Rice, and J. K. Lynas. 1986. Clinical pathology of myodegeneration (pancreas disease) in Atlantic salmon (*Salmo salar*). *Vet. Rec.* 119:297–99.

Ferguson, H. W., R. J. Roberts, R. H. Richards, R. O. Collins, and D. A. Rice. 1986. Severe degenerative cardiomyopathy associated with pancreas disease in Atlantic salmon, *Salmo salar* L. *J. Fish Dis.* 20:95–98.

Fjolstad, M., and A. L. Heyeraas. 1985. Muscular and myocardial degeneration in cultured Atlantic salmon, *Salmo salar* L., suffering from 'Hitra disease'. *J. Fish Dis.* 8:367–72.

Hamilton, A. J., and E. U. Canning. 1987. Studies on the proposed role of *Tubifex tubifex* (Muller) as an intermediate host in the lifecycle of *Myxosoma cerebralis* (Hofer 1903). *J. Fish Dis.* 10, 145–51.

Herman, R. L., and F. W. Kircheis. 1985. Steatitis in Sunapee trout, *Salvelinus alpinus oquassa* Girard. *J. Fish Dis.* 8:237–39.

Hine, P. M. 1980. A review of some species of *Myxidium* Butschli, 1882 (Myxosporea) from eels (*Anguilla* spp.). *J. Protozool.* 27:260–67.

Hoffman, G. L., and J. A. Hutcheson. 1970. Unusual pathogenicity of a common metacercaria of fish. *J. Wildl. Dis.* 6:109–14.

Holloway, H. L., Jr., and C. E. Smith. 1982. A myopathy in North Dakota walleye, *Stizostedion vitreum* (Mitchill). *J. Fish Dis.* 5:527–30.

Johnston, I. A. 1982. Physiology of muscle in hatchery raised fish. *Comp. Biochem. Physiol.* 73:105–24.

Kilarski, W., and M. Kozlowska. 1985. Histochemical and electronmicroscopical analysis of muscle fiber in myotomes of teleost fish (*Noemacheilus barbatulus* L.). *Gegenbaurs Morph. Jahrb.* (Leipz.) 131:55–72.

Lester, R. J. G., and W. R. Kelly. 1984. Ankylosing spondylosis in the giant perch, *Lates calcarifer* (Bloch). *J. Fish Dis.* 7:193–97.

Lim, C., and R. T. Lovell. 1978. Pathology of the Vitamin C deficiency syndrome in Channel catfish (*Ictalurus punctatus*). *J. Nutr.* 108:1137–46.

Lom, J., and E. R. Noble. 1984. Revised classification on the class Myxosporea Butschli, 1881. *Folia Parasitol.* (Praha) 31:193–205.

Markiw, M. E., and K. Wolf. 1978. *Myxosoma cerebralis*: fluorescent antibody techniques for antigen recognition. *J. Fish. Res. Board Can.* 35:828–32.

———. 1983. *Myxosoma cerebralis* (Myxozoa:Myxosporea) etiologic agent of salmonid whirling disease requires tubificid worm (Annelida:Oligochaeta) in its life cycle. *J. Protozool.* 30:561–64.

Matthews, R. A., and B. F. Matthews. 1980. Cell and tissue reactions of turbot *Scophthalmus maximus* (L.) to *Tetramicra brevifilum* gen. n., sp. n. (Microspora). *J. Fish Dis.* 3:495–515.

Mitchell, L. G. 1977. Myxosporida. In *Parasitic Protozoa: Babesia, Theileria, Myxosporida, Microsporida, Bartonellaceae, Anaplasmataceae, Ehrlichia, and Pneumocystis,* ed. J. P. Kreier, vol. 4, 115–54.

Miyazaki, T. 1986. A histopathological study on the carp fed α-tocopherol deficient diets including oxidized methyl linolate. *Fish Pathol.* 21:73–79.

Miyazaki, T., and J. A. Plumb. 1986. A histopathological study on a natural case of broken-back syndrome of the channel catfish. *Bull. Fac. Fish. Mie Univ.* 13:11–16.

Moss, M. L. 1965. Studies on the acellular bone of teleost fish. V. Histology and mineral homeostasis of freshwater species. *Acta Anat.* 60:262–76.

Munro, A. L. S., A. E. Ellis, A. H. McVicar, H. A. McLay, and E. A. Needham. 1984. An exocrine pancreas disease of farmed Atlantic salmon in Scotland. *Helgolander Meeresunters* 37:571–86.

Murthy, V. K., P. Reddanna, M. Bhaskar, and S. Govindappa. 1981. Muscle metabolism of freshwater fish, *Tilapia mossambica* (Peters), during acute exposure and acclimation to sublethal acidic water. *Can. J. Zool.* 59:1909–15.

Newsome, C. S., and R. D. Piron. 1982. Aetiology of skeletal deformities in the Zebra Danio fish (*Brachydanio rerio,* Hamilton-Buchanan). *J. Fish Biol.* 21:231–37.

Poppe, T. T., T. Hastein, A. Froslie, N. Koppang, and G. Norheim. 1986. Nutritional aspects of haemorrhagic syndrome ('Hitra Disease') in farmed Atlantic salmon *Salmo salar. Dis. Aquat. Org.* 1:155–62.

Siau, Y. 1980. Observation immunologique sur des poissons du genre *Mugil* parasites par la Myxosporidie *Myxobolus exiguus* Thelohan, 1895. *Z. Parasitenkd.* 62:1–6.

Sinclair, N. R. 1972. Studies on the heterophyid trematode *Apophallus brevis,* the "sandgrain grub" of yellow perch (*Perca flavescens*). II. The metacercaria: position, structure, and composition of the cyst; hosts; geographical distribution and variation. *Can. J. Zool.* 50:577–84.

Sprague, V., and K. L. Hussey. 1980. Observations on *Ichthyosporidium giganteum* (Microsporida) with particular reference to the host-parasite relations during merogony. *J. Protozool.* 27:169–75.

Tave, D., J. E. Bartels, and R. O. Smitherman. 1983. Saddleback: a dominant, lethal gene in *Sarotherodon aureus* (Steindachner) (*Tilapia aurea*). *J. Fish Dis.* 6:59–73.

Weatherley, A. H., H. S. Gill, and S. C. Rogers. 1980. Growth dynamics of mosaic muscle fibres in fingerling rainbow trout (*Salmo gairdneri*) in relation to somatic growth rate. *Can. J. Zool.* 58:1535–41.

Wieser, W., R. Lackner, S. Hinterleitner, and U. Platzer. 1987. Distribution and properties of lactate dehydrogenase isoenzymes in red and white muscle of freshwater fish. *Fish Physiol. Biochem.* 3:151–62.

Wobeser, G., and N. W. Smith. 1975. Proliferative bone lesions in Lake Whitefish (*Coregonus clupeaformis*). *J. Fish. Res. Board Can.* 32:2065–68.

Zimmer, M. A. 1984. Idiopathic skeletal muscle necrosis in bluegills. *J. Am. Vet. Med. Assoc.* 185:1371–73.

Zinn, J. L., K. A. Johnson, J. E. Sanders, and J. L. Fryer. 1977. Susceptibility of salmonid species and hatchery strains of chinook salmon (*Oncorhynchus tshawytscha*) to infections by *Ceratomyxa shasta*. *J. Fish. Res. Board. Can.* 34:933–36.

13 Neoplasia in Fish

By M. A. Hayes and H. W. Ferguson

Introduction

Teleosts generally resemble other vertebrates in their basic susceptibility to develop neoplastic lesions. As diseases of fish have been studied, characterized, and categorized, a large variety of neoplasms of various tissues have been recognized and described in a broad range of freshwater and marine species of fish (elasmobranchs are a notable exception). A limited number have been studied in considerable detail, especially those relating to tumors found at relatively high frequencies in important commercial food species or other species inhabiting industrially polluted regions.

An ever-increasing number of experimental models for chemically or virally induced neoplasms of various species of captive fish have been developed as a means of elucidating the causes and mechanisms of carcinogenesis. Part of the rationale for the development of these experimental models is the potential use of wild fish as sentinels for environmental carcinogenic agents that are potentially harmful to humans or other terrestrial vertebrates. Also, comparative studies on the biology of neoplastic diseases of various animal species provide an important means for understanding the causes, epidemiology, and biological behavior of neoplasms in man. Recently recognized alterations in highly conserved genes (i.e., protooncogenes) for growth regulatory factors and receptors now explain many of the abnormal growth characteristics of neoplasms. Accordingly, a broad comparative analysis of the biology of protooncogenes and their counterparts (i.e., oncogenes) in neoplasms in all species, including fish, should provide a conceptual basis from which to understand normal and neoplastic growth in all species.

Classification of Neoplasms

Most systems of classification and interpretation of neoplasms have been developed from clinical oncology in human medicine. These systems have been developed over many years to deal with neoplastic lesions that present clinical health problems in human patients. They are based on recognizable phenomena that characterize the identity and behavior of neoplasms and attempt to orient clinical management practices in directions indicated by the important salient features of the neoplasm presented. The

M. A. Hayes, B.V. Sc., Ph.D., is an associate professor in the Department of Pathology, Ontario Veterinary College, University of Guelph.

general objective of pathological examination of neoplasms of human patients is to enable clinical oncologists to select an appropriate therapeutic protocol from those demonstrated to be the most beneficial for various established categories of neoplasms. Because many advanced neoplasms have few adequately successful established therapeutic protocols, the classification systems are in a state of constant flux, with an ever-increasing number of phenomena (morphological, immunological, biochemical, etc.) being identified. The resulting catalogs of types and subtypes of human neoplasms are large, with confusing nomenclature systems that are probably too complex for systematic classification of fish neoplasms.

There are, however, a number of useful conceptual, biological, and mechanistic classifications of neoplasms developed from experimental models of carcinogenesis that relate well to the multistep processes of cancer development. Presently, these nonclinical classifications do not provide much direction in the clinical management of neoplasms, but they do provide a useful conceptual basis from which to understand neoplastic disease at the population level. They are, therefore, probably better for an epidemiological assessment of neoplasms in fish, or other animal populations, in which therapeutic management of an individual's cancer problem is not the highest priority. Accordingly, complex classification systems that attempt to accommodate the full breadth of phenotypic diversity in neoplasms of many species of fish could complicate rather than clarify the higher-priority efforts to characterize and prevent the major disease processes leading to important neoplasms in fish. As a prelude to defining appropriate procedures for pathologic interpretation and classification of fish neoplasms, it is necessary to briefly review the processes by which neoplasms develop.

Pathogenesis of Neoplastic Development

Neoplasms develop in response to a series of changes in cells. These changes can be associated with different types of harmful effects on tissues exposed to oncogenic insults. It is important to realize that many, but not all, frequent neoplasms in older individuals

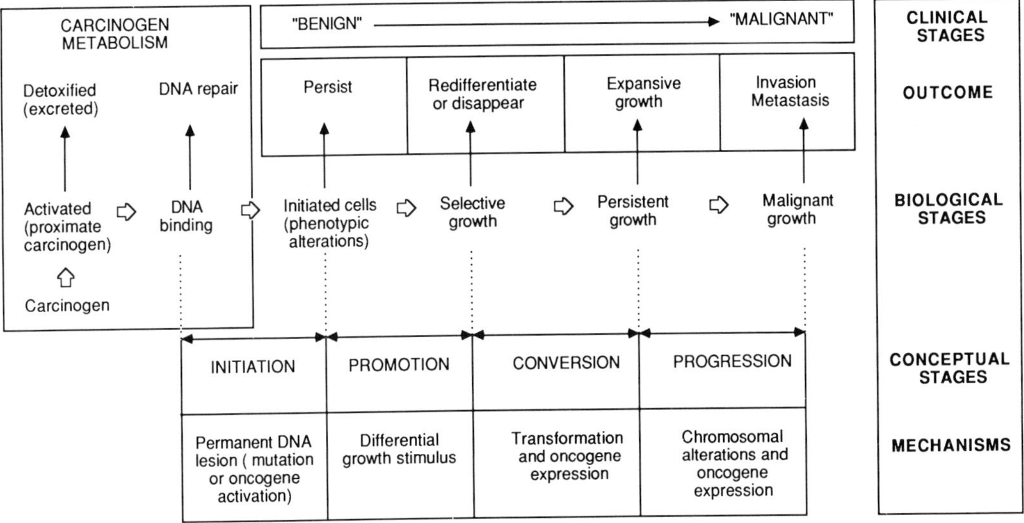

Fig. 13.1. Schematic representation of biological, conceptual, and mechanistic stages in chemical carcinogenesis.

develop in organs or tissues subjected to chronic insult (viral, chemical, parasitic, nutritional, etc.), the nature of which is often manifest by early and ongoing changes in the surrounding non-neoplastic tissue, rather than in the tumor itself. The multistep pathogenesis of neoplasia as a consequence of chronic insult has been demonstrated in many major classes of naturally occurring and experimentally induced neoplasms in terrestrial vertebrates and, to a lesser extent, in some fish. However, neoplasms sometimes develop later in life in animals exposed early in life to a single dose of a genotoxic carcinogen, for example in trout exposed as embryos to aflatoxin. All neoplasms develop through a series of stages that have been reasonably well characterized in biological terms, as populations of cells with sequentially developing novel phenotypic properties and proliferative behavior. These new populations of cells are frequently generated by genotoxic agents but, more importantly, are stimulated to enlarge and progress toward neoplasia by virtue of their ability to *withstand* the harmful effects of the chronic insults that either generated them or enhanced their survival (Fig. 13.1).

The combined influences of *selection, diversification,* and *growth* explain the biology of neoplastic disease. The majority of "early" atypical lesions are reversible or nonprogressive (i.e., benign) but have changes that predispose them to subsequent changes, some of which lead to harmful (malignant) growth behavior. Progression through the various stages depends on the *intensity of selection* and the *rate of diversification,* each of which is influenced in different ways by the intensity of oncogenic insults on the target tissue. The biology of a particular neoplasm depends on its phenotypic changes, its native behavioral growth patterns, and its acquired novel characteristics that are the result of diversification and heterogeneity. Some neoplasms such as those of hematopoietic tissue may be malignant early in the sequence because the cells are normally able to spread to other tissues. Others may not even be noticed during the early stages and are recognized as harmful neoplasms only when they are very advanced, for example neoplasms of parenchymal or structural tissues.

Evaluating Stages of Neoplastic Progression

In clinical oncology, neoplasms are evaluated prognostically as either benign (nonaggressive, innocuous) or malignant (aggressive, harmful). While this is very useful when making decisions for surgical or chemotherapeutic management of tumors in individual patients, such a division is not always easily achieved by histopathological assessment, or even understood in molecular biological terms. Moreover, in experimental or population oncology of fish where individual therapy is not performed, the terms benign and malignant are possibly even inappropriate especially when it is appreciated that, with relatively few exceptions (lymphosarcoma of northern pike or aflatoxin-induced hepatic carcinomas of rainbow trout that metastasize) fish tumors do not often appear to behave in a particularly aggressive manner. Prognostication (i.e., deciding a neoplasm is malignant because eventually it would have become harmful if the animal had continued to survive) is imprecise, especially when the criteria for making such judgments are not reported. The problems that can arise from imprecise prognostic assessment of fish neoplasms are obvious when we consider how regulatory agencies might deal with malignant rather than benign neoplasms in fish in polluted environments or in chemical carcinogenicity assays.

Since the stage of tumor progression is an important measure of the cumulative effects of carcinogenic insults over time, it is especially important for pathologists evaluating animals in carcinogenicity assays to precisely assess the stage of a neoplasm. This should be based on factors such as growth rate, invasiveness, heterogeneity, location, and metastatic behavior. By so doing, the present biological behavior (i.e., stage) of a particular tumor is implied.

Assessment of Neoplasms in Fish Populations

From the foregoing it should have become obvious that the objectives for pathologists required to identify and interpret neoplasms in fish are not all the same (Table 13.1). For example, neoplasms that occur frequently in wild fish need to be considered either with an epidemiological perspective towards possible causes and prevention of the disease or with a public-health perspective towards reducing potential risk for anyone eating the fish or for the coinhabitants of their extended natural environment. By comparison, pathologists evaluating experimentally produced neoplasms in fish need a perspective towards the causes and pathogenetic mechanisms of carcinogenesis. From a broader comparative perspective of the differences in behavior of neoplasms in all species, pathologists can characterize unusual aspects of fish neoplasms that might eventually lead to new preventive and therapeutic strategies applicable to human neoplasms.

From the basis of known mechanisms of carcinogenesis, initial evaluation of populations of fish with neoplasms has been conducted in our laboratory with the following rationale in mind.

1. *Target organ* (topographical occurrence). This aids in attempts to identify classes of carcinogenic influences with known capacity to induce neoplasms in this tissue, and directs further attention to the pathological processes that might be present in the surrounding non-neoplastic tissue.

2. *Stage of neoplastic progression.* This enables an assessment of the duration and intensity of carcinogenic influences in relation to age, bearing in mind the age-related mechanisms of proliferation required for initiation and promotion. The major stages are altered focal (preneoplastic) proliferations and established neoplasms with expansive, diversified, invasive, and metastatic behavior (see Fig. 13.1). Despite possessing histological criteria for malignancy, fish tumors rarely metastasize, although the reasons for this are unknown. Fish cells may

Table 13.1. Objectives for pathological evaluation of neoplasms in fish

Problem	Major Objectives
Neoplasms in individual "valuable" fish	Prognosis and treatment; prevention
Neoplasms in captive fish populations	
Ornamental (pets, exhibits, etc.)	Treatment and prevention; fish health significance
Aquaculture (food fish)	Public health and significance; effects on production; prevention
Experimental (carcinogenesis models)	Pathogenesis of neoplasms; mechanisms and interactions
Experimental (other studies)	Interactions of neoplasms on experimental objectives
Neoplasms in wild fish populations	
Commercial fishery (food)	Public health significance: food safety, water safety, and environmental safety; effects on production; prevention
Noncommercial species	Water and environmental quality; fish health significance; prevention
All neoplasms (comparative oncology)	Mechanisms: biological and molecular; characterization and classification

lack the ability to express secretory proteases or membrane glycoproteins required for successful metastatic spread. Alternatively, the hypothesis that immune surveillance is involved in defense against some neoplasms needs to be assessed in the light of differences between fish and mammalian immune systems.

3. *Epidemiology.* Sporadic tumors in old or otherwise normal populations of fish are put aside as rare accidents due to cumulative molecular events in a single population of cells. Clustered tumors, on the other hand, draw attention to the likelihood that one or more distinct categories of tissue insults has given rise to the increased incidence. For example, the increased occurrence of liver tumors in rainbow trout originally led to the identification of aflatoxin as a food-borne liver carcinogen. Also, the frequent occurrence of skin and liver tumors in fish in polluted rivers or marine environments has stimulated intensive efforts to identify and eliminate carcinogens in industrial effluents.

4. *Host aspects.* The age at which neoplasms develop, particularly in younger fish, is an important lead to the further identification of possible pathogenetic and etiological influences. Neoplasms in young fish usually reflect exposure to potent viral or chemical carcinogens during developmental growth, whereas tumors in very old fish usually reflect a weaker and possibly multifactorial chronic carcinogenic insult. Comparing bottom-dwelling species (such as brown bullheads or white suckers) with free-swimming ones (such as salmonids or carp) may help identify effects due to habitat or due to species differences in, for example, levels of detoxifying enzyme systems. Similarly, different strains of fish may have differing sensitivities to carcinogens, so that tumor incidence in one location is higher than in another. Hybridization may also enhance susceptibility. This is seen in the carp-goldfish hybrids that develop gonadal tumors or in the swordtail-platy hybrids that develop melanomas.

5. *Pathogenetic mechanisms.* The nonneoplastic tissues are examined carefully for evidence of processes that may have contributed to the pathogenesis of the neoplasm observed. For example, if there are many other small preneoplastic lesions in neighboring tissue, there is a strong possibility that the animal has been exposed to a major genotoxic insult. Also the size of these other preneoplastic lesions and the degree of normal cell proliferation may indicate the intensity of a mitogenic promotional influence. Most importantly, the tissue is examined for evidence of ongoing necrosis or atrophy, each of which can contribute to a reciprocal promotional effect on preneoplastic lesions. In this case, the effects on the surrounding tissue may then indicate the nature of the major carcinogenic insult. Similarly, chronic regenerative proliferation, fibroblastic or angiogenic repair, or lymphoproliferative inflammatory reactions may be relevant in explaining the pathogenesis of neoplasms of fibroblasts, endothelial cells, or hematopoietic cells. Some chemically induced neoplasms have an altered biochemical phenotype such as elevations in various glutathione-dependent or other detoxification enzyme systems. This phenotype can be detected by appropriate immuno- or enzyme-histochemical methods and its presence may suggest that the tumor has arisen because it is resistant to a toxic insult. Obviously, such interpretations give some lead towards defining etiological agents (viruses, parasites, chemicals, etc.) and, more important, the mechanisms by which they contribute to the process as a whole.

6. *Etiological mechanisms.* The previous assessments can, as explained, provide an idea of what might be involved in causing the tumors. This needs to be confirmed by a process of identifying, if possible, the nature of the major etiological agents involved, in conjunction with the previously established pathogenetic mechanisms they elicit. It is important to remember that the multiple stages may be influenced by separate etiological influences such that many cancers

have a multifactorial etiology. While this complicates the search for etiological agents, it is a reality that should not be overlooked by searching for a single agent that explains everything in the tissue. A further complication is that the agent may not be present in the tissues, especially in the neoplasms, by virtue of the fact that many neoplasms arise as populations *resistant* to the agent. Accordingly, the search for the agent should focus on the environment, or on the pathologically altered surrounding tissue. For example, it is possible to detect DNA adducts in target tissues of fish exposed to some carcinogens in polluted environments.

7. *Molecular mechanisms.* Some chemical and viral carcinogens generate characteristic alterations in genes for growth regulation. These altered genes are termed oncogenes, which explain to a large extent the different growth behavior of neoplasms. Accordingly, molecular biological characterization of the genetic alteration in a group of tumors may sometimes help to explain the type of virus or chemical involved in their generation. For example, the point mutations in the ras oncogene in liver and skin neoplasms of mammals differ for different classes of genotoxic carcinogens. Also, for most viruses, it is relatively easy to identify evidence of persistent viral infection, either by immunological or nucleic acid hybridization techniques, if the appropriate probes are available. This strategy is presently unsuited to studies of a single neoplasm; rather it should be reserved for detailed evaluation of idiopathic tumors that occur with unusually high frequencies.

8. *Histogenetic and phenotypic characterization.* These studies, which use microscopic features and probes for specific differentiated functions of histological cell types, help to identify the cell of origin in the target organ. However, this approach is prone to error when used on advanced neoplasms because during phenotypic diversification they may have lost certain markers, or they may newly express proteins that are not characteristically expressed in the parent cell type. For example, some advanced hepatic neoplasms have differentiated features of hepatocytes and also of bile ducts making it impossible to determine if such a tumor is of hepatocytic or bile epithelial origin. However, if this approach is reserved for groups of neoplasms, as explained in item 7, and interpreted in conjunction with recognition of early preneoplastic lesions that have not changed substantially from the parent cell type, the major target cell types can be identified more reliably.

Common Neoplasms in Fish

As already stated, classification of fish neoplasms is largely based on mammalian criteria, and there seems little point therefore in merely reiterating these with detailed histological descriptions. Accordingly the reader is referred to the standard texts such as Moulton (1978). A brief overview of some of the more common neoplasms found in fish is nevertheless considered worthwhile, if for no other reason than to underline the fact that they are morphologically similar to their counterparts in other vertebrates. Differences can arise, however, because of cells or tissue organizations unique to teleosts. In keeping with the rest of the book, a systemic approach has been adopted. For reviews on fish tumors, see Lucke and Schlumberger (1949), Mawdesley-Thomas (1971), Budd and Roberts (1978), and Peters (1984).

Skin. In general, skin tumors are quite commonly encountered and diagnosed, partly of course because they are easily seen. The most common skin tumors are generally termed *papillomas,* although their morphology is not always papillary (wartlike) with supporting ingrowing fingers of dermal fibrous tissue typical of classical papillomas of mammals. Many are flatter and sometimes called hyperplastic plaques. These lesions are discrete expansive growths of altered epidermis (malpighian cells) (Fig. 13.2)

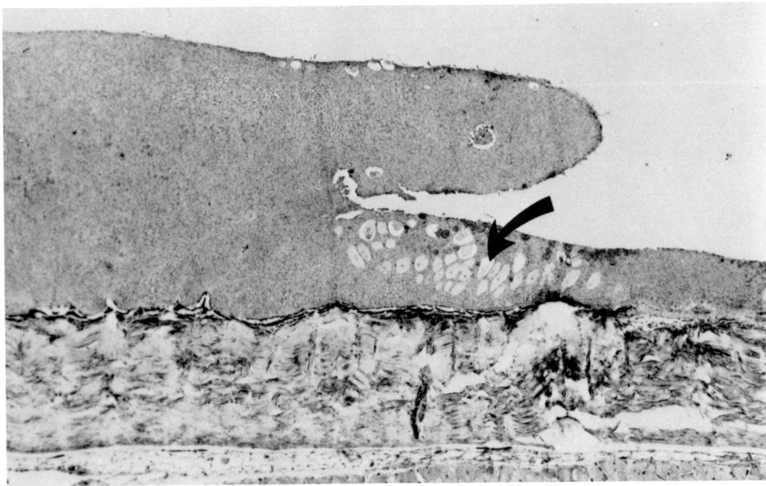

Fig. 13.2. Papilloma from brown bullhead showing plaque-type configuration. The more normal epidermis still retains club cells (*arrow*) and goblet cells.

that tend to obliterate the other goblet cells and club cells by displacement or necrosis. The proliferative epithelium is often spongiotic with focally intense lymphoid cell infiltrates. Papillomas, especially those on the lips, frequently ulcerate due to constant abrasion, supporting the view that physical injury may have a role in promoting their growth. Papillomas are common on Great Lakes fish, especially on the lips of brown bullheads and virtually anywhere on the body surface of white suckers. Viruslike particles have been seen ultrastructurally in papillomas from both of these species of fish, but no viral agent has been firmly implicated or identified. The frequent occurrence of these lesions on fish in industrially polluted locations has led to suspicions that chemicals may have a role in initiating or promoting these lesions in conjunction with a viral agent. Some papillomas in bullheads may progress towards locally invasive lesions, more appropriately termed squamous cell carcinomas.

Papillomatosis ("cauliflower disease") of the European eel is a condition producing particularly florid lesions, most commonly around the mouth. These may be large enough to interfere with feeding and respiration. They are most frequently seen (up to 30% of the population) in young fish in the summer in brackish water, and it is interesting that the lesions regress with lower water temperatures or higher salinities. Once again, a viral etiology is suspected but not confirmed. Hyperplastic epidermal plaques are seen in "papillomatosis" of Atlantic salmon. These lesions are most common in young fish in fresh water in August or September, and although they usually spontaneously regress with no apparent damage, some may become secondarily infected.

Oncorhynchus masou virus (yamame tumor virus) is a herpesvirus that naturally causes disease in young masu salmon (yamame) populations in Japan. Survivors show a remarkably high incidence (60%) of skin tumors including papillomas and basal cell tumors that are most commonly located around the mouth, then the caudal fin, operculum, and eye.

In the Pacific flatfish papillomas, there is a proliferation of malpighian cells that often become replaced by large, clear, rounded cells, x-cells, which lack the epidermal features of desmosomes and tonofilaments and which have the ultrastructural hallmarks of degeneration, e.g., swollen mitochondria. The identity of these x-cells is unknown, and a debate centers even on whether they are host cells or not (protozoa?), similar to the perplexing identity of rodlet cells (see page

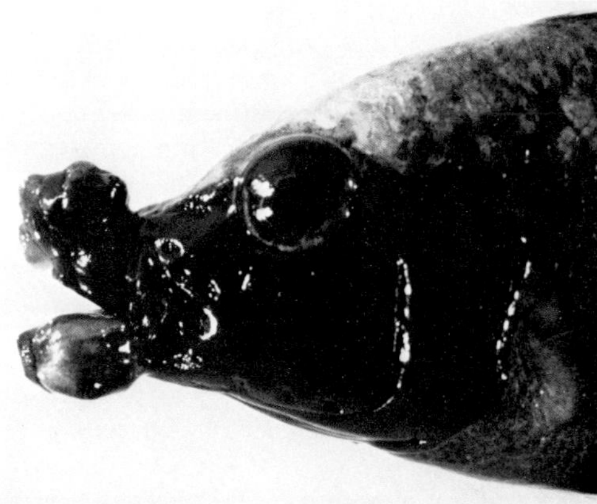

Fig. 13.3. Cichlid with squamous cell carcinoma of lips. (*Courtesy of R. D. Moccia*)

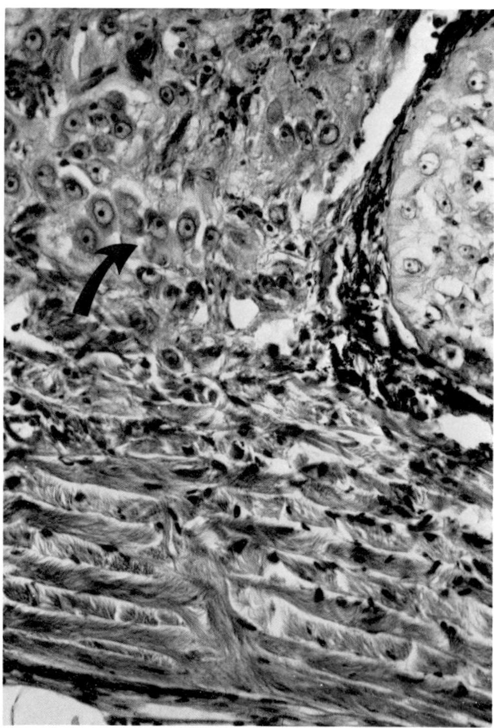

Fig. 13.4. Brown bullhead squamous cell carcinoma showing loss of basal lamina (*arrow*) and early migration of cells down into dermis.

16). In some areas the incidence of these papillomas correlates with heavy pollution, while in others (Queen Charlotte Islands in British Columbia) sand sole may have infection levels of 30% despite inhabiting apparently pollutant-free water. Nevertheless, there has to date been no successful transmission of the disease.

Squamous cell carcinomas are less common (Fig. 13.3), but in one survey of fish from Hamilton Harbour, Lake Ontario, out of 25 brown bullheads necropsied because of grossly visible skin lesions, 10 were subsequently diagnosed as having early squamous cell carcinoma. The relevant criteria were an unusually proliferative basal epithelium with loss of basal lamina, and invasion of basal cells down into the underlying dermis and muscle and up into the overlying epidermis (Fig. 13.4). In most cases, these arose from coexisting papillomas. In only one animal did the carcinoma appear to arise from a moderately hyperplastic rather than papillomatous area of epidermis. *Pigment cell tumors* include melanomas, erythrophoromas, xanthophoromas, and guanophoromas. Of these, melanomas are by far the commonest and may show a greatly increased incidence in hybrids (as do other tumors) such as the swordtail-platy hybrid, in which there appears to be repression of the gene controlling melanophore production. As in mammals, some of the melanomas in fish may be amelanotic, and there may be considerable variation in cellular morphology from round to fusiform or spindle-shaped with a distinct "neural" appearance. Erythrophoromas are also composed of spindle-shaped cells and often are multinucleate with a fine granular cytoplasm.

Fibromas and *fibrosarcomas* are relatively common tumors affecting a wide variety of different species (Fig. 13.5). They are often loosely organized (and some of these may be termed *myxomas*) but occasionally may be hard, or extremely hard, as is the case with the specific entity dermal fibrosarcoma of walleye, associated with a retrovirus. These latter tumors arise from dermal elements and cause multifocal hard nodular lesions on the body surface. They

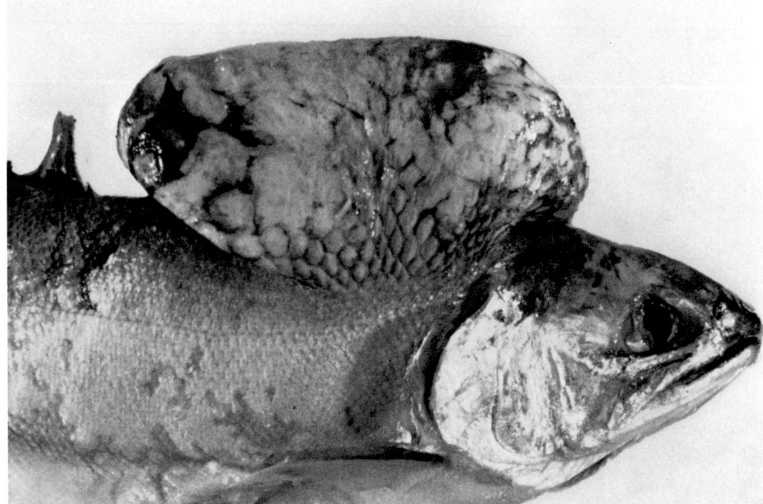

Fig. 13.5. Coho salmon with dermal fibroma.

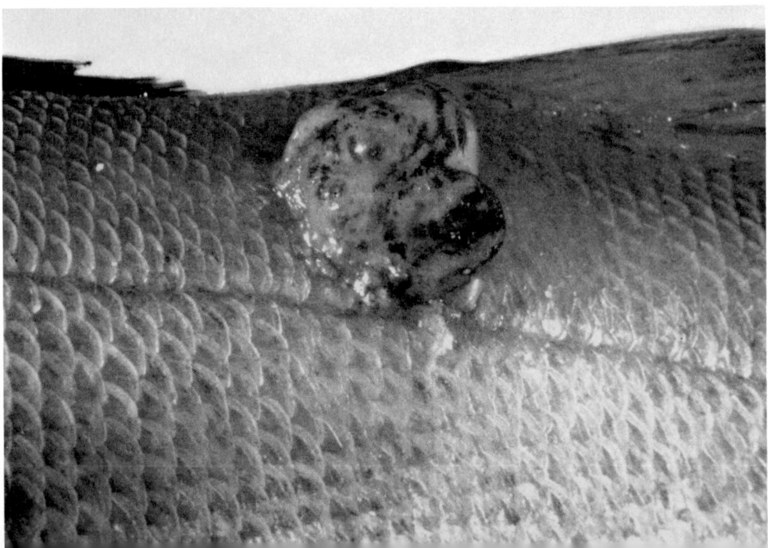

Fig. 13.6. "Dermal fibrosarcoma of walleye." **A.** Locally invasive tumor tending to obliterate many cranial features. **B.** Hard nodular lesion adjacent to dorsal fin showing superficial congestion.

may be extremely disfiguring, locally invasive, and may ulcerate (Fig. 13.6). They are not infrequently found in association with a herpes-virally induced epidermal hyperplasia and/or lymphocystis disease.

Lipomas are occasionally encountered (Fig. 13.7), but they are much less common than fibromas. They too may cause dermal ulceration. *Hemangiomas,* composed of benign proliferations of small blood vessels, are sometimes encountered in the dermis, although occasionally in the trunk musculature.

Gills and Pseudobranchs. Gill tumors are rare, which is interesting, considering their often much greater surface area (and hence greater exposure to carcinogens) than the skin, in which tumors are common. A chondroma arising from filamental or arch cartilage (Fig. 13.8), a fibroma of the arch, and one possible adenoma represent some of the mere handful of spontaneous neoplasms. Branchial lamellar papillomas have been experimentally produced in medaka using N-methyl-N'-nitro-N-nitrosoguanidine.

Pseudobranchial "tumors" are not uncommon, especially in Atlantic and Pacific cod and other gadoids. Histologically they are composed of large pale cells that appear virtually identical to the "*x*-cells" seen in the Pacific flatfish "papillomas". Inevitably the question arises once again as to whether these cells are of host or parasitic origin (indeed a proliferation of similar cells is re-

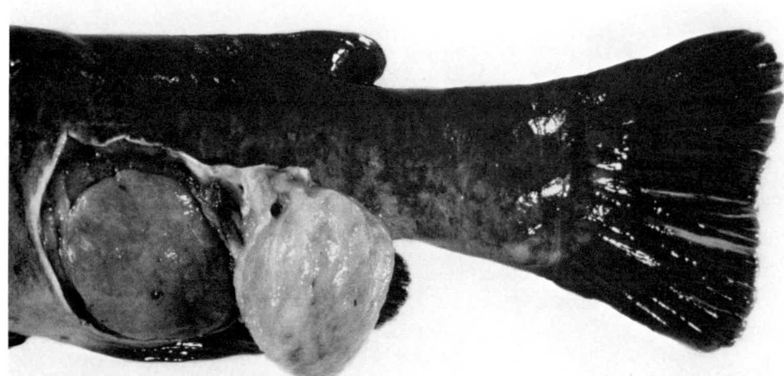

Fig. 13.7. Brown bullhead with subcutaneous lipoma. (*Courtesy of R. D. Moccia*)

Fig. 13.8. Brown trout with branchial chondroma.

ported in the *gills* of some marine fish, just to confound the situation!). The incidence rate in Atlantic cod may be 2% around Nova Scotia and confined mainly to 1-, 2-, or 3-year-old fish. Incidence in Pacific fish tends to be higher.

Kidney. In general, renal neoplasia in fish is relatively uncommon although nephroblastomas and adenocarcinomas are both reported. In the former, there is proliferation of pluripotential blast-type embryonic cells to form most of the cellular elements normally encountered in the posterior excretory portion (Fig. 13.9). Thus there is usually a proliferation of epithelial cells that in places try to form tubules or glomeruli. Connective tissue is present in variable amounts, and either fibrous tissue or cartilage may dominate from one location to the next.

We have seen nephroblastomas both in wild fish (smelt) and in farmed rainbow trout. The renal adenocarcinomas we have encountered have had a papillary appearance.

Hematopoietic Tissue. Lymphosarcoma is well described from northern pike in Scandinavia, Ireland, and North America. In the latter location, muskellunge are also affected. As with similar disease in mammals, a retrovirus is associated, and there has been successful experimental transmission using cell-free extracts. Many fish with lymphosarcoma have hyperplastic epidermal plaques in which it is relatively easy to see C-

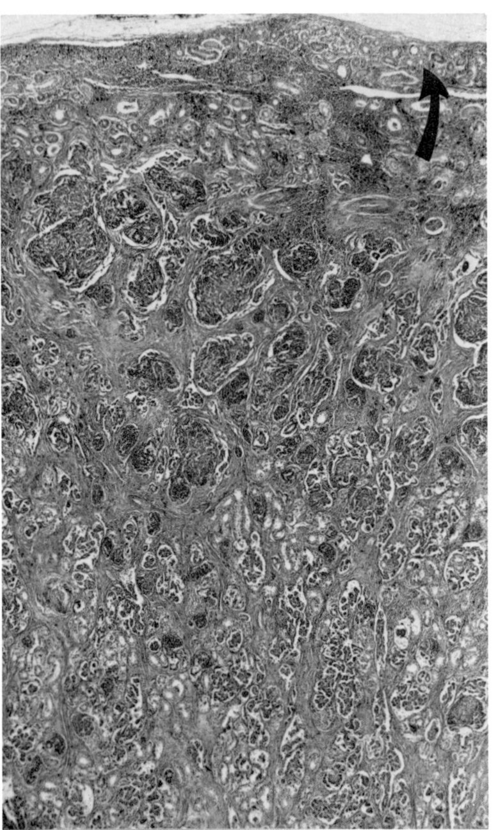

A

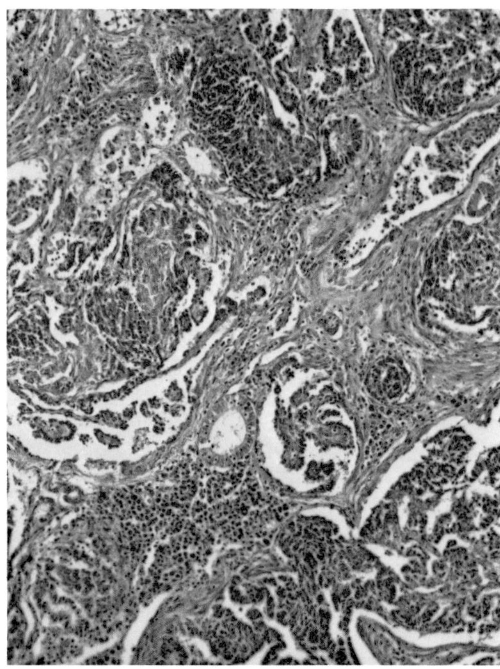

B

Fig. 13.9. Nephroblastoma in rainbow trout. **A.** Only a remnant of relatively normal renal tissue can be seen (*arrow*). **B.** Higher power of **A** showing abortive attempts at glomerular and tubular formation.

A

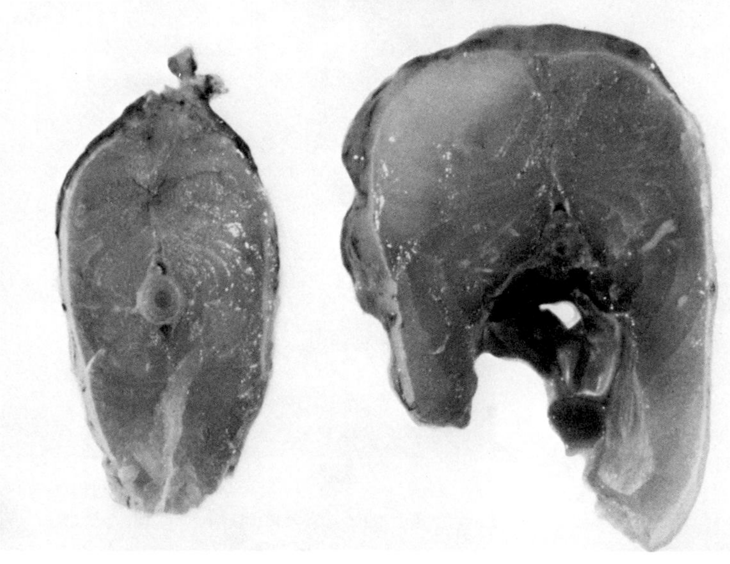

B

Fig. 13.10. A and **B.** Lymphosarcoma of northern pike.

type particles. The lesions are widely disseminated and may be leukemic. They often start as small nodules in the skin, may ulcerate, and then invade the underlying muscle (Fig. 13.10). In late stages, spleen, liver, and kidney are affected. The disease is approximately 99% fatal to muskellunge but less so with pike, and in this species, recovery may even be the norm. Disease incidence in muskellunge is highest in the spring, and by summer many fish have died. In the Baltic, the disease was not recorded until the 1950s and there is a suggestion that the incidence has increased in line with pollution, especially with chlorinated hydrocarbons.

Lymphosarcoma of thymic origin (*thymoma*) is also found and has been reported several times from salmonids (Fig. 13.11). In most instances, the tumor is leukemic. Occasionally there is diffuse involvement of the

Fig. 13.11. Speckled trout with thymoma.

kidney, and such lesions must be carefully differentiated from the granulomatous nephritides so common in salmonids and other species. Granulomas comprised of a monomorphic cell population have been mistakenly identified as lymphoma (hemogregarine infection; see Fig. 5.6).

Liver. Hepatomas and hepatic carcinomas are common tumors in fish. Rainbow trout, like ducks and turkeys, are exquisitely sensitive to aflatoxin, a product of the mold *Aspergillus flavus,* which grows on peanuts and other oil seeds under warm humid conditions. In the 1960s, aflatoxin caused large numbers of farmed rainbow trout to develop both hepatocellular and bile duct carcinomas. This is one of the few examples of a fish tumor that has been shown to *metastasize*—often to the kidney in this particular case. Because of the susceptibility of the liver of various small fish (e.g., medaka) to chemical carcinogens, this organ is a major focus in short-term bioassays for carcinogens. These fish resemble mammals in their susceptibility to hepatocarcinogenic nitrosamines.

A variety of hepatic tumors are seen in wild populations of fish, often in areas of high industrial pollution. Recent examination of a population of white suckers on a spawning run up Oakville Creek in Lake Ontario showed roughly 17% to have advanced hepatic tumors resembling both hepatocellular and bile duct types. In some cases, as might be expected, both were found together in the same liver. Early changes were often represented by small nodules of morphologically altered cells. An increase in nuclear to cytoplasmic ratio, or staining intensity, was common. More advanced lesions were discrete nodules or multiloculated hepatocellular carcinomas, some of which were extensively invasive within the liver. The bile duct epithelium frequently showed inflammatory, degenerative, and metaplastic change before presumably progressing to discrete cholangiomas or invasive carcinomas. In general, the hepatocellular carcinomas were very cystic and hemorrhagic (Fig. 13.12) while the bile duct carcinomas were solid and often contained retained bile (Fig. 13.13), giving them a grossly distinct color. Despite locally aggressive behavior and cytological pleomorphism and heterogeneity, only one of these liver tumors was seen to metastasize, in this case to the heart.

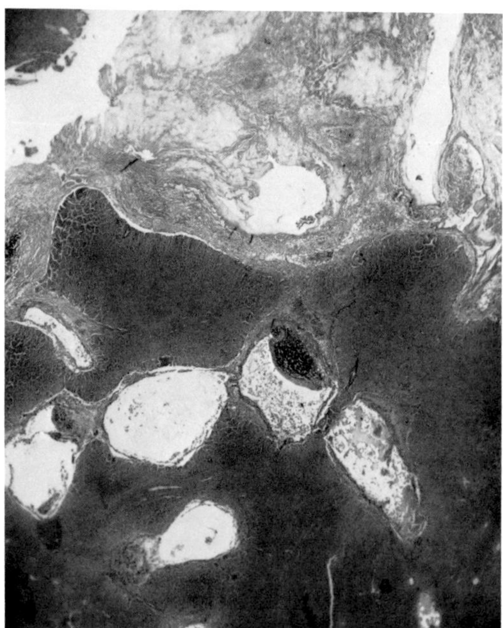

Fig. 13.12. White sucker with hepatocellular carcinoma showing characteristic blood and proteinaceous fluid-filled cystic arrangement.

Another component of the liver in some species is the melanomacrophage center. We have once seen a *melano-histiocytoma,* characterized by large aggregations of the cells tending to obliterate normal parenchyma, and by the presence of multinucleated syncitial giant cells.

Gastrointestinal Tract and Swimbladder. Teeth tumors include ameloblastomas, invasive tumors of the odontogenic epithelium (syn. adamantinoma and enameloblastoma). Odontomas are considered to be

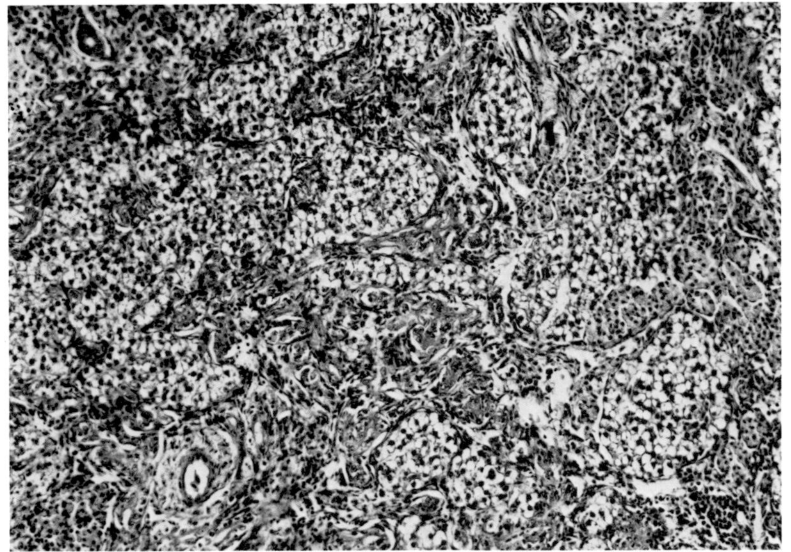

A

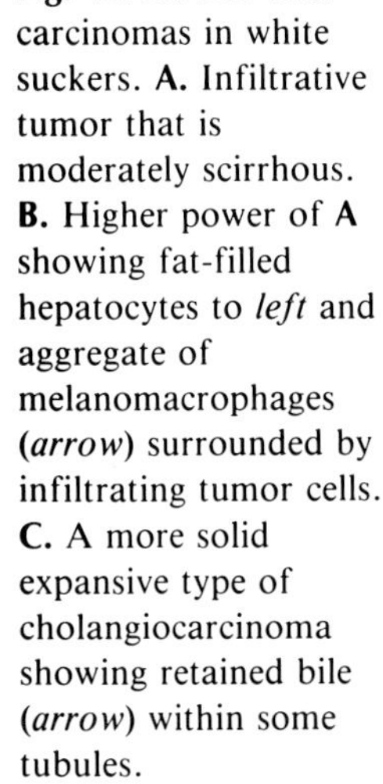

Fig. 13.13. Bile duct carcinomas in white suckers. **A.** Infiltrative tumor that is moderately scirrhous. **B.** Higher power of **A** showing fat-filled hepatocytes to *left* and aggregate of melanomacrophages (*arrow*) surrounded by infiltrating tumor cells. **C.** A more solid expansive type of cholangiocarcinoma showing retained bile (*arrow*) within some tubules.

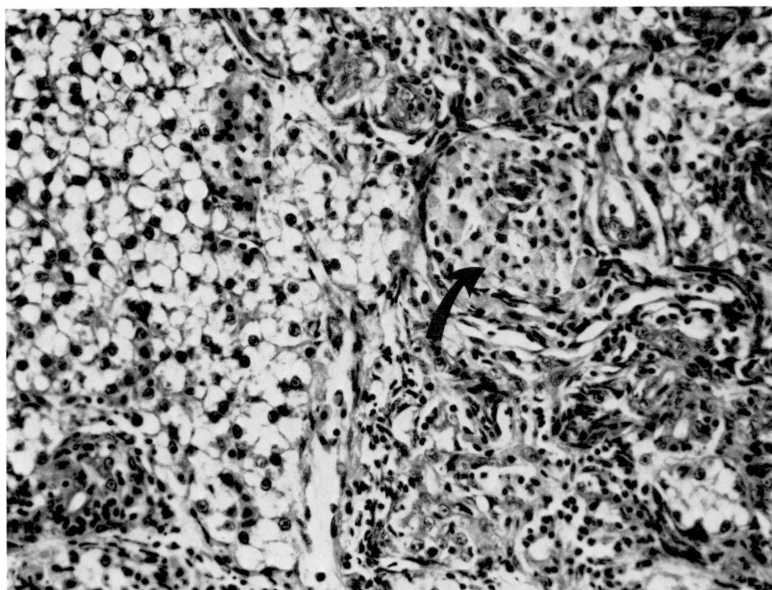

B

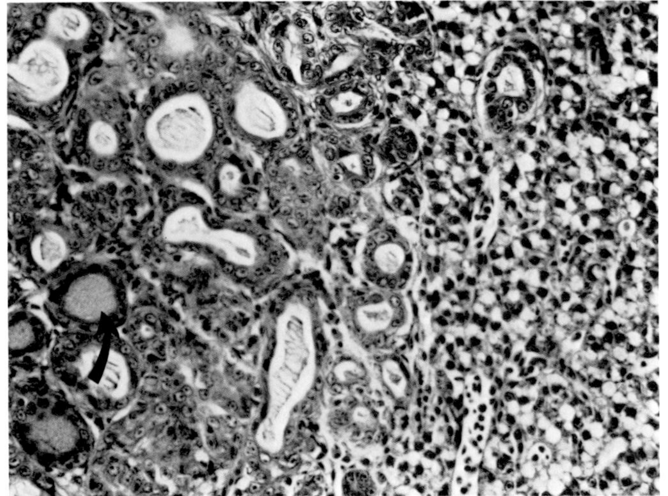

C

malformations. Ameloblastomas are not uncommon, especially in salmonids, and are also reported in the Atlantic cunner (*Tautogolabrus adspersus*), a fish that grazes barnacles and clams and that may therefore suffer mechanical damage. They are locally invasive and may ulcerate. Other than the retrovirus-associated fibrosarcoma of Atlantic salmon that occurred in Scotland in caged fish in epizootic proportion (McKnight 1978) there are only a few reports of swimbladder tumors in other species, notably cod.

Tumors in the stomach and intestine are uncommon, although a tumor of the cod stomach was one of the first recorded (mid-eighteenth century). So-called adenomatous polyps are recognized in cultured yellowtails, sea bream, and eels in Japan, although the significance of these pinhead-sized proliferations is unknown. A few peritoneal mesotheliomas are also reported.

Endocrine and Reproductive Systems

THYROID. Differentiating physiological hyperplasia (goiter; see Fig. 11.4) from neoplasia becomes very difficult in the thyroid, especially when it is appreciated that in most teleost species the thyroid is a diffuse organ with follicles normally scattered in often remote sites such as kidney, splenic capsule, or epicardial surface. Thus the assumption of metastasis is usually mistaken, and reports of thyroid "carcinoma" should be viewed in this light. Nevertheless, the extremely aggressive and invasive behavior that some of these goiters demonstrate, infiltrating muscle, bone, and even myocardium (see Fig. 11.4), means that in a biological sense they are indeed acting as malignancies.

GONADS. Leiomyomas and fibroleiomyomas are quite common tumors of the testis of the yellow perch in the Great Lakes and indeed one study showed 8% of susceptible fish to be affected. A lesser number of ovaries have similar tumors.

Carp-goldfish hybrids in the Great Lakes and elsewhere also have a very high incidence of gonadal tumors (Fig 13.14). As with the tumors in yellow perch, these may grow to a large size, causing space-occupying problems.

Sertoli cell tumors, interstitial cell tumors, and seminomas are also reported as individual cases, again often in hybrids. So

Fig. 13.14. Carp-goldfish hybrid with gonadal tumor. The lesion in this fish was large enough to cause compression atrophy of adjacent tissues. (*Courtesy of E. Downs*)

too are ovarian papillary adenocarcinomas.

Musculoskeletal System. Rhabdomyomas are usually reported as individual cases, although an epizootic has been described in cultured ayu in Japan. The tumors were seen in approximately 1% of fish, affecting mainly the trunk and tail muscles and grossly distorting the fish. In the absence of obvious virus involvement, chemotherapeutics were suspected as the cause.

Other than hyperostoses (reactive bone), which are not infrequently found (an example being in the mandible of pike in the Great Lakes), bone tumors are not common. They are sometimes encountered, however, in salmonid vertebrae, usually at the fulcrum of flexure beneath the dorsal fin, and a traumatic etiology should be suspected. Other individually reported cases occur in a variety of other species and indeed represent some of the earliest recorded tumors.

Nervous Tissue. There are a few reports of central nervous system tumors (ependymomas). By contrast, peripheral nerve tumors are quite common, especially those associated with the skin. Neurofibromas (formed from endo- and epineurial connective tissue) and neurilemmomas (derived from Schwann cells) are the two commonly reported, although they are difficult to differentiate from fibromas or sometimes melanomas, especially the amelanotic-melanomas. While usually seen as isolated cases, they may occur in high frequency in some populations. Such is the case reported for several *Lutianus* spp. (snappers) along the Florida coast, affecting the nerve branches of roughly 1% of the population. The possibility of granuloma should always be borne in mind however. Ganglioneuromas are rarely reported.

An epizootic (12 of 100,000) of neuroblastomas was reported in coho salmon raised in chlorinated-dechlorinated water; they were located in the skeletal muscle adjacent to the dorsal fin. The suggestion is that the halogenation of the water may have produced carcinogens.

Eye. A variety of tumors involving the eye have been reported. They include ocular fibromas, neurofibromas of the limbus, lymphosarcoma, and retinoblastoma. The best-known ocular tumor, however, is the melanoma of the swordtail-platy hybrids. Neoplastic cells may be seen extending from the choroid through the sclera to form usually densely pigmented retrobulbar masses (Levine and Gordon 1946).

References

Bernstein, J. W. 1984. Leukaemic lymphosarcoma in a hatchery-reared rainbow trout, *Salmo gairdneri* Richardson. *J. Fish Dis.* 7:83–86.

Black, J. J. 1983. Field and laboratory studies of environmental carcinogenesis in Niagara River fish. *J. Great Lakes Res.* 9:326–34.

Brittelli, M. R., H. H. C. Chen, and C. F. Muska. 1985. Induction of branchial (gill) neoplasms in the Medaka fish (*Oryzias latipes*) by N-methyl-N′-nitro-N-nitrosoguanidine. *Cancer Res.* 45:3209–14.

Budd, J., and R. J. Roberts. 1978. Neoplasia of teleosts. In *Fish Pathology,* ed. R. J. Roberts. London: Bailliere Tindall.

Budd, J., J. D. Schroder, and K. Davey Dukes. 1975. Tumors of the yellow perch. In *The Pathology of Fishes,* ed. W. E. Ribelin and G. Migaki. Madison: The University of Wisconsin Press.

Bylund, G., E. T. Valtonen, and E. Niemela. 1980. Observations on epidermal papillomata in wild and cultured Atlantic salmon *Salmo salar* L. in Finland. *J. Fish Dis.* 3:525–28.

Coulombe, R. A., Jr., G. S. Bailey, and J. E. Nixon. 1984. Comparative activation of aflatoxin B_1 to mutagens by isolated hepatocytes from rainbow trout (*Salmo gairdneri*) and coho salmon (*Oncorhynchus kisutch*). *Carcinog.* 5:29–33.

Duncan, I. B. 1978. Evidence for an oncovirus in swimbladder fibrosarcoma of Atlantic salmon *Salmo salar* L. *J. Fish Dis.* 1:127–31.

Egidius, E. C., J. V. Johannessen, and E. Lange. 1981. Pseudobranchial tumours in Atlantic cod, *Gadus morhua* L., from the Barents Sea. *J. Fish Dis.* 4:527–32.

Fujimoto, Y., H. Madarame, H. Yoshida, R. Moriguchi, H. Kodama, and H. Iizawa. 1986. Pathomorphological observations on epidermal papilloma of flatfish (*Liopsetta obscura*). *Jpn. J. Vet. Res.* 34:81–103.

Gorlin, R. J. 1972. Odontogenic tumors in mammals and fish. *Oral Surg.* 33:86–90.

Grizzle, J. M., and E. H. Williams, Jr. 1983. Dermal fibroma in a redband parrotfish, *Sparisoma aurofrenatum* (Valenciennes). *J. Fish Dis.* 6:205–9.

Grizzle, J. M., T. E. Schwedler, and A. L. Scott. 1981. Papillomas of black bullheads, *Ictalurus melas* (Rafinesque), living in a chlorinated sewage pond. *J. Fish Dis.* 4:345–51.

Harshbarger, J. C., and C. J. Dawe. 1973. Hematopoietic neoplasms in invertebrate and poikilothermic vertebrate animals. I. Comparative studies on animal species. In *Unifying Concepts of Leukemia,* Bibl. haemat. no. 39, ed. R. M. Dutcher and L. Chieco-Bianchi, 1–25. Basel: Karger.

Harshbarger, J. C., S. E. Shumway, and G. W. Bane. 1976. Variably differentiating oral neoplasms, ranging from epidermal papilloma to odontogenic ameloblastoma, in cunners [(*Tautogolabrus adspersus*) Osteichthyes; Perciformes:Labridae]. *Prog. Exp. Tumor Res.* 20:113–28.

Huizinga, H. W., and J. Budd. 1983. Nephroblastoma in the smelt, *Osmerus mordax* (Mitchill). *J. Fish Dis.* 6:389–91.

Hussein, S. A., and D. H. Mills. 1982. The prevalence of 'cauliflower disease' of the eel, *Anguilla anguilla* L., in tributaries of the River Tweed, Scotland. *J. Fish Dis.* 5:161–65.

Ishikawa, T., and S. Takayama. 1979. Importance of hepatic neoplasms in lower vertebrate animals as a tool in cancer research. *J. Toxicol. Environ. Health* 5:537–50.

Kimura, I., S. Morikawa, T. Kiriyama, and H. Kitaori. 1983. An epizootic occurrence of rhabdomyoma and a case of ganglioneuroma in hatchery-reared ayu, *Plecoglossus altivelis* Temminck & Schlegel. *J. Fish Dis.* 6:195–200.

Kranz, H., N. Peters, G. Bresching, and H. F. Stich. 1980. On cell kinetics in skin tumours of the Pacific English sole *Parophrys vetulus* Girard. *J. Fish Dis.* 3:125–32.

Landolt, M. L., and R. M. Kocan. 1984. Lethal and sublethal effects of marine sediment extracts on fish cells and chromosomes. *Helgolander Meeresunters.* 37:479–91.

Levine, M., and M. Gordon. 1946. Ocular tumours with exophthalmia in Xiphophorin fishes. *Cancer Res.* 6:197–204.

Lucke, B., and H. G. Schlumberger. 1949. Neoplasia in cold-blooded vertebrates. *Am. Physiol. Soc.* 29:91–126.

McCain, B. B., W. D. Gronlund, M. S. Myers, and S. R. Wellings. 1979. Tumours and microbial diseases of marine fishes in Alaskan waters. *J. Fish Dis.* 2:111–30.

McKnight, I. J. 1978. Sarcoma of the swimbladder of Atlantic salmon (*Salmo salar* L.). *Aquaculture* 13:55–60.

Majeed, S. K., D. W. Jolly, and C. Gopinath. 1984. An outbreak of liver cell carcinoma in rainbow trout, *Salmo gairdneri* Richardson, in the U.K. *J. Fish Dis.* 7:165–68.

Manier, J. F., A. Raibaut, A. Lopez, and J.-A. Rioux. 1984. A calcified fibroma in the common carp, *Cyrpinus carpio* L. *J. Fish Dis.* 7:283–92.

Masahito, P., T. Ishikawa, and S. Takayama. 1984. Spontaneous spermatocytic seminoma in African lungfish, *Protopterus aethiopicus* Heckel. *J. Fish Dis.* 7:169–72.

Mawdesley-Thomas, L. E. 1971. Neoplasia in fish: a review. *Curr. Top. Comp. Pathobiol.* 1:87–170.

Metcalfe, C. D., and R. A. Sonstegard. 1984. Microinjection of carcinogens into rainbow trout embryos: an in vivo carcinogenesis assay. *J. Natl. Cancer Inst.* 73:1125–32.

Moulton, J.E., ed. 1978. Tumors in domestic animals. 2d ed. Berkeley: University of California Press.

Murchelano, R. A., and R. L. Edwards. 1981. An erythrophoroma in ornamental carp, *Cyprinus carpio* L. *J. Fish Dis.* 4:265–68.

Nishimoto, M., and U. Varanasi. 1985. Benzo[a]pyrene metabolism and DNA adduct formation mediated by English sole liver enzymes. *Biochem. Pharmacol.* 34:263–68.

Park, E-H., and D. S. Kim. 1984. Hepatocarcinogenicity of diethylnitrosamine to the self-fertilizing hermaphroditic fish *Rivulus marmoratus* (Teleostomi:Cyprinodontidae). *J. Natl. Cancer Inst.* 73:871–76.

Peters, G. 1976. The papillomatosis of the European eel (*Anguilla anguilla* L.): Analysis of seasonal fluctuations in the tumor incidence. *Arch. Fisch Wiss.* 27:251–63.

Peters, G., and N. Peters. 1979. The influence of salinity on growth and structure of epidermal papillomas of the European eel *Anguilla anguilla* L. *J. Fish Dis.* 2:13–26.

Peters, N. 1984. Diseases caused by neoplasia. In *Diseases of Marine Animals,* vol. 4, pt. 1, ed. O. Kinne. Hamburg: Biologische Anstalt Helgoland.

Peters, N., G. Peters, H. F. Stich, A. B. Acton, and G. Bresching. 1978. On differences in skin tumours of Pacific and Atlantic flatfish. *J. Fish Dis.* 1:3–25.

Sano, T., H. Fukuda, N. Okamoto, and F. Kaneko. 1983. Yamame tumor virus: lethality and oncogenicity. *Bull. Jpn. Soc. Sci. Fish.* 49:1159–63.

Scarpelli, D. G. 1976. Neoplasia in poikilotherms. In *Cancer: A Comprehensive Treatise,* ed. F. F. Becker, vol. 4. New York: Plenum Press.

Schultz, M. E., and R. J. Schultz. 1982. Diethylnitrosamine-induced hepatic tumors in wild vs. inbred strains of a viviparous fish. *J. Hered.* 73:43–48.

Smith, C. E., T. H. Peck, R. J. Klauda, and J. B. McLaren. 1979. Hepatomas in Atlantic tomcod *Microgadus tomcod* (Walbaum) collected in the Hudson River estuary in New York. *J. Fish Dis.* 2:313–19.

Swain, L., and P. Melius. 1984. Characterization of benzo[a]pyrene metabolites formed by 3-methylcholanthrene-induced goldfish, black bullhead and brown bullhead. *Comp. Biochem. Physiol.* 79:151–58.

Thornton, S. C., L. Diamond, and W. M. Baird. 1982. Metabolism of benzo[a]pyrene by fish cells in culture. *J. Tox. Environ. Health* 10:157–67.

Walker, R. 1969. Virus associated with epidermal hyperplasia in fish. *Natl. Cancer Inst. Monogr.* 31:195–207.

Warr, G. W., B. R. Griffin, D. P. Anderson, P. E. McAllister, B. Lidgerding, and C. E. Smith. 1984. A lymphosarcoma of thymic origin in the rainbow trout, *Salmo gairdneri* Richardson. *J. Fish Dis.* 7:73–82.

Williams, D. E., and D. R. Buhler. 1983. Purified form of cytochrome P-450 from rainbow trout with high activity toward conversion of aflatoxin B_1 to aflatoxin B_1-2,3-epoxide. *Cancer Res.* 43:4752–56.

Winqvist, G., O. Ljungberg, and B. Hellstroem. 1968. Skin tumours of northern pike (*Esox lucius* L.). II. Viral particles in epidermal proliferations. *Bull. Off. Int. Epiz.* 69:1023–31.

Yamamoto, T., R. D. MacDonald, D. C. Gillespie, and R. K. Kelly. 1976. Viruses associated with lymphocystis disease and dermal sarcoma of walleye (*Stizostedion vitreum vitreum*). *J. Fish. Res. Board Can.* 33:2408–19.

Bibliography

I. Books

1. GENERAL AND VETERINARY PATHOLOGY

Cheville, N. F. 1983. *Cell Pathology.* 2d ed. Ames: Iowa State University Press.

Cohen, N., and M. M. Sigel, eds. 1982. *The Reticuloendothelial System.* Vol. 3, *Ontogeny and Phylogeny.* New York: Plenum Press.

Ghadially, F. N. 1982. *Ultrastructural Pathology of the Cell and Matrix: A Text and Atlas of Physiological and Pathological Alterations in the Fine Structure of Cellular and Extracellular Components.* 2d ed. London: Butterworths.

Hill, R. B., and M. F. LaVia, eds. 1980. *Principles of Pathobiology.* 3d ed. Oxford: Oxford University Press.

Jubb, K. V. F., P. C. Kennedy, and N. Palmer. 1985. *Pathology of Domestic Animals.* 3d ed. Vols. 1–3. New York: Academic Press.

Manning, M. J., ed. 1980. *Phylogeny of Immunological Memory.* Amsterdam: Elsevier.

Mims, C. A. 1982. *The Pathogenesis of Infectious Disease.* 2d ed. New York: Academic Press.

Moulton, J. E., ed. 1978. *Tumors in Domestic Animals.* Berkeley: University of California Press.

Roitt, I., J. Brostoff, and D. Male. 1986. *Immunology.* St. Louis: C. V. Mosby Co.

Slauson, D. O., and B. J. Cooper. 1982. *Mechanisms of Disease: A Textbook of Comparative General Pathology.* Baltimore: Williams and Wilkins.

Thomson, R. G. 1984. *General Veterinary Pathology.* 2d ed. Philadelphia: W. B. Saunders Co.

Tizard, I. 1987. *Veterinary Immunology: An Introduction.* 3d ed. Philadelphia: W. B. Saunders Co.

2. WATER QUALITY

Alabaster, J. S., and R. Lloyd. 1982. *Water Quality Criteria for Freshwater Fish.* 2d ed. London: Butterworth Scientific.

Allen, L. J., and E. C. Kinney, eds. 1979. *Proceedings of the Bio-Engineering Symposium for Fish Culture.* Published by the Fish Culture Section of the American Fisheries Society and the Northeast Society of Conservation Engineers. Bethesda, Md.

American Society for Testing and Materials 75th Annual Meeting Symposium. 1973. *Biological Methods for the Assessment of Water Quality.* ASTM Special Technical Publication no. 528. Philadelphia.

Boyd, C. E. 1979. *Water Quality in Warmwater Fish Ponds.* Auburn, Ala.: Auburn University.

Clark, J. R., and R. L. Clark, eds. *Sea-Water Systems for Experimental Aquariums.* Neptune City, N.J.: TFH Publications Inc.

European Inland Fisheries Advisory Commission, Food and Agriculture Organization of the United Nations. 1980. *Water Quality Criteria for European Freshwater Fish.* EIFAC Technical Paper no. 37.

Hawkins, A. D., ed. 1981. *Aquarium Systems.* London: Academic Press.

Kraybill, H. F., C. J. Dawe, J. C. Harshbarger, and R. G. Tardiff, eds. 1977. *Aquatic Pollutants and Biologic Effects with Emphasis on Neoplasia.* Vol. 298. New York Academy of Sciences.

McLean, M. P., R. E. McNicol, and E. Scherer. 1980. *Bibliography of Toxicity Test Methods for the Aquatic Environment.* Canadian Special Publication of Fisheries and Aquatic Sciences no. 50.

Rand, G. M., and S. R. Petrocelli, eds. 1985. *Fundamentals of Aquatic Toxicology.* Washington, D.C.: Hemisphere Publ. Corp.

Stumm, W., and J. J. Morgan. 1981. *Aquatic Chemistry: An Introduction Emphasizing Chemical Equilibria in Natural Waters.* 2d ed. New York: John Wiley & Sons.

Weber, L. J., ed. 1984. *Aquatic Toxicology.* Vol. 2. New York: Raven Press.

Wheaton, F. W. 1985. *Aquacultural Engineering.* Malabar, Fla: Robert E. Krieger Publ. Co. Inc.

3. FISH AND FISH DISEASES

Alexander, R. McN. 1974. *Functional Design in Fishes.* London: Hutchinson & Co. Ltd.

Ali, M. A., and M. Anctil. 1976. *Retinas of Fishes.* Berlin: Springer-Verlag.

Anderson, D. P., M. Dorson, and P. H. Dubourget, eds. 1982. *Les antigenes des microorganismes pathogenes des poissons.* Symposium International de Talloires, May 10, 11, 12, 1982. Collection Fondation Marcel Merieux.

Austin, B., and D. A. Austin. 1987. *Bacterial Fish Pathogens: Diseases in Farmed and Wild Fish.* Chichester, U.K.: Ellis Horwood.

Berka, R. 1986. *The Transport of Live Fish: A Review.* European Inland Fisheries Advisory Commission, Food and Agriculture Organization of the United Nations. EIFAC Technical Paper no. 48.

Bullock, G. L., D. A. Conroy, and S. F. Snieszko. *Diseases of Fishes.* Book 2A: *Bacterial Diseases of Fishes,* and Bullock, G. L. Book 2B: *Identification of Fish Pathogenic Bacteria.* Neptune City, N.J.: TFH Publications.

Davy, F. B., and A. Chouinard, eds. 1982. *Fish Quarantine and Fish Diseases in Southeast Asia.* Report of a Workshop Held in Jakarta, Indonesia. Ottawa: IDRC.

De Kinkelin, P., C. H. Michel, and P. Ghittino. 1985. *Precis de Pathologie des Poissons.* Institut National de la Recherche Agronomique.

Ellis, A. E., ed. 1985. *Fish and Shellfish Pathology.* London: Academic Press.

Groman, D. B. 1982. *Histology of the Striped Bass.* American Fisheries Society Monograph no. 3. Bethesda, Md.

Herwig, N. 1979. *Handbook of Drugs and Chemicals Used in the Treatment of Fish Diseases: A Manual of Fish Pharmacology and Materia Medica.* Springfield, Ill.: C. C. Thomas Publisher.

Hoar, W. S., and D. J. Randall, eds. 1984. *Fish Physiology.* Vol. 10, *Gills.* Orlando, Fla.: Academic Press. (See also volumes 1–9 for coverage of most other systems.)

Hoffman, G. L. 1967. *Parasites of North American Freshwater Fishes.* Berkeley: University of California Press.

Hoover, K. L. 1981. *Use of Small Fish Species in Carcinogenicity Testing.* U.S. Dept. of Health and Human Services, Public Health Service, National Institutes of Health Monograph no. 65. Bethesda, Md.

Humphrey, J. D., and J. S. Langdon, eds. 1985. *Proceedings of the Workshop on Diseases of Australian Fish and Shellfish.* Australian Fish Health Reference Laboratory. Regional Veterinary Laboratory, Benalla, Victoria.

Kabata, Z. 1985. *Parasites and Diseases of Fish Cultured in the Tropics.* London and Philadelphia: Taylor & Francis.

Kinne, O., ed. 1984. *Diseases of Marine Animals.* Vol. 4, pt. 1, *Introduction, Pisces.* Hamburg: Biologische Anstalt Helgoland.

Lagler, K. F., J. E. Bardach, R. R. Miller, and D. R. M. Passino, 1977. *Ichthyology.* 2d ed. New York: John Wiley & Sons.

Liewes, E. W. 1984. *Culture, Feeding and Diseases of Commercial Flatfish Species.* Rotterdam: A. A. Balkema.

Margolis, L., and J. R. Arthur. 1979. *Synopsis of the Parasites of Fishes of Canada.* Fisheries Research Board of Canada Bulletin no. 199. Ottawa.

Matty, A. J. 1985. *Fish Endocrinology.* Portland, Ore.: Croom Helm, London and Timber Press.

Meyer, F. P., J. W. Warren, and T. G. Carey, eds. 1983. *A Guide to Integrated Fish Health Management in the Great Lakes Basin.* Great Lakes Fishery Commission Special Publication no. 83-2. Ann Arbor, Mich.

Moller, H., and K. Anders. 1986. *Diseases and Parasites of Marine Fishes.* Kiel: Verlag Moller.

National Research Council. 1977. *Nutrient Requirements of Warmwater Fishes.* National Academy of Sciences. Washington.

______. 1982. *Nutrient Requirements of Coldwater Fishes.* National Academy of Sciences. Washington.

Neish, G. A., and G. C. Hughes. 1980. *Diseases of Fishes.* Bk. 6, *Fungal Diseases of Fishes.* Neptune City, N.J.: TFH Publications Inc.

Northcutt, R. G., and R. E. Davis, eds. 1983. *Fish Neurobiology.* Vol. 1, *Brain Stem and Sense Organs.* Ann Arbor, Mich.: University of Michigan Press.

Patt, D. I., and G. R. Patt. 1969. *Comparative Vertebrate Histology.* New York: Harper & Row Publishers.

Plumb, J. A., ed. 1985. *Principal Diseases of Farm Raised Catfish.* Southern Cooperative Series Bulletin no. 225. Auburn University, Ala.

Reddacliff, G. L. 1985. *Diseases of Aquarium Fishes: A Practical Guide for the Australian Veterinarian.* Published by the University of Sydney, The Post-Graduate Foundation in Veterinary Science, New South Wales.

Ribelin, W. E., and G. Migaki, eds. 1975. *The Pathology of Fishes.* Madison: University of Wisconsin Press.

Rickards, W. L., ed. 1978. *A Diagnostic Manual of Eel Diseases Occurring under Culture Conditions in Japan.* UNC Sea Grant Publication. Raleigh.

Roberts, R. J., ed. 1978. *Fish Pathology.* London: Bailliere Tindall.

______, ed. 1982. *Microbial Diseases of Fish.* Special Publications of the Society for General Microbiology. London: Academic Press.

Ross, L. G., and B. Ross. 1984. *Anaesthetic and Sedative Techniques for Fish.* Institute of Aquaculture. University of Stirling.

Sindermann, C. J. 1970. *Principal Diseases of Marine Fish and Shellfish.* New York: Academic Press.

Snieszko, S. F., ed. 1970. *A Symposium of the American Fisheries Society on Diseases of Fishes and Shellfishes.* American Fisheries Society. Washington.

Vivares, C. P., J.-R. Bonami, and E. Jaspers, eds. 1986. *Pathology in Marine Aquaculture.* European Aquaculture Society Special Publication no. 9. Belgium.

Wedemeyer, G. A., F. P. Meyer, and L. Smith. 1976. *Diseases of Fishes.* Bk. 5, *Environmental Stress and Fish Diseases.* Neptune City, N.J.: TFH Publications Inc.

Yasutake, W. T., and J. H. Wales. 1983. *Microscopic Anatomy of Salmonids: An Atlas.* U.S. Dept. of the Interior, Fish and Wildlife Service Resource Publication no. 150, Washington.

II. Articles

1. IMMUNOLOGY

Ambrosius, H., and D. Hadge. 1983. Phylogeny of low molecular weight immunoglobulins. *Dev. Comp. Immunol.* 7:721–24.

Anderson, D. P., W. B. van Muiswinkel, and B. S. Roberson. 1984. Effects of chemically induced immune modulation on infectious diseases of fish. In *Chemical Regulation of Immunity in Veterinary Medicine,* 187–211. New York: Alan Liss Inc.

Avtalion, R. R. 1981. Environmental control of the immune response in fish. In *Critical Reviews in Environmental Control,* vol. 11, 163–88. Boca Raton, Fla.: CRC Press, Inc.

Avtalion, R. R., A. Wojdani, Z. Malik, R. Shahrabani, and M. Duczyminer. 1973. Influence of environmental temperature on the immune response in fish. *Curr. Top. Microbiol. Immunol.* 61:1–35.

Badet, M. T., P. Chateaureynaud, and G. Mayer. 1983. Immunitary relations between mother and embryo during gestation: a comparative study in lower and higher vertebrates. *Dev. Comp. Immunol.* 7:731–34.

Baldo, B. A., and T. C. Fletcher. 1973. C-reactive protein-like precipitins in plaice. *Nature* 246:145–46.

Bisset, K. A. 1946. The effect of temperature on non-specific infections of fish. *J. Path. Bact.* 58:251–58.

Blazer, V. S., and R. E. Wolke. 1984. Effect of

diet on the immune response of rainbow trout (*Salmo gairdneri*). *Can. J. Fish. Aquat. Sci.* 41:1244–47.

______. 1984. The effects of a-tocopherol on the immune response and non-specific resistance factors of rainbow trout (*Salmo gairdneri* Richardson). *Aquaculture* 37:1–9.

Cenini, P., and R. J. Turner. 1983. *In vitro* effects of zinc on lymphoid cells of the carp, *Cyprinus carpio* L. *J. Fish Biol.* 23:579–83.

Chilmonczyk, S. 1978. *In vitro* stimulation by mitogens of peripheral blood lymphocytes from rainbow trout (*Salmo gairdneri*). *Ann. Immunol.* 129:3–12.

Clawson, C. C., J. Finstad, and R. A. Good. 1966. Evolution of the immune response. V. Electron microscopy of plasma cells and lymphoid tissue of the paddlefish. *Lab. Invest.* 15:1830–47.

Clelland, G. B., and R. J. Sonstegard. 1987. Natural killer cell activity in rainbow trout (*Salmo gairdneri*): effect of dietary exposure to Aroclor 1254 and/or Mirex. *Can. J. Fish. Aquat. Sci.* 44:636–38.

Clerx, J. P. M., A. Castel, J. F. Bol, and G. J. Gerwig. 1980. Isolation and characterization of the immunoglobin of pike (*Esox lucius* L.). *Vet. Immunol. Immunopathol.* 1:125–44.

Collins, M. T., D. L. Daw, and J. B. Gratzek. 1976. Immune responsiveness of channel catfish under different experimental conditions. *J. Am. Vet. Med. Assoc.* 169:991–94.

Corbel, M. J. 1975. The immune response in fish: a review. *J. Fish Biol.* 7:539–63.

Ellis, A. E. 1977. Ontogeny of the immune response in *Salmo salar.* Histogenesis of the lymphoid organs and appearance of membrane immunoglobulin and mixed leucocyte reactivity. In *Developmental Immunobiology,* ed. J. B. Solomon and J. D. Horton, 225–31. Amsterdam: Elsevier/North Holland Biomedical Press.

______. 1982. Difference between the immune mechanisms of fish and higher vertebrates. In *Microbial Diseases of Fish,* ed. R. J. Roberts, 1–30. Society for General Microbiology, Academic Press.

Ellis, A. E., and M. de Sousa. 1974. Phylogeny of the lymphoid system. 1. A study of the fate of circulating lymphocytes in plaice. *Eur. J. Immunol.* 4:338–43.

Evans, D. L., R. L. Carlson, S. S. Graves, and K. T. Hogan. 1984. Nonspecific cytotoxic cells in fish (*Ictalurus punctatus*). IV. Target cell binding and recycling capacity. *Dev. Comp. Immunol.* 8:823–33.

Evans, D. L., S. S. Graves, D. Cobb, and D. L. Dawe. 1984. Nonspecific cytotoxic cells in fish (*Ictalurus punctatus*). II. Parameters of target cell lysis and specificity. *Dev. Comp. Immunol.* 8:303–12.

Faulmann, E., M. A. Cuchens, C. J. Lobb, N. W. Miller, and L. W. Clem. 1983. An effective culture system for studying in vitro mitogenic responses of channel catfish lymphocytes. *Trans. Am. Fish. Soc.* 112:673–79.

Fletcher, T. C. 1986. Modulation of nonspecific host defenses in fish. *Vet. Immunol. Immunopathol.* 12:59–67.

Fletcher, T. C., and A. White. 1973. Antibody production in the plaice (*Pleuronectes platessa* L.) after oral and parenteral immunization with *Vibrio anguillarum* antigens. *Aquaculture* 1:417–28.

Fryer, J. L., K. S. Pilcher, J. E. Sanders, J. S. Rohovec, J. L. Zinn, W. J. Groberg, and R. H. McCoy. 1976. Temperature, infectious diseases, and the immune response in salmonid fish. 1–70. U.S. Dept. of Commerce, National Technical Information Service. Bull. no. EPA-600/3-76-021. Springfield, Va.

Graves, S. S., D. L. Evans, D. Cobb, and D. L. Dawe. 1984. Nonspecific cytotoxic cells in fish (*Ictalurus punctatus*). I. Optimum requirements for target cell lysis. *Dev. Comp. Immunol.* 8:293–302.

Groberg, W. J., Jr., R. H. McCoy, K. S. Pilcher, and J. L. Fryer. 1978. Relation of water temperature to infections of coho salmon (*Oncorhynchus kisutch*), chinook salmon (*O. tshawytscha*), and steelhead trout (*Salmo gairdneri*) with *Aeromonas salmonicida* and *A. hydrophila. J. Fish. Res. Board Can.* 35:1–7.

Grondel, J. L., and E. G. M. Harmsen. 1984. Phylogeny of interleukins: growth factors produced by leucocytes of the cyprinid fish, *Cyprinus carpio* L. *Immunology* 52:477–82.

Henderson-Arzapalo, A., R. R. Stickney, and D. H. Lewis. 1980. Immune hypersensitivity in intensively cultured *Tilapia* species. *Trans. Am. Fish. Soc.* 109:244–47.

Hildemann, W. H. 1974. Some new concepts in

immunological phylogeny. *Nature* 250:116–20.

Hodgins, H. O., R. S. Weiser, and G. J. Ridgway. 1967. The nature of antibodies and the immune response in rainbow trout (*Salmo gairdneri*). *J. Immunol.* 99:534–44.

Hodgins, H. O., F. L. Wendling, B. A. Braaten, and R. S. Weiser. 1973. Two molecular species of agglutinins in rainbow trout (*Salmo gairdneri*) serum and their relation to antigenic exposure. *Comp. Biochem. Physiol.* 45:975–77.

Horne, M. T., M. Tatner, S. McDerment, C. Agius, and P. Ward. 1982. Vaccination of rainbow trout, *Salmo gairdneri* Richardson, at low temperatures and the long-term persistence of protection. *J. Fish Dis.* 5:343–45.

Ingram, G. A. 1980. Substances involved in the natural resistance of fish to infection—a review. *J. Fish Biol.* 16:23–60.

Ingram, G. A., and J. B. Alexander. 1981. A comparison of the methods used to detect the cellular and humoral immune response of brown trout. *Dev. Biol. Stand.* 49:295–99.

Koppenheffer, T. L. 1987. Serum complement systems of ectothermic vertebrates. *Dev. Comp. Immunol.* 11:279–86.

Lamers, C. H. J. 1986. Histophysiology of a primary immune response against *Aeromonas hydrophila* in carp (*Cyprinus carpio* L.). *J. Exp. Zool.* 238:71–80.

Lamers, C. H. J., and M. J. M. de Haas. 1983. The development of immunological memory in carp (*Cyprinus caprio* L.) to a bacterial antigen. *Dev. Comp. Immunol.* 7:713–14.

Manning, M. J., M. F. Grace, and C. J. Secombes. 1982. Developmental aspects of immunity and tolerance in fish. In *Microbial Diseases of Fish,* ed. R. J. Roberts, 31–46. London: Academic Press.

Miller, N. W., and M. R. Tripp. 1982. The effect of captivity on the immune response of the killifish, *Fundulus heteroclitus* L. *J. Fish Biol.* 20:301–8.

______. 1982. An immunoinhibitory substance in the serum of laboratory held killifish, *Fundulus heteroclitus* L. *J. Fish Biol.* 20:309–16.

Moody, C. E., D. V. Serreze, and P. W. Reno. 1985. Non-specific cytotoxic activity of teleost leukocytes. *Dev. Comp. Immunol.* 9:51–64.

Murray, C. K., and T. C. Fletcher. 1976. The immunohistochemical localization of lysozyme in plaice (*Pleuronectes platessa* L.) tissues. *J. Fish Biol.* 9:329–34.

O'Leary, P. J., J. O. Cisar, and J. L. Fryer. 1978. The effect of temperature on agglutination activity of coho salmon *Oncorhynchus kisutch* (Walbaum) antiserum. *J. Fish Dis.* 1:123–25.

O'Neill, J. G. 1979. The immune response of the brown trout, *Salmo trutta,* L. to *MS2* bacteriophage: immunogen concentration and adjuvants. *J. Fish Biol.* 15:237–48.

______. 1980. Temperature and the primary and secondary immune responses of three teleosts, *Salmo trutta, Cyrpinus carpio* and *Notothenia rossii,* to *MS2* bacteriophage. In *Phylogeny of Immunological Memory,* ed. M. J. Manning, 123–30. Amsterdam: Elsevier/North Holland Biomedical Press.

______. 1981. Effects of intraperitoneal lead and cadmium on the humoral immune response of *Salmo trutta. Bull. Environ. Contam. Toxicol.* 27:42–48.

______. 1981. The humoral immune response of *Salmo trutta* L. and *Cyprinus carpio* L. exposed to heavy metals. *J. Fish Biol.* 19:297–306.

Ourth, D. O., and L. M. Bachinski. 1987. Bacterial sialic acid modulates activation of the alternative complement pathway of channel catfish (*Ictalurus punctatus*). *Dev. Comp. Immunol.* 11:551–64.

Post, G. 1966. Serum proteins and antibody production in rainbow trout (*Salmo gairdneri*). *J. Fish. Res. Board Canada* 23:1957–63.

Prendergast, R. A., G. A. Lutty, and A. L. Scott. 1983. Directed inflammation: The phylogeny of lymphokines. *Dev. Comp. Immunol.* 7:629–32.

Reynolds, W. W., J. B. Covert, and M. E. Casterlin. 1978. Febrile responses of goldfish *Carassius auratus* (L.) to *Aeromonas hydrophila* and to *Escherichia coli* endotoxin. *J. Fish. Dis.* 1:271–73.

Rijkers, G. T., and W. B. van Muiswinkel. 1977. The immune system of cyprinid fish. The development of cellular and humoral responsiveness in the rosy barb (*Barbus conchonius*). In *Developmental Immunobiology,* ed. J. B. Solomon and J. D. Horton, 233–40. Amsterdam: Elsevier/North Holland Biomedical Press.

Stolen, J. S., T. Gahn, V. Kasper, and J. J. Nagle. 1984. The effect of environmental temperature on the immune response of a marine teleost (*Paralichthys dentatus*). *Dev. Comp. Immunol.* 8:89–98.

Tatner, M. F. 1986. The ontogeny of humoral immunity in rainbow trout, *Salmo gairdneri. Vet. Immunol. Immunopathol.* 12:93–105.

Tatner, M. F., and M. J. Manning. 1983. Growth of the lymphoid organs in rainbow trout, *Salmo gairdneri* from one to fifteen months of age. *J. Zool. Lond.* 199:503–20.

van Muiswinkel, W. B., G. T. Rijkers, and R. van Oosterom. 1978. The origin of vertebrate immunity. The lymphoid system in fish. *Proc. Zodiac Symp. on Adaptation,* 84–87. Wageningen, Netherlands: Pudoc Agric. Publ.

Warr, G. W., and R. C. Simon. 1983. The mitogen response potential of lymphocytes from the rainbow trout (*Salmo gairdneri*) re-examined. *Dev. Comp. Immunol.* 7:379–84.

Weiss, E., and R. R. Avtalion. 1977. Regulatory effect of temperature and antigen upon immunity in ectothermic vertebrates. II. Primary enhancement of anti-hapten antibody response at high and low temperatures. *Dev. Comp. Immunol.* 1:93–104.

Wetzel, M. C., C. Vilain, and J. Charlemagne. 1983. Antibody diversity in cyprinid fish. *Dev. Comp. Immunol.* 7:729.

Zeeman, M. 1986. Modulation of the immune response in fish. *Vet. Immunol. Immunopathol.* 12:235–41.

2. THE RETICULOENDOTHELIAL SYSTEM

Agius, C. 1979. The role of melano-macrophage centres in iron storage in normal and diseased fish. *J. Fish Dis.* 2:337–43.

Agius, C., and S. A. Agbede. 1984. An electron microscopical study on the genesis of lipofuscin, melanin and haemosiderin in the haemopoietic tissues of fish. *J. Fish Biol.* 24:471–88.

Alexander, J. B., G. A. Ingram, S. M. Shamshoom, and A. Bowers. 1983. Antigen clearance in trout. *Dev. Comp. Immunol.* 7:707–8.

Avtalion, R. R., and R. Shahrabani. 1975. Studies on phagocytosis in fish. I. *In vitro* uptake and killing of living *Staphylococcus aureus* by peripheral leucocytes of carp (*Cyprinus carpio*). *Immunology* 29:1181–87.

Braun-Nesje, R., G. Kaplan, and R. Seljelid. 1982. Rainbow trout macrophages in vitro: Morphology and phagocytic activity. *Dev. Comp. Immunol.* 6:281–91.

Braun-Nesje, R., K. Bertheussen, G. Kaplan, and R. Seljelid. 1981. Salmonid macrophages: separation, *in vitro* culture and characterization. *J. Fish Dis.* 4:141–51.

Brown, C. L., and C. J. George. 1985. Age-dependent accumulation of macrophage aggregates in the yellow perch, *Perca flavescens* (Mitchill). *J. Fish Dis.* 8:135–38.

Ellis, A. E. 1980. Antigen-trapping in the spleen and kidney of the plaice *Pleuronectes platessa* L. *J. Fish Dis.* 3:413–26.

Ellis, A. E., A. L. S. Munroe, and R. J. Roberts. 1976. Defence mechanisms in fish. 1. A study of the phagocytic system and the fate of intraperitoneally injected particulate material in the plaice (*Pleuronectes platessa* L.). *J. Fish Biol.* 8:67–78.

Ferguson, H. W. 1984. Renal portal phagocytosis of bacteria in rainbow trout (*Salmo gairdneri* Richardson): ultrastructural observations. *Can. J. Zool.* 62:2505–11.

Ferguson, H. W., M. J. Claxton, and J. Lesperance. 1984. The effect of temperature and starvation on the clearance of bacteria from the bloodstream of rainbow trout (*Salmo gairdneri* Richardson). *J. Leuk. Biol.* 35:209–16.

Ferguson, H. W., M. J. Claxton, R. D. Moccia, and E. J. Wilkie. 1982. The quantitative clearance of bacteria from the bloodstream of rainbow trout (*Salmo gairdneri*). *Vet. Pathol.* 19:687–99.

Griffin, B. R. 1983. Opsonic effect of rainbow trout (*Salmo gairdneri*) antibody on phagocytosis of *Yersinia ruckeri* by trout leukocytes. *Dev. Comp. Immunol.* 7:253–59.

Lamers, C. H. J., and M. J. H. De Haas. 1985. Antigen localization in the lymphoid organs of carp (*Cyprinus carpio*). *Cell Tissue Res.* 242:491–98.

Lamers, C. H. J., and H. K. Parmentier. 1985. The fate of intraperitoneally injected carbon particles in cyprinid fish. *Cell Tissue Res.* 242:499–503.

MacArthur, J. I., T. C. Fletcher, and A. W. Thomson. 1983. Distribution of radiolabeled erythrocytes and the effect of temperature on clearance in the plaice (*Pleuronectes platessa*

L.). *J. Reticuloendothel. Soc.* 34:13–21.

McCumber, L. J., M. M. Sigel, R. J. Trauger, and M. A. Cuchens. 1983. RES structure and function of the fishes. In *The Reticuloendothelial System.* Vol. 3, *Phylogeny and Ontogeny,* ed. Cohen and Sigel, 393–422. New York: Plenum Publ. Corp.

McKinney, E. C., S. B. Smith, H. G. Haines, and M. M. Sigel. 1977. Phagocytosis by fish cells. *J. Reticuloendothel. Soc.* 21:89–95.

O'Neill, J. G. 1980. Blood clearance of *MS2* bacteriophage in *Salmo trutta*: a paradoxon. *Experientia* 36:1226–27.

Ruben, L. N. 1984. Some aspects of the phylogeny of macrophage-lymphocyte immune regulation. *Dev. Comp. Immunol.* 8:247–56.

Russell, W. J., S. A. Taylor, and M. M. Sigel. 1976. Clearance of bacteriophage in poikilothermic vertebrates and the effect of temperature. *J. Reticuloendothel. Soc.* 19:91–96.

Sakai, D. K. 1984. Opsonization by fish antibody and complement in the immune phagocytosis by peritoneal exudate cells isolated from salmonid fishes. *J. Fish Dis.* 7:29–38.

Secombes, C. J., and M. J. Manning. 1980. Comparative studies on the immune system of fishes and amphibians: antigen localization in the carp *Cyprinus carpio* L. *J. Fish Dis.* 3:399–412.

Stave, J. W., B. S. Roberson, and F. M. Hetrick. 1983. Chemiluminescence of phagocytic cells isolated from the pronephros of striped bass. *Dev. Comp. Immunol.* 7:269–76.

______. 1984. Factors affecting the chemiluminescent response of fish phagocytes. *J. Fish Biol.* 25:197–206.

Suzuki, K. 1984. A light and electron microscope study on the phagocytosis of leucocytes in rockfish and rainbow trout. *Bull. Jpn. Soc. Sci. Fish.* 50:1305–15.

Tatner, M. F., and M. J. Manning. 1985. The ontogenetic development of the reticulo-endothelial system in the rainbow trout, *Salmo gairdneri* Richardson. *J. Fish Dis.* 8:189–95.

3. STRESS

Hazel, J. R. 1984. Effects of temperature on the structure and metabolism of cell membranes in fish. *Am. J. Physiol.* 246:R460–R470.

Kindle, K. R., and D. H. Whitmore. 1986. Biochemical indicators of thermal stress in *Tilapia aurea* (Steindachner). *J. Fish Biol.* 29:243–55.

Leach, G. J., and M. H. Taylor. 1980. The role of cortisol in stress-induced metabolic changes in *Fundulus heteroclitus. Gen. Comp. Endocrinol.* 42:219–27.

Mazeaud, M. M., F. Mazeaud, and E. M. Donaldson. 1977. Primary and secondary effects of stress in fish: some new data with a general review. *Trans. Am. Fish. Soc.* 106:201–12.

Peters, G., and R. Schwarzer. 1985. Changes in hemopoietic tissue of rainbow trout under influence of stress. *Dis. Aquat. Org.* 1:1–10.

Pickering, A. D. 1984. Cortisol-induced lymphocytopenia in brown trout, *Salmo trutta* L. *Gen. Comp. Endocrinol.* 53:252–59.

Pickering, A. D., and T. G. Pottinger. 1987. Crowding causes prolonged leucopenia in salmonid fish, despite interrenal acclimation. *J. Fish Biol.* 30:701–12.

Robertson, O. H., S. Hane, B. C. Wexler, and A. P. Rinfret. 1963. The effect of hydrocortisone on immature rainbow trout (*Salmo gairdneri*). *Gen. Comp. Endocrinol.* 3:422–36.

Swift, D. J. 1982. Changes in selected blood component values of rainbow trout, *Salmo gairdneri* Richardson, following the blocking of the cortisol stress response with betamethasone and subsequent exposure to phenol or hypoxia. *J. Fish Biol.* 21:269–77.

White, A., and T. C. Fletcher. 1986. Serum cortisol, glucose, and lipids in plaice (*Pleuronectes platessa* L.) exposed to starvation and aquarium stress. *Comp. Biochem. Physiol.* 84A:649–53.

Willemse, J. J., L. Markus-Silvis, and G. H. Ketting. 1984. Morphological effects of stress in cultured elvers, *Anguilla anguilla* (L.). *Aquaculture* 36:193–201.

4. INFLAMMATION

Cannon, M. S., H. H. Mollenhauer, A. M. Cannon, T. E. Eurell, and D. H. Lewis. 1980. Ultrastructural localization of peroxidase activity in neutrophil leukocytes of *Ictalurus punctatus. Can. J. Zool.* 58:1139–43.

Ellis, A. E. 1986. The function of teleost fish lymphocytes in relation to inflammation.

Int. J. Tissue React. 8:263–70.

Finn, J. P., and N. O. Nielsen. 1971. The effect of temperature variation on the inflammatory response of rainbow trout. *J. Pathol.* 105:257–68.

Fishman, J. A., R. P. Daniele, and G. G. Pietra. 1979. Lung defenses in the African lungfish (*Protopterus*): Cellular responses to irritant stimuli. *J. Reticuloendothel. Soc.* 25:179–95.

Hunt, T. C., and A. F. Rowley. 1986. Leukotriene B4 induces enhanced migration of fish leucocytes *in vitro. Immunology* 59:563–68.

MacArthur, J. I., A. W. Thomson, and T. C. Fletcher. 1985. Aspects of leucocyte migration in the plaice, *Pleuronectes platessa* L. *J. Fish Biol.* 27:667–76.

MacArthur, J. I., T. C. Fletcher, B. J. S. Pirie, R. J. L. Davidson, and A. W. Thomson. 1984. Peritoneal inflammatory cells in plaice, *Pleuronectes platessa* L.: effects of stress and endotoxin. J. *Fish Biol.* 25:69–81.

Marx, J., R. Hilbig, and H. Rahmann. 1984. Endotoxin and prostaglandin E_1 fail to induce fever in a teleost fish. *Comp. Biochem. Physiol.* 77:483–87.

Nash, K. A., T. C. Fletcher, and A. W. Thomson. 1986. Migration of fish leucocytes in vitro: the effect of factors which may be involved in mediating inflammation. *Vet. Immunol. Immunopathol.* 12:83–92.

______. 1987. Effect of opsonization on oxidative metabolism of plaice (*Pleuronectes platessa* L.) neutrophils. *Comp. Biochem. Physiol.* 86(B):31–36.

Piomelli, D., and B. Tota. 1983. Different distribution of serotonin in an elasmobranch (*Scyliorhinus stellaris*) and in a teleost (*Conger conger*) fish. *Comp. Biochem. Physiol.* 74:139–42.

Secombes, C. J. 1985. The *in vitro* formation of teleost multinucleate giant cells. *J. Fish Dis.* 8:461–64.

Sohnle, P. G., and M. J. Chusid. 1983. The effect of temperature on the chemiluminescence response of neutrophils from rainbow trout and man. *J. Comp. Pathol.* 93:493–97.

Sommer, C. V., and J. M. Bartos. 1981. In vivo leukocyte migration assay in rainbow trout with a flexible silicone coverslip. *J. Comp. Pathol.* 91:443–45.

Yasuda, T., M. Endo, T. Sakai, and M. Kimura. 1984. Histochemical study on the granulocytes in the inflammation of carp. *Bull. Jpn. Soc. Sci. Fish.* 50:1375–80.

Index

Page numbers in **bold type** indicate figures or tables.